Creating Powerful Brands

Creating Powerful Brands

Fourth Edition

Leslie de Chernatony, Malcolm McDonald and Elaine Wallace

AMSTERDAM • BOSTON • HEIDELBERG • LONDON • NEW YORK • OXFORD • PARIS
SAN DIEGOSAN FRANCISCO • SINGAPORE • SYDNEY • TOKYO

Butterworth-Heinemann is an imprint of Elsevier

Butterworth-Heinemann is an Imprint of Elsevier
The Boulevard, Langford Lane, Kidlington, Oxford OX5 1GB, UK
30 Corporate Drive, Suite 400, Burlington, MA 01803, USA

First published 1992
Reprinted 1993, 1994, 1995, 1996, 1997
Second edition 1998
Third edition 2003
Fourth Edition 2011
Reprinted 2011

British Library Cataloguing-in-Publication Data
A catalogue record for this book is available from the British Library.

Library of Congress Cataloging-in-Publication Data
A catalog record for this book is available from the Library of Congress.

ISBN: 978-1-85617-849-5

For information on all Butterworth-Heinemann publications
visit our Web site at elsevierdierct.com

Printed and bound in China

11 12 13   10 9 8 7 6 5 4 3 2

Working together to grow
libraries in developing countries

www.elsevier.com | www.bookaid.org | www.sabre.org

ELSEVIER   BOOK AID International   Sabre Foundation

Dedication

This book is dedicated to my wife, Carolyn, daughter, Gemma and son, Russell, with love

- Leslie de Chernatony

Dedicated to all my academic colleagues who have helped me over the years, with thanks

- Malcolm H.B. McDonald

Dedicated to Seán and Ann Wallace, with love

- Elaine Wallace

Contents

Part 2   **Brand Management in Different Sectors**

Part 3   Winning the Brands Battle

About the Authors

Professor Leslie de Chernatony

Professor of Brand Marketing, Università della Svizzera Italiana, Lugano, Switzerland and Aston Business School, Birmingham, UK. Managing Partner, Brands Box Marketing and Research Consultancy.

The Chartered Institute of Marketing cites Prof Leslie de Chernatony as one of the 50 gurus who have shaped the future of marketing. This is because of his pioneering work on Brand Management for which he has an international reputation. His cutting edge work on strategically building brands has helped many organisations to develop more effective brand strategies.

With a doctorate in brand marketing, Leslie has a substantial number of publications in American and European journals, in addition to frequent presentations at international conferences. Several of his papers have won best paper awards. His books have been translated into Chinese, Russian, Polish, Czech and Slovenian. He has several books on Brand Management, a recent one being 'From Brand Vision to Brand Evaluation', published by Butterworth-Heinemann and backed by online videos from Oxford Learning Lab. Winning several major research grants helped him and his team progress projects on factors associated with brand success, services branding and the future of brand management.

Leslie is Visiting Professor at Thammasat University in Bangkok. He is a Fellow of the Chartered Institute of Marketing, Fellow of the Market Research Society and Liveryman of the Worshipful Company of Marketors.

Leslie's work has resulted in TV programmes and radio broadcasts. He is a frequent speaker at management conferences.

A firm advocate of the need for managers to benefit from his work on brand marketing, he has run many highly acclaimed management development workshops throughout Europe, the USA, the Middle East, the Far East, Asia and Australia. His advice has been sought by numerous organisations throughout the world on developing more effective brand strategies. On several occasions he has acted as an Expert Witness in court cases over branding issues.

Leslie can be contacted at dechernatony@btinternet.com

Emeritus Professor Malcolm H.B. McDonald, MA (Oxon), MSc, PhD, D.Litt. FCIM, FRSA

Malcolm, until recently, Professor of Marketing and Deputy Director, Cranfield School of Management with special responsibility for E-Business, is a graduate in English Language and Literature from Oxford University and in Business Studies from Bradford University Management Centre. He has a PhD from Cranfield University. He also has an Honorary Doctorate from Bradford University. He has extensive industrial experience, including a number of years as Marketing Director of Canada Dry.

He is Chairman of Brand Finance plc and spends much of his time working with the operating boards of the world's biggest multinational companies, such as IBM, Xerox, BP and the like in many countries in the world, including Japan, the USA, Europe, South America, ASEAN and Australasia.

He has written forty three books, including the best seller 'Marketing Plans; how to prepare them; how to use them' and many of his papers have been published.

His current interests centre around the measurement of the financial impact of marketing expenditure and global best practice key account management.

He is Professor at Cranfield, Henley, Warwick, Aston and Bradford Business Schools.

Malcolm can be contacted at: m.mcdonald@cranfield.ac.uk

Dr Elaine Wallace

J.E. Cairnes School of Business & Economics, National University of Ireland Galway.

Dr. Elaine Wallace is Lecturer in Marketing and Associate Head of Development and Promotion at the J.E. Cairnes School of Business & Economics, National University of Ireland Galway. Elaine has developed programmes in Brand Management and Branding Strategy at undergraduate and post-graduate level.

Having graduated with an honours degree in Commerce, Elaine worked initially with Siemens Limited on corporate marketing and then with Unilever as a trade marketer for brands including Persil and Timotei. She subsequently worked as Panadol Product Manager for Glaxo SmithKline and was a member of the European launch team for NiQuitin CQ. Elaine also worked as Nurofen Brand Manager in Boots Healthcare and managed the launch of Nurofen for Children into the Irish market.

Elaine completed her doctorate at the University of Birmingham, working with Professor de Chernatony. Her subsequent research has explored the anteced-ents and components of service employee performance, the role of the front line employee as brand champion and the nature and management of brand sabotage. Elaine adopts qualitative and quantitative methodologies in her research, has published in international journals and presented at international conferences. Her research activities to date, have been supported in part by funding from the Irish Research Council. Elaine is also interested in brand building, as well as exploring the relationship between consumers and brands and the impact of controls and compliance requirements on brand orientation.

Elaine can be contacted at elaine.wallace@nuigalway.ie

Preface

As we look forward with confidence that this new edition significantly benefits students and managers, we look back to how this book became a well established core text in many business schools internationally and amongst a lot of managers. It seems not that long ago when we embarked on this project, yet as our publisher attests, the first edition of this book appeared in 1992. That was the year when the European Union offered the potential for developing brands that could cross frontiers without too much hindrance. Cries were heard about identities being lost as corporations sought to consider how more standardised approaches could be followed in their brand strategies. Yet today we see a loose-tight approach being followed that helps different communities to celebrate their diversity, but at the same time allows them to conform to societal norms. Successful brands did not adopt the "one shape fits all" mantra. Rather, they evolved using a loose-tight approach. They are tight in so far as there are a core set of brand values that are not transgressed, yet loose since country managers have the latitude to enact the brand values in a manner more suited to their country.

The changing environment makes this new edition even more relevant. Having a good foundation in the principles of brand management is essential if resources are to be wisely used to grow brand value. One only needs to look at league tables to appreciate the significant financial value of brands. It is becoming more common to see organisations monitoring brand equity and brand value and then to assess how different brand strategies are contributing to changes in brand equity and brand value .

A passive view about the role of customers in branding used to be the norm. In other words customers were viewed as the end of the value chain consuming the value of brands. Brand managers who believe in this notion are cutting their brands' life expectancies. Customers are part of real or virtual communities and they expect to be able to shape the nature of brands. The result is that customers are increasingly both co-creators of brands and brand publics. The internet not only levels the playing fields between large corporates and SMEs, but also empowers customers to make brands better fit their needs.

The growth of the services sector, at the expense of the manufacturing sector, necessitates the need to understand the principles of services branding. Appreciating the differences between services and goods branding is but part of the process of managing services brands. Capitalising on the benefits of committed and passionate staff whose personal values are aligned with their brand values is the hallmark of many successful services brands. Thriving services brands do not enjoy such vitality just because they identify gaps in markets. Rather, it is because they then take new brand propositions inside the organisation, engaging staff to be involved in developing integrated processes that deliver unique and welcome brand experiences.

Retailer brands no longer play a game of catch-up behind leading brand manufacturers. Rather, they are innovators, reshaping markets. Well devised brand promises, coherently delivered through very sophisticated processes, belittle the term 'private label'. Impressive though retailers' brand strategies are, there are those who question whether there are some retailer brands that are excessively adopting category conventions resulting from years of investment by category leaders, making it more difficult to appreciate the difference between them and the brand leaders.

These changing conditions are but a few of the factors that prompted us to thoroughly revise this book. In this new edition we have stayed true to the enduring objective of previous editions of this successful book, i.e. to clarify the concept of brands and to help students and managers appreciate issues that are germane to their growth.

Frequent interactions with managers in management development work-shops made it clear that they wanted grounded, pragmatic frameworks that enables them to better characterise their brands and develop sound strategies for growth. The use of this text on MBA, MSc and undergraduate pro-grammes showed that they needed to have a solid foundation in brand management, which gives them insights to powerful models. This book bridges the requirements of managers and students. Powerful branding theories are not just well grounded, but are also easily applied in a variety of situations. Our concern has therefore been to provide a solid grounding for models and frameworks, while at the same time indicating how readers can take advantage of these through examples and through the exercises that we have provided. Furthermore, we sought to make the ideas applicable across consumer, service and business markets, to both corporate and SME managers and to those operating in either the for profit or the not for profit sectors.

As part of our intent of thoroughly revising the book, we also wanted to bring in a new perspective from someone who is very well versed in brand management. We were delighted when Dr Elaine Wallace accepted our invitation to become a co-author. Elaine has been undertaking research into brand management with us over many years. Her experience as a Brand Manager and an academic at the National University of Ireland Galway are ideal for this book. Furthermore Elaine's insightful perspective and considerable breadth and depth of knowledge have significantly contributed to this new edition. We are also pleased that Elaine, from her Irish location, has helped bring a more international orientation.

How to Use this Book to Achieve the Best Results

Each chapter of this book has undergone significant revision. In addition to the exercises in the Marketing Action Checklists, Student Based Enquiry exercises are set enabling students to explore some of the topics in more detail and to engage them in the application of frameworks. A new selection of advertisements is used to bring the material to life, along with excerpts from websites. A unique enhancement of this book is the availability of two and a half hours of video produced by Oxford Learning Lab (http://bit.ly/oxlbrnd2). This video streaming provides more insights to some of the topics in this book.

We asked ourselves whether the structure of the text is still appropriate for the changing environment. Our conclusion is that we have a structure that is still sound. It logically moves from laying the foundations of brand management, which can then be applied in different sectors, enabling organisations to beat competitors in the branding battle. This enhanced book is structured in three parts that logically enables students and managers to appreciate the nature of brands and how to help grow and nurture these valuable assets. The constituent parts are:

PART ONE. FOUNDATIONS OF BRAND MANAGEMENT

The first two chapters address the core characteristics of brands and the factors that influence their growth.

Chapter 1 lays the foundations for this book, summarising the latest thinking and best practice in marketing and takes a fresh look at organisations' assets, which are represented by the brand.

Chapter 2 provides an overview of the key issues in planning for brand success by explaining the nature of brands, reviewing their evolution, identifying different types of brands and highlighting the forces that shape brands.

PART TWO. BRAND MANAGEMENT IN DIFFERENT SECTORS

The next six chapters explore the characteristics of effective brand management in diverse sectors.

Chapter 3 addresses the key question of how consumers choose brands and therefore how managers can influence brand choice.

Chapter 4 concentrates on the psychological and social aspects of consumer brands, exploring their symbolic nature, and investigating the importance of values, their expression through brand personality and enactment through relationships.

Chapter 5 focuses on business-to-business brands, appreciating how they are bought, the importance of value, brands as relationship builders, the role of emotion and the importance of corporate identity.

Chapter 6 discusses the importance and characteristics of service brands, addressing the numerous issues critical to building them, including the role of the service employee as brand ambassador.

Chapter 7 first explores the concept of the retail brand, store loyalty and brand image then it considers store brands and examines aspects such as shelf space allocation, power dynamics between retailers and manufacturers, customer loyalty and consumer response to store brands in specific product categories.

Chapter 8 considers how branding principles can be used to help grow brands on the internet, exploring the role of online communities and the opportunities arising from their having a greater involvement shaping brands.

PART THREE. WINNING THE BRANDS BATTLE

The remaining four chapters address techniques to win the branding battle and evaluate brand success.

Chapter 9 adopts a strategic perspective on positioning, identifying sources of sustainable competitive advantage to beat competitors through building, buying or extending brands and then exploring factors associated with brand success.

Chapter 10 explains the critical role of added values, suggesting ways of identifying these and ensuring their sustainability in challenging environments.

Chapter 11 considers planning issues that ensure consistent brand values over time through adopting a holistic perspective on brands.

Chapter 12 explores the concept of brand equity and its dimensions, looks at ways of valuing brands and then considers the contribution from employing brand scorecards.

GETTING THE MOST FROM THE ADDITIONAL TEACHING RESOURCES

This new edition of Creating Powerful Brands is complemented by a package of free additional teaching resources. These valuable resources contain a series of slides that summarise the key concepts introduced during each chapter. Lecturers can gain access to this material by registering at www.textbooks. elsevier.com.

The slides have been created to help lecturers who intend to use Creating Powerful Brands as their core teaching text. The materials enable lecturers to identify key areas of the text when preparing lectures whilst the annotated diagrams enhance the presentation of material. Each chapter concludes with 'Talking Points' which facilitate student engagement through the exploration, application and evaluation of key concepts each chapter introduces.

It is my hope these additional teaching resources will help with lecture preparation and enrich the student learning experience.

Dr. Darren Coleman, Managing Consultant, Wavelength Marketing.

Darren has worked at a number of blue chip organisations in brand, proposition and strategic marketing. As the Managing Consultant of Wavelength Marketing (www.wavelengthmarketing.co.uk) Darren now works with organisations that look to differentiate their brand through service and expect measurable marketing returns. Darren holds a brand marketing PhD. He also frequently presents at academic and practitioner conferences on brand and other marketing-related issues.

Darren can be contacted at darren@wavelengthmarketing.co.uk

Acknowledgements

Every effort has been made to locate the copyright owners of material used. The authors would like to thank all of the organizations who have granted permission to include material copyrighted to them in the book.

PART

1

Foundations of Brand Management

CHAPTER 1

Why it is Crucial to Create Powerful Brands

SUMMARY

This introductory chapter lays the foundations for the remaining chapters of this book. It summarises the latest thinking and best practice in the domain of marketing and takes a fresh look at the real nature of an organisation's assets, such as market share and supplier and customer relationships, all of which are represented by the brand. It also questions traditional thinking and practice in asset accounting and suggests alternative approaches designed to focus attention on the core purpose of this book — how to create powerful brands.

After almost a century of marketing, it is sad to note that as recently as 2005, the Harvard Business Review reported that of 30,000 new products launched in the USA, 90 percent of them failed because of poor marketing. The other 10 percent went on to become successful brands.

If this seems like a somewhat dramatic way to start a book on branding by three professional, experienced practitioners, researchers, teachers and writers, let us say at once that the reason is that, wherever we go in the world, chief executive officers immediately accost us with questions about either their corporate or product/service brands. Alas, we have to reign in their obvious enthusiasm for branding with some questions relating to the markets they serve, to the segments within those markets and to where they are positioned in these segments vis-à-vis their competitors. Only then can a sensible discussion take place about their brands.

Indeed, even a cursory glance at the work of gurus like Phillip Kotler, Tom Peters, the ex-chairman of Unilever Sir Michael Parry and the like, reveals a very broad agreement that the main components of world class marketing are:

1. A deep understanding of the market place
2. Correct needs-based segmentation and prioritisation
3. Segment-specific propositions
4. Powerful differentiation, positioning and branding
5. Effective strategic marketing planning processes
6. Long-term integrated marketing strategies

CONTENTS

Creating Powerful Brands. DOI: 10.1016/B978-1-85617-849-5.10001-7

7. A deep understanding of the needs of major customers
8. Market/customer-driven organisation structures
9. Professionally-qualified marketing people
10. Institutionalised creativity and innovation

The order is important and justifies our belief that it is impossible to discuss branding in isolation from the context to which it belongs, because everything an organisation does, from production through to eventual consumption, all adapt to and converge on the business value proposition that is offered to the customer. This value proposition has to have a name attached to it, so the brand name comes to represent everything a company does or strives to do. Hence the crucial importance of brands.

This chapter begins by reminding readers what marketing is and how it works, then goes on to spell out the growing importance of intangible assets and how success is measured. Finally, it describes what product management is and concludes by spelling out this difference between a brand and a successful brand.

CONFUSION ABOUT WHAT MARKETING IS

We all know that one of the stumbling blocks to those of us in marketing is the cacophony of definitions of marketing that exists. It doesn't help when one of CIM's ex Presidents, Diane Thompson declared: "Marketing isn't a function. It is an attitude of mind". Many will wonder how an attitude of mind can be measured, researched, developed, protected, examined, etc. Of course she was correct in one sense, because marketing as a function can never be effective in any organisation that does not put the customer at the core of its operations. Add to this the hundreds of different definitions of marketing to be found in books and papers on marketing and the confusion is complete. A selection of 30 such definitions are to be found in McDonald's 6th edition of Marketing Plans, most of which involve doing things to customers (2007).

Whilst definitions such as CIM's are admirable and correct, they provide little guidance on what should be included and excluded, with the result that they are difficult to use for a research exercise on what should be measured in marketing. Therefore, let us be unequivocal about marketing. Just like finance, or HR, or IT, it is a function, a specific business activity that fulfils a fundamental business purpose. The following describes marketing in terms of what it actually entails.

Marketing is a process for:

- Defining markets in terms of needs
- Quantifying the needs of the customer groups (segments) within these markets

- Putting together the value propositions to meet these needs and communicating these value propositions to all those people in the organisation responsible for delivering them
- Playing an appropriate part in delivering these value propositions (usually only communications)
- Monitoring the value actually delivered

For this process to be effective, organisations need to be consumer/customer-driven (McDonald M., 2007). This consolidated summary of the marketing process is shown diagrammatically in Fig. 1.1.

The map of the process in Fig. 1.1 works to simplify what is a complex process, into a series of manageable steps. It provides a practical framework for understanding and tackling the multitude of issues that comprise marketing, leading to sustainable competitive advantage; but in particular, it helps to determine the parameters of measurement and accountability.

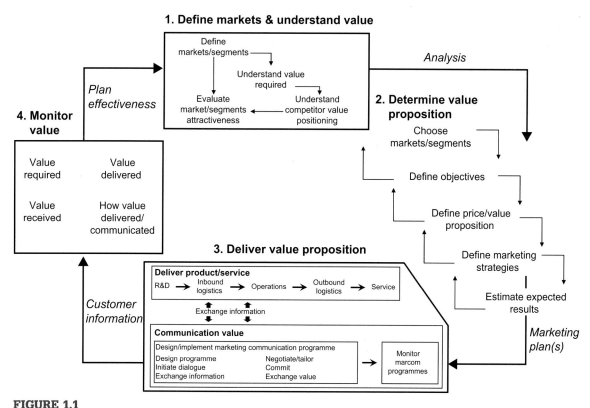

FIGURE 1.1

Summary of marketing map

Steps 1 and 2 are about strategy determination, while Steps 3 and 4 are about tactical implementation and measurement. It is these latter two that have come to represent marketing as a function that is still principally seen as sales support and promotion.

We have used the term 'Determine value proposition', to make plain that we are here referring to the decision-making process of deciding what the offering to the customer is to be — what value the customer will receive and what value (typically the purchase price and ongoing revenues) the organisation will receive in return. The process of delivering this value, such as by making and delivering a physical product or by delivering a service, is covered by 'Deliver value proposition'.

It is well known that not all of these marketing activities will be under the control of the marketing department, whose role varies considerably between organisations. The marketing department should be in charge of the first two sub processes, 'define markets and customer value' and 'determine value proposition', although even these need to involve numerous functions, albeit coordinated by specialist marketing personnel. However, the responsibility for delivering value is the shared domain of the whole company, requiring cross-functional expertise and collaboration. It will include, for example, product development, manufacturing, purchasing, sales promotion, direct mail, distribution, sales and customer service.

The marketing process is clearly cyclical, in that monitoring the value delivered will update the organisation's understanding of the value that is required by its customers. The cycle may be predominantly an annual one, with a marketing plan documenting the output from steps 1 and 2; but equally, changes throughout the year may involve fast iterations around the cycle to respond to particular opportunities or problems.

Choices may be influenced by physical assets and/or the less tangible but substantial value afforded by the organisation's people, brands, financial status and information technology.

The authors make a plea here that rather than arguing incessantly about a suitable definition of marketing, we at least take this one as a starting point for considering the role of brands and how powerful brands create success.

THE GROWING IMPORTANCE OF INTANGIBLE ASSETS

In 2006, Proctor and Gamble paid £31 billion for Gillette, of which only £4 billion was accounted for by tangible assets, as Table 1.1 shows.

Table 1.1 Intangibles

▪ Gillette brand	£ 4.0 billion
▪ Duracell brand	£ 2.5 billion
▪ Oral B	£ 2.0 billion
▪ Braun	£ 1.5 billion
▪ Retail and supplier network	£10.0 billion
▪ Gillette innovative capability	£ 7.0 billion
Total	£27.0 billion

(David Haigh, Brand Finance, Marketing Magazine, 1st April 2005)

Recent estimates of companies in the USA and in the UK show that over 80 percent of the value of companies resides in intangibles. Whilst this has reduced during the recession of 2008/9, intangibles will nonetheless remain a substantial percentage of corporate wealth. Table 1.2 and Fig. 1.2 show some of this research. Fig. 1.3 shows a typical breakdown of intangibles, whilst Table 1.1 above is an example of the breakdown of intangibles in a recent acquisition. Yet very little is known about intangibles by shareholders and the investment community. Traditional accounting methods are biased towards tangible assets, for this is where the wealth used to reside.

Generalising from this, what typically appears in a balance sheet can be seen from Fig. 1.4 below. However, when a predator bids for such a company, it is often forced to pay substantially more than the £100 million shown in this balance sheet.

Table 1.2 Some Market Definitions (Personal Market)

Invisible Business: Some Research Findings

•Brand Finance analysis of top 25 stock markets – $31.6 trillion (99% of global market value)

•62% of global market value is intangible – $19.5 trillion

•Technology is the most intangible sector (91%)

•The technology sector in the USA is 98% intangible

Source: Brand Finance, 2005

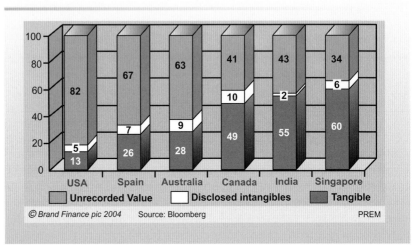

FIGURE 1.2
Asset split across selected economies

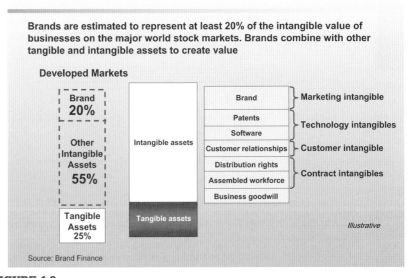

FIGURE 1.3
Brands are key intangibles in most businesses

In this hypothetical example shown in Fig. 1.5, it can be seen that in this case, it is £900 million—£800 million more than is shown in the balance sheet in Fig. 1.4.

The problem is that it leaves a balance sheet that doesn't balance, so this is corrected in Fig. 1.6, which shows a balancing figure of £800 million.

A critic of accounting procedures might be justified in pointing out that this £800 million entry is the mistake made by accountants in valuing this company and that it takes an acquisition (or the threat of an acquisition) to work out how big this mistake is.

Assets	Liabilities
- Land - Buildings - Plant - Vehicles etc.	- Shares - Loans - Overdrafts etc.
£100 million	£100 million

FIGURE 1.4
Balance sheet

Of course this is not true and in any case, the share price of a company is usually a good guide to its worth. There are also clear rules agreed internationally concerning how such intangibles should be recorded and treated, following an acquisition. But this isn't the point.

The point is that incongruously, most large companies have formally-constituted audit committees doing financial due diligence on major investments such as plant and machinery, using discounted cash flows, probability theory, real option analysis and the like; yet few have anything even remotely rigorous to evaluate the real value of the company's intangibles. There is a massive body of research over the past 50 years on how companies carry out strategic planning and much of it verifies that a lot of what passes for strategy amounts to little more than forecasting and budgeting, which are of little value to the investment community in estimating risk; with the result that it uses its own methods and frequently downgrades the capital value of shares, even when the earnings per share have been raised and when forecasts appear to look good.

As readers of this book will know, in capital markets, success is measured in terms of Shareholder Value Added (SVA), having taken account of the risks associated with

Assets	Liabilities
- Land - Buildings - Plant - Vehicles etc.	- Shares - Loans - Overdrafts etc.
£100 million	£900 million

FIGURE 1.5
Balance sheet

Assets	Liabilities
- Land - Buildings - Plant - Vehicles Goodwill £800m	- Shares - Loans - Overdrafts etc.
£900 million	£900 million

FIGURE 1.6
Balance sheet

future strategies, the time value of money and the cost of capital. The role of powerful brands in this assessment of risk is crucial.

First, however, there are some basic concepts relating to risk and return and stock markets all over the world that are best explained here. Fig. 1.7 shows a simple matrix encompassing financial risk and business risk. A combination of high business and financial risks can be fatal.

For example, although there were other factors at play, Sir Freddie Laker's airline in the 1970s involved a high financial gearing. He then chose to compete on the busy, high risk London/North Atlantic route, employing a low price strategy. His high financial gearing/breakeven model subsequently left him open to tactical low price promotions from more global, established airlines, such as British Airways. The result was financial disaster.

Compare this with Virgin's low financial risk entry in the same market, with a highly differentiated marketing strategy. Virgin is now an established and profitable international airline.

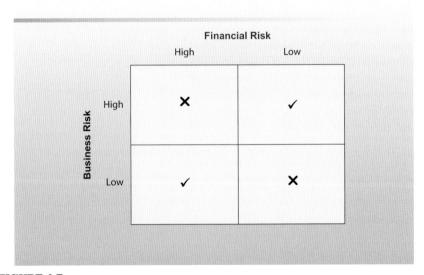

FIGURE 1.7

Fig. 1.8 shows a typical stock exchange, with shares plotted against return and risk. From this it can be seen that a Beta is drawn (the diagonal line).

At the low end, investors do not mind a lower return for a low risk investment, whilst at the high end, investors expect a high return for a high risk investment. At any point on the line (take the middle point for example), the point of intersection represents the minimum that any investor would be prepared to accept from an investment in this sector. This weighted average return on investment is referred to as the cost of capital. Any player in such a sector returning the weighted average cost of capital is neither creating nor destroying shareholder value—to return more is creating shareholder value; to return less is destroying shareholder value.

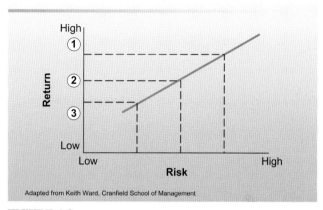

Adapted from Keith Ward, Cranfield School of Management

FIGURE 1.8
Financial risk and return

It is interesting to note, however, that the reason the capital value of shares is often marked down after a company has created shareholder value is that the investment community does not believe that such a performance is sustainable. This is often because they have observed that the source of profit growth has been cost cutting, which is, of course, finite; whereas customer value creation is infinite and is only limited by a company's creativity and imagination.

A good example of this is a major British retailer in the mid-90s, shown in Fig. 1.9, from which it can be seen that whilst underlying customer service was steadily declining, the share price was rising.

The inevitable almost terminal decline of this retailer was only reversed after a customer-oriented Chief Executive began to focus again on creating value for consumers, rather than boosting the share price by cost cutting. Shareholders, in the meantime, suffered almost a decade of poor returns.

It is, of course, not as simplistic as this and those readers who would like a more detailed explanation of the technical aspects of stock market risk and return, together with the relevant financial formulae, are directed to Chapter 3 of 'Marketing Due Diligence; reconnecting strategy to share price' (McDonald, Smith and Ward, 2007).

We have gone to so much trouble to explain how stock markets work in relation to risk and return because, as this book will make clear, the most established way to reduce risk is through powerful brands—which is why Proctor

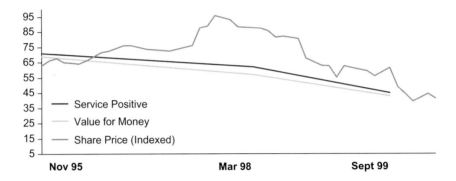

Base: Marks & Spencer Customers

FIGURE 1.9

A major UK retailer

and Gamble paid £27 billion pounds for the intangible assets of Gillette. Also, although brand names per se account for only around 30 percent of intangible assets, they are without doubt the most valuable and − according to Rita Clifton, Chairman of Interbrand (2009) − account for a quarter of all global corporate wealth.

Indeed, in the case of Proctor and Gamble's acquisition of Gillette (Table 1), it is clear that their excellent relationship with their channels is only possible because of the strength of their brand names. Equally, their innovation capability can only be effective if it has an outlet through powerful brands.

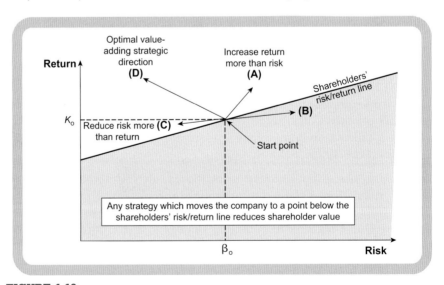

FIGURE 1.10

Shareholder value-adding strategies

Most marketing strategies are aimed at generating growth in sales revenues and profits but, for many mature products and markets, such strategies increase the risk profile of the business; indeed, the word 'growth' can normally be taken to indicate a risk-increasing strategy. This does not automatically mean that these strategies cannot be shareholder value-enhancing; but it does mean, as can be seen from directions A and B in Fig. 1.10, that the return from the more risky strategy must increase proportionately more than does the risk profile of the company. Direction B shows an increase in return less than the increase in risk, thus reducing shareholder value. Remember that merely moving along the shareholders' indifference line does not create shareholder value; this is achieved only by moving to a position above the line.

SHAREHOLDER VALUE-ADDING STRATEGIES

More interestingly, direction C in Fig. 1.10 highlights another type of shareholder value-enhancing strategy that is often ignored in marketing plans. A reduction from the current risk profile of the business (diagrammatically shown as a move to the left) means that shareholder value can be created even if the rate of return is reduced slightly. This time, the reduction in return must be proportionately less than the reduction in the risk profile. Since risk is associated with volatility in returns, this means that marketing strategies that make the future returns more stable and predictable can be shareholder value-enhancing, even if these less volatile future returns are slightly reduced. Thus, marketing strategies designed to increase customer loyalty through long-term discounts and so on can, if properly designed, be shareholder value-enhancing, even though the discounts given actually reduce profit levels.

Obviously, the optimal marketing strategy seeks to increase returns while reducing the associated risk levels (i.e., the volatility of these increased returns), see direction D of Fig. 1.3.

Any such strategy must leverage some already established, sustainable competitive advantages or first seek to develop a new, sustainable, competitive advantage, as the overall purpose and focus of strategic marketing is the identification and creation of such sustainable, competitive advantages.

A good example of this, was BMW's launch of the Mini—low-risk, because all the upmarket, hot hatches indicated that a market existed and low share risk because the proposition was highly differentiated and well positioned.

WHAT IS A PRODUCT?

Let us preface our introduction to the topic of brands by asking ourselves 'what is a product?' or 'what is a service?' The central role that the product plays in business management makes it such an important subject and mismanagement in this area is unlikely to be compensated by good management in other areas. Misunderstanding in relation to the nature of *product* management is also the root of whatever subsequent misunderstanding there is about *brand* management.

A product or a service is a problem solver

It should hardly be necessary to explain that a product or a service is a **problem solver**, in the sense that it solves the customer's problems and is also the means by which the organisation achieves its own objectives. And since it is what actually changes hands, it is clearly a subject of great importance.

The clue to what constitutes a product can be found in an examination of what it is that customers appear to buy. Theodore Levitt (1960) in what is perhaps one of the best-known articles on marketing, said that what customers want when they buy 1/4 inch drills is 1/4 inch holes. In other words the drill itself is only a means to an end. The lesson here for the drill manufacturer is that if they really believe their business is the manufacture of drills rather than, say, the manufacture of the means of making holes in materials, they are in grave danger of going out of business as soon as a better means of making holes is invented—such as, say, a pocket laser.

The important point about this somewhat simplistic example is that a company, which fails to think of its business in terms of customer benefits rather than in terms of physical products, is in danger of losing its competitive position in the market.

We can now begin to see that when customers buy a product, even if they are industrial buyers purchasing a piece of equipment for their company, they are still buying a particular bundle of benefits, which they perceive as satisfying their own particular needs and wants.

More serious real-world examples with disastrous consequences include Gestetner, who genuinely believed they were in the duplicator business, IBM, who thought they were in the mainframe business and Encyclopedia Britannica, whose business was badly affected by new information channels. Unless a company defines its products in terms of *needs,* not only will history repeat itself but any hope for successful branding will not be fulfilled.

Take the insurance industry, for example. A pension is clearly a problem solver yet many insurance companies continue to say that they are not in the pensions business.

In financial services, one very successful company uses the definitions as shown in Table 1.3.

Table 1.3

Market	Need (On-line)
Emergency cash ('rainy day')	Cash to cover an undesired and unexpected event, often the loss of/damage to property
Future event planning	Schemes to protect and grow money which are for anticipated and unanticipated cash calling events (e.g. car replacement/repairs, education, weddings, funerals, health care)
Asset purchase	Cash to buy assets they require (e.g. car purchase, house purchase, once-in-a-lifetime holiday)
Welfare contingency	The ability to maintain a desired standard of living (for self and/or dependants) in times of unplanned cessation of salary
Retirement income	The ability to maintain a desired standard of living (for self and/or dependants) once the salary cheques have eased
Wealth care and building	The care and growth of assets (with various risk levels and liquidity levels).
Day-to-day money management	Ability to store and readily access cash for day-to-day requirements

An international publisher until recently classified its business as books—in this particular case, Marketing book. Having drawn a market map of the market for Marketing books, the authors then challenged them on their market definition, i.e., "books". They eventually realised that they were in the knowledge promulgation business. The total difference between the market maps shown in Figs. 1.11 and 1.12 indicates how this transformed their entire thinking and strategy making.

What this different kind of logic and thinking leads to is a firm foundation, not only for brands but for the whole of marketing, in terms of measuring market share, market size, market growth, the listing of relevant competitors and the delineation of marketing strategies.

We can now begin to appreciate the danger of leaving product decisions entirely to engineers, actuaries, R&D people and the like. If we do, technicians will sometimes assume that the only point in product management is the actual technical performance or the functional features of the product itself. These ideas are incorporated in Fig. 1.13 and appear right at the very centre of the circle.

We can go even further than this and depict two outer circles as the 'product surround'. This product surround can account for as much as 80 percent of the added value and impact of a product or service. Often, these only account for about 20 percent of costs, whereas the reverse is often true of the inner circle.

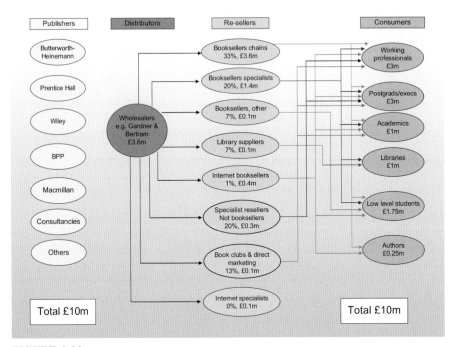

FIGURE 1.11
Original product-based market map

So far we have said little about service products, such as banking, consultancy, insurance and so on. This is because the marketing of services is not very different from the marketing of goods—the greater difference being that services cannot be stored. Thus, an airline seat, if not utilised at the time of the flight, is gone forever.

THE IMPORTANCE OF THE BRAND

It will be clear that here we are talking about not just a physical product but a *relationship* with the customer—a relationship that is personified either by the company's name or by the **brand name** on the product itself. IBM, BMW and Shell are excellent examples of company brand names. Persil, Coca-Cola, Fosters Lager, Dulux Paint and Castrol GTX are excellent examples of product brand names.

Most people are aware of the Coca-Cola/Pepsi-Cola blind taste tests, in which little difference was perceived when the colas were drunk 'blind'. On revealing the labels, however, 65 percent of consumers claimed to prefer Coca-Cola. This is one of the best indications of the value of what we have referred to as the 'product surround'. There can be little doubt that it is a major determinant of commercial

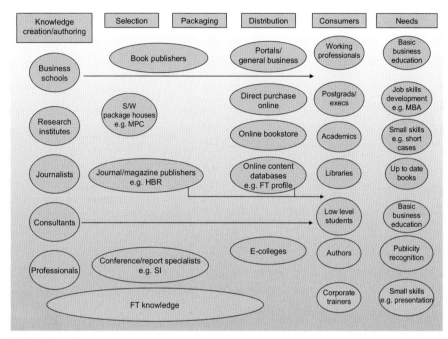

FIGURE 1.12
New needs-based market map

success. When one company buys another, as in the case of Nestle and Rowntree, it is abundantly clear that the purpose of the acquisition is not to buy the tangible assets which appear on the balance sheet (such as factories, plants, vehicles and so on) but the *brand names* owned by the company to be acquired.

This is because it is not factories which make profits but relationships with customers; and it is company and brand names which secure these relationships.

The example shown in Table 1.1 is Proctor and Gamble's purchase of Gillette for £31 billion, of which only £4 billion was for tangible assets. It is only one of thousands of examples of the value of brands and the associated link with strong customer relationships.

Philip Morris bought Kraft for £12.9 billion—four times the value of their tangible assets. Grand Metropolitan bought Pillsbury for £5.5 billion—a 50 percent premium on Pillsbury's pre-bid premium. RHM, taking its cue from this, more than trebled its asset value when it voluntarily incorporated its own brands into its balance sheet.

It is also a fact that whenever brand names are neglected, what is known as 'the commodity slide' begins. This is because the physical characteristics of products are becoming increasingly difficult to differentiate and easy to emulate. In

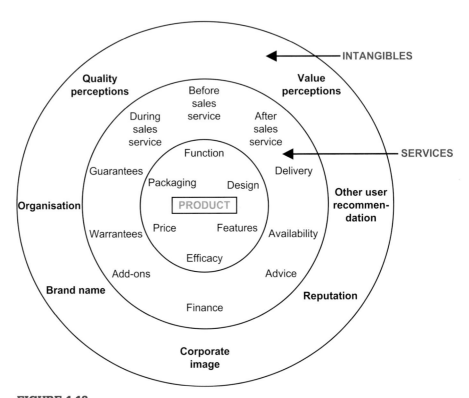

FIGURE 1.13
The components of a product

situations like these, one finds that purchasing decisions tend to be made on the basis of price or availability in the absence of strong brands.

Business history is replete with examples of strong brand names which have been allowed to decay through lack of attention, often because of a lack of both promotion and continuous product improvement programmes.

The fruit squash drink market is typical of this. The reverse can be seen in the case of Intel, which is a fantastic branding success story in a highly competitive global market.

Fig. 1.14 depicts the process of decay from brand to commodity as the distinctive values of the brand are eroded over time, with a consequent reduction in the ability to command a premium price.

The difference between a brand and a commodity can be summed up in the term 'added values', which are the additional attributes or intangibles that the consumer perceives as being embodied in the product. Thus, a product with a strong brand name is more than just the sum of its component parts. The

Coca-Cola example is only one of thousands of examples of the phenomenon.

Research has shown that perceived product quality, as explained above, is a major determinant of profitability.

It is the difference between successful and unsuccessful brands.

Successful brand building helps profitability by adding values that entice customers to buy. They also provide a firm base for expansion into product improvements, variants, added services, new countries and so on. They also protect companies against the growing power of intermediaries. Last but not least, they help transform organisations from being faceless bureaucracies to entities that are attractive to work for and deal with.

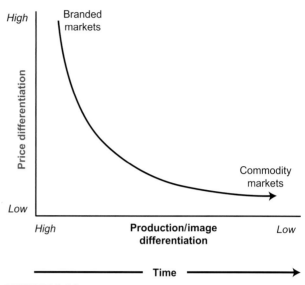

FIGURE 1.14

We must not, however, make the mistake of confusing successful and unsuccessful 'brands'. The world is full of products and services that have brand names but which are not successful brands. They fall down on other important criteria.

A successful brand has a name, symbol or design (or some combination of these) that identifies the 'product' of an organisation as having a sustainable, competitive advantage; for example, Coca-Cola, IBM, Tesco. A successful brand invariably results in superior profit and market performance (PIMS). Brands are only assets if they have sustainable, competitive advantage. Like other assets, brands depreciate without further investment; for example, Hoover, Singer, MG, Marks & Spencer and so on.

There are many 'products' that pretend to be brands but are not the genuine article. As the Director of Marketing at Tesco said, 'Pseudo brands are not brands. They are manufacturer's labels. They are "me-toos" and have poor positioning, poor quality and poor support. Such manufacturers no longer understand the consumer and see retailers solely as a channel for distribution' (reported in *Marketing Globe*, Vol. 2, No. 10, 1992).

Seen in this light, pseudo brands can never be mistaken for the real thing, because the genuine brand provides added brand values. Customers believe that the product:

- will be reliable
- is the best

- is something that will suit them better than product X
- is designed with them in mind.

These beliefs are based not only on perceptions of the brand itself relative to others, but also on customers' perceptions of the supplying company and their beliefs about its reputation and integrity.

The title 'successful brand' has to be earned. The company has to invest in everything it does, so that the product meets the physical needs of customers, as well as having an image to match their emotional needs. Thus, it must provide concrete and rational benefits that are sustained by a marketing mix that is compatible, believable and relevant.

By dint of considerable effort, IBM, despite its trials and tribulations over a decade ago, has succeeded in building a substantial world market share and that three lettered logo is still very powerful.

Often, the added values of brands are emotional values that customers might find difficult to articulate—for example, the prestige someone might feel in using their American Express Platinum card. These added values result from well thought through marketing strategies, which develop a distinctive position for the brand in the customers' mental map of the market. In commodity markets, competing brands, because they are undifferentiated, are seen by the customer as occupying virtually identical positions; and thus to all intents and purposes they are substitutable. The more distinctive a **brand** position with favourable attributes that the customer considers important, the less likelihood that a customer will accept a substitute.

Fig. 1.15 is a simplified version of Fig. 1.13, which illustrates that it is successful brands and the 'product surround' that create 80 percent of the market impact—hence the substantial premiums paid over tangible assets. It might be argued that if it is possible to value a company for sale, then surely it should be possible to do so on an ongoing basis and specifically to recognise the worth of marketing assets as represented by brands.

The question of asset protection and development is in a sense what marketing is all about. The 'stewardship' of marketing assets is a key responsibility which is recognised in many companies by, for example, the organisational concept of brand management. Here, an executive is given the responsibility for a brand or brands and acts as its 'champion', competing

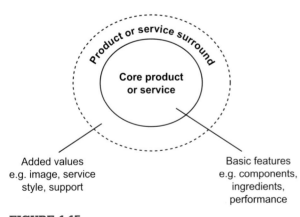

FIGURE 1.15

The importance of added values

internally for resources and externally for market position. It is but a short step from this organisational concept to a system of 'brand accounting', which would seek to identify the net present value of a brand, based upon the prospect of future cash inflows compared with outgoings; indeed a whole industry has grown round this concept in recent years, led by companies such as Brand Finance and Interbrand.

One advantage of such an approach is that it forces the manager to acknowledge that money spent on developing the market position of a brand, is in fact an investment which is made in order to generate future revenues. There is a strong argument for suggesting that for internal decision making and on questions of resource allocation, a 'shadow' set of management accounts be used; this is not the traditional approach whereby marketing costs are treated as expenditure in the period in which they are incurred— but an approach which recognises such expenditure as investments.

There is some interesting evidence from the IPA's analysis of almost 900 promotional campaigns, presented in a report (Binet. L, 2007). In one experimental scenario, the promotional budget was cut to zero for a year, then returned to normal; whilst in another, the budget was cut by 50 percent. Sales

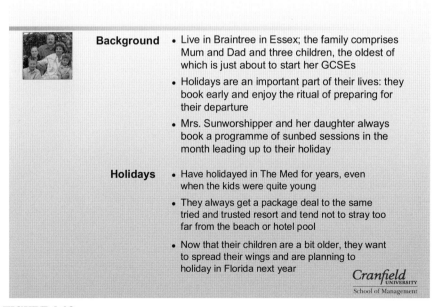

FIGURE 1.16
The Sunworshippers

recovery to pre-cut levels took five years and three years respectively, with cumulative negative impacts on net profits of £1.7 million and £0.8 million respectively.

Buying a major brand via acquisition nowadays often makes more sense to organisations than launching a new brand, with all the risk and uncertainty that this entails. This is just one of the reasons why brand valuation has emerged as a major issue in recent times and why brands are increasingly sought after as assets.

With the advent of the Internet, particularly in the light of online auctions, it is easier for customers to establish the lowest price of any product. However, the destabilising effect this has on prices affects commodities more than brands. Indeed, it is interesting to note that it is the major brand leading organisations who have embraced e-commerce, not just to reduce transaction costs, but also to create added values for customers. For example, Thomas Cook, the travel

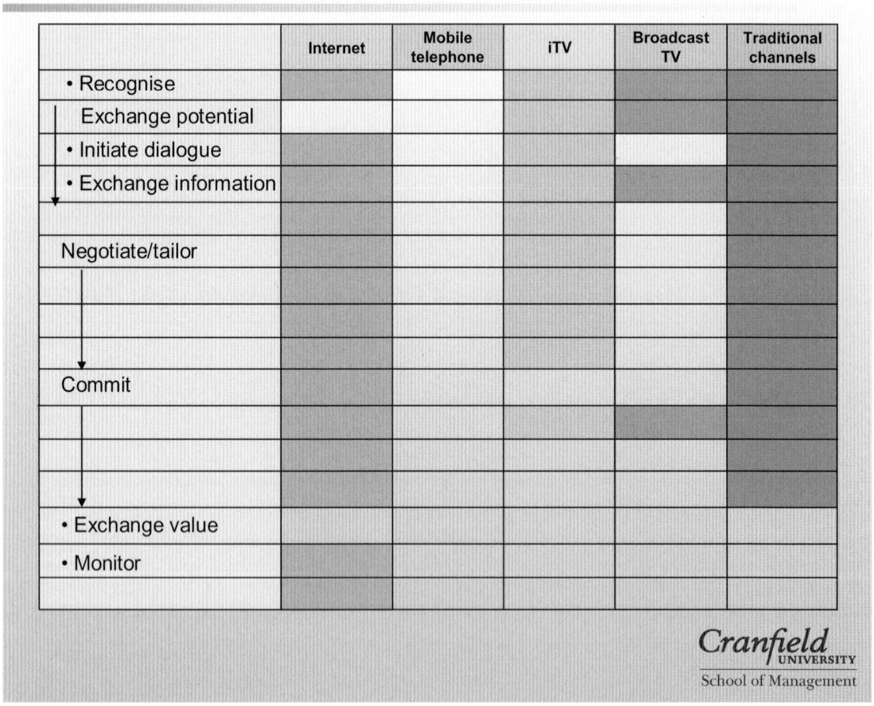

	Internet	Mobile telephone	iTV	Broadcast TV	Traditional channels
• Recognise					
Exchange potential					
• Initiate dialogue					
• Exchange information					
Negotiate/tailor					
Commit					
• Exchange value					
• Monitor					

Cranfield UNIVERSITY
School of Management

FIGURE 1.17
The Sunworshippers

specialists, have used the Internet to make it easier for customers to contact them, using whatever medium suits them best. This is a far cry from the traditional approach of viewing the customer as a passive entity to absorb whatever the supplier decides to do to them. Indeed, in the electronic age, given that customers today have as much information about suppliers as suppliers have about them, the most powerful brands will be customer-centric. Successful companies will know the customer and will be the customer's advocate.

The following four figures (Figs. 1.16, 1.17, 1.18 and 1.19) heavily disguised for the purpose of confidentiality, shows but two segments and demonstrates how the preferences for media and channels of each segment in relation to the booking of holidays, are totally different in each case.

So, it is wise to remember that e-commerce has given the customer speed, choice, control, and comparability. It is for this reason that branding has become more important than it has ever been. In the past, all marketers thought that the customer was part of a herd. Marketing was more about broadcasting than about building relationships.

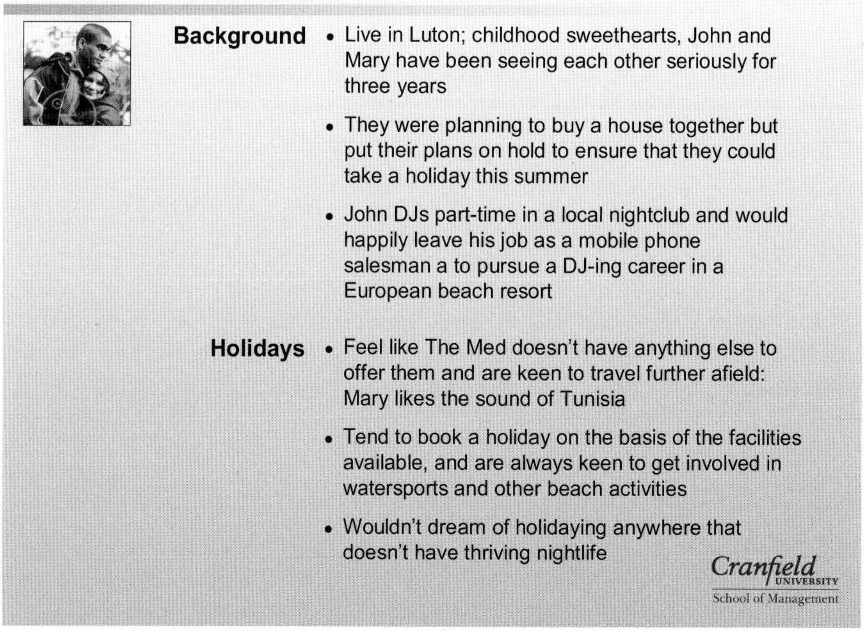

Background
- Live in Luton; childhood sweethearts, John and Mary have been seeing each other seriously for three years
- They were planning to buy a house together but put their plans on hold to ensure that they could take a holiday this summer
- John DJs part-time in a local nightclub and would happily leave his job as a mobile phone salesman a to pursue a DJ-ing career in a European beach resort

Holidays
- Feel like The Med doesn't have anything else to offer them and are keen to travel further afield: Mary likes the sound of Tunisia
- Tend to book a holiday on the basis of the facilities available, and are always keen to get involved in watersports and other beach activities
- Wouldn't dream of holidaying anywhere that doesn't have thriving nightlife

Cranfield
UNIVERSITY
School of Management

FIGURE 1.18
John and Mary Lively

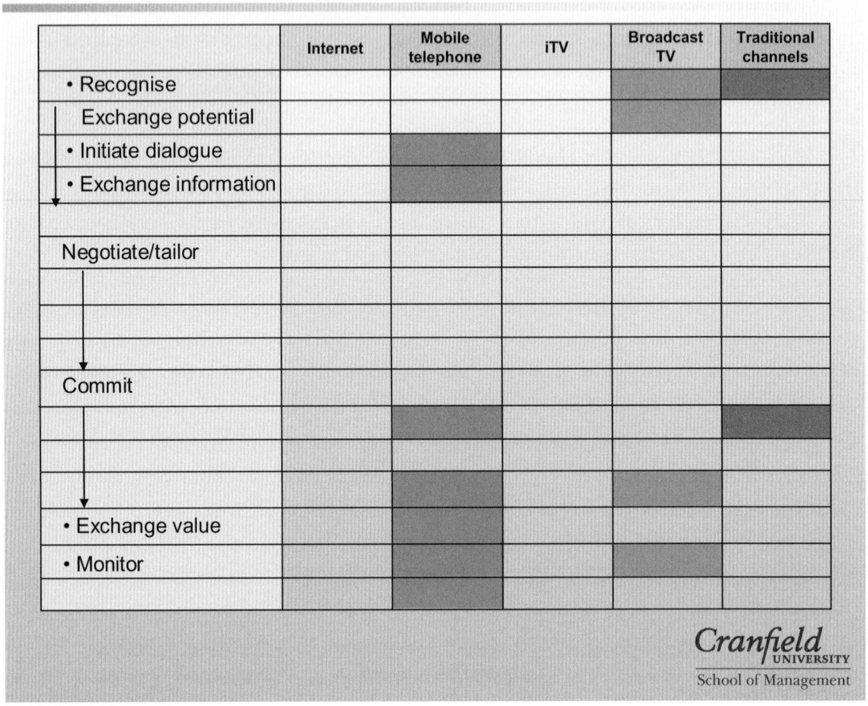

	Internet	Mobile telephone	iTV	Broadcast TV	Traditional channels
• Recognise					
Exchange potential					
• Initiate dialogue					
• Exchange information					
Negotiate/tailor					
Commit					
• Exchange value					
• Monitor					

Cranfield
UNIVERSITY
School of Management

FIGURE 1.19
John and Mary Lively

CONCLUSION

To summarise this section, a successful brand delivers sustainable competitive advantage and invariably results in superior profitability and market performance.

This introductory chapter began by asserting that the brand is the ultimate manifestation of a company's relationship with its market. This is represented in Fig. 1.20.

This is a model of the value-creating process. It shows a number of cross-functional management processes, two of which relate to understanding markets (the traditional marketing process) and to managing the relationship with markets (the traditional selling and service processes). The three remaining vertical lines represent other key processes involved in creating

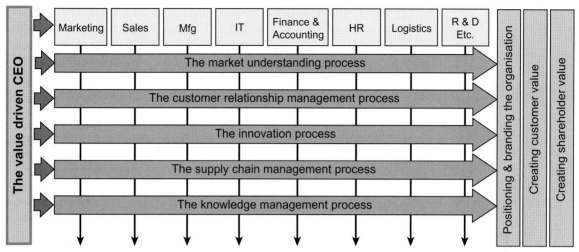

FIGURE 1.20

shareholder and stakeholder value. What is notable, however, in the ultimate creation of shareholder value, is the penultimate vertical box 'positioning and branding the organisation'; for it is this, above all else, that creates value from assets.

BUILDING SUCCESSFUL BRANDS

How brands encapsulate the value-creating capabilities of an organisation

We hope that we have, by now, been able to give some initial signals about the increasing importance of brands in business success. Later in this book, we refer to the PIMS database (Profit Impact of Market Strategies), which, along with other databases, show conclusively that strong, successful brands enable organisations to build stable, long-term demand and enable them to build and hold better margins than either commodities or unsuccessful brands.

Successful brand building helps profitability by adding value that entices customers to buy. They provide a firm base for expansion into product improvements, variants, added services, new countries and so on. They protect organisations against the growing power of intermediaries. And last, but not least, they help transform organisations from being faceless bureaucracies to ones that are attractive to work for and to deal with.

The following chapters of this book contain an in-depth treatment of aspects relevant to successful brand building. How to create powerful brands is a major challenge facing all organisations today. It is unlikely that this challenge will be met unless a more rigorous approach is taken to the issues surrounding branding.

We urge you to read on!

BOOK MODUS OPERANDI

Each of the following chapters covers a number of vital aspects of brand management and concludes with an action checklist. Finally, for the convenience of our readers, we have included a further reading list on the more important aspects covered in each chapter.

Further Reading

Binet, L. & Field, P. (2007) Marketing in an Era of Accountability. *IPA DataMINE*.

Christenson, C., Cook, S., & Hall, T. (December 2005). Marketing Malpractice — the Cause and the Cure. *Harvard Business Review*.

Clifton, R. (March 2009). Brand Valuation from Market to Boardroom. *Market Leader*.

Interbrand (2009). Best Global Brands List (online). Available at http://www.interbrand.com/best_global_brands.aspx (Accessed 21st July 2010).

Levitt, T. (July–August 1960). Marketing Myopia. *Harvard Business Review*, 45–56.

McDonald, M. (2007). McDonald on Marketing Planning: Understanding Marketing Plans and Strategy. London: Kogan Page.

McDonald, M., Smith, B., & Ward, K. (2006). Marketing Due Diligence: Reconnecting Strategy to Share Price. Oxford: Butterworth-Heinemann.

CHAPTER 2

Understanding the Branding Process

OBJECTIVES

After reading this chapter, you will be able to:

- Understand the meaning of brands and explain their strategic significance for manufacturers, distributors and consumers.
- Describe the characteristics that add value to brands.
- Trace the historical evolution of brands.
- Discuss the eight ways managers can emphasise aspects of brands.
- Recognise the issues which impede brand potential.

CONTENTS

SUMMARY

The purpose of this chapter is to provide an overview of the key issues involved in planning for brand success. It begins by explaining that successful branding is more than merely the use of names, then goes on to discuss the concept of the brand, brand characteristics and the role of brands in relationship marketing. A historical review of the evolution of brands, distributors' brands and generics is presented along with a consideration of ways of categorising brands. The value of brands to manufacturers, distributors and buyers is addressed along with the importance of brand planning and issues influencing the potential of a brand.

BRAND SUCCESS THROUGH INTEGRATING MARKETING RESOURCES

When BMW drivers proudly turn the ignition keys for the first time in 'the ultimate driving machine', they are not only benefiting from a highly engineered car with excellent performance, but also taking ownership of a symbol that signifies the core values of exclusivity, performance, quality and technical innovation. British Airways Executive Club customers who collect air miles are not just availing themselves of membership benefits such as executive lounge

27

Creating Powerful Brands. DOI: 10.1016/B978-1-85617-849-5.10002-9

access, or the option to upgrade their flights. They are also repeatedly using the service to feel that they belong to an exclusive club of successful travelers. Likewise a managing director outsourcing the IT function of a company chooses a partner not just because of the large resources and the expertise offered. He or she is also buying a name that stands for commitment to service excellence, global recognition and sustained credibility.

While these purchasers in the consumer, service and industrial markets have bought solutions for their individual problems, they have also paid a price premium for the added values provided by buying brands. In addition to satisfying their core purchase requirements, they have bought an augmented solution to their problem, for which they perceive sufficient added value to warrant paying a premium over other alternatives that might have satisfied their buying needs.

The added values they sought, however, were not just those provided through the presence of a brand name as a differentiating device, nor through the use of brand names to recall powerful advertising. Instead, they perceived a total entity, the **brand**, which is the result of a coherent organisational and marketing approach that uses all elements of the marketing mix. A man does not give a woman a box of branded chocolates because she is hungry. Instead, he selects a brand that communicates something about his relationship with her. This, he hopes, will be recognised through the pack design, her recall of a relevant advertising message, the quality of the contents, her chiding of him for the price he paid and her appreciation of the effort he took to find a retailer specialising in stocking such an exclusive brand. The same goes for a woman buying a man a special bottle of wine.

These examples show that thinking of branding as 'being to do with naming products', or 'about getting the right promotion with the name prominently displayed', or 'getting the design right', is too myopic. In the 1990s when VW purchased Skoda, this brand had to compete on price as it was perceived as being akin to a poor quality commodity. Following significant investment in Skoda in areas such as product development, production and marketing, managers worked hard to develop a more favourable identity for this brand. An integrated approach to building the brand resulted in its being able to charge a higher price. One TV advertising campaign majored on a Skoda production worker even wondering whether he should put the Skoda badge on the newly produced cars!

Thus, branding is a powerful marketing concept that does not just focus on one element of the marketing mix, but represents the result of a carefully conceived array of activities across the whole spectrum of the marketing mix, directed towards making the buyer recognise relevant added values that are unique when compared with competing products and services and which are difficult for competitors to emulate. One purpose of branding is to facilitate the organisation's task of getting and maintaining a loyal customer base in a cost-effective manner to achieve the highest possible return on investment. In other words, branding should not be regarded as a tactical tool directed towards one element of the marketing mix, but rather should be seen as the result of strategic thinking, integrating a marketing programme across the complete marketing mix.

Neither is this a concept that should be regarded as appropriate primarily for consumer markets. As this book will explore, it is of value in consumer and business-to-business markets, in goods and services sectors, in the for profit as well as the not for profit sectors and in the real and virtual worlds. Indeed, the concept of branding is increasingly being applied to people and places, such as politicians, pop stars, cities and the like, whilst it has always been equally relevant to the marketing of products and services. Strategic branding is concerned with identifying a vision for a brand and evaluating how to achieve the highest return on investment from brands, through analysing, formulating and implementing a strategy that best satisfies stakeholders.

THE CONCEPT OF THE BRAND

Successful brands, that is, those which are the focus of a coherent blending of marketing resources, represent valuable marketing assets. During the 1980s, the value of brands was brought to the attention of marketers by the financial community. For example, in 1985 Reckitt & Colman acquired Airwick Industries and put on its balance sheet £127 m as the financial value resulting from the intangible benefits of goodwill, heritage and loyalty conveyed by the newly acquired brands. While this may have been one of the opening shots to make organisations aware of the financial value of brands, it was Rank Hovis McDougal who really brought the brand debate to life. They announced in 1988 that they had put £678 m on their balance sheet as the valuation of their brand names. In the same year Jacobs Suchard and Nestlé fought for the ownership of Rowntree. At the time of the takeover battle it was estimated that Rowntree's tangible net assets were worth around £300 million, yet Nestlé won control by paying £2.5 billion. This difference of £2.2 billion represented the value that Nestlé saw in the potential earnings of strong brands such as KitKat, Polo, Quality Street and After Eight Mints!

A 2009 analysis by Interbrand calculated the value of the world's top brands. As Table 2.1 shows, the value of these assets are considerable.

Successful brands are valuable because they guarantee future income streams. It is notable that new online service brands such as Google are replacing traditional product brands in the Top 10. Companies recognise that loyal customers will repeatedly buy their brands, trust their brands, and are also willing to support them during crises, for example when people maliciously tamper with

Table 2.1 The World's Most Valuable Brands in 2009	
Brand	**Value ($bn)**
Coca Cola	68.73
IBM	60.21
Microsoft	56.65
GE	47.78
Nokia	34.86
McDonalds	32.28
Google	31.98
Toyota	31.33
Intel	30.64
Disney	28.45
Source: Interbrand	

brands. In such instances the rapid response of management and their commitment to communicating developments to the brands' stakeholders helps to rapidly restore normality. Such actions, following the tampering with Tylenol tablets enabled the management of Johnson & Johnson to quickly regain public confidence. The ultimate assessor of the value of a brand, however, is not the manufacturer or the distributor, but the buyer or the user. Marketers are able to develop strategies to communicate added values to purchasers, but because of the 'perceptual process', the target audience may focus on only a part of the available information and 'twist' some of the messages to make them congruent with their prior beliefs. For example, should a wallpaper paste manufacturer show an apparently incompetent DIY house-holder mixing paste in a television commercial in an attempt to communicate the smoothness and ease of application of their brand of wallpaper paste, they run the risk of some consumers interpreting the brand as being more 'suitable for idiots'. This is one example of the perceptual process.

People interpret messages and images through their own perceptions, often with very different results.

It is imperative to recognise that while marketers instigate the branding process (branding as an *input*), it is the buyer or the user who forms a mental picture of the brand (branding as an *output*), which may be different from the intended marketing thrust. While marketers talk about the branding effort they are undertaking, they should never lose sight of the fact that the final form of the brand is the mental evaluation held by the purchasers or users. Branding, then, needs to be appreciated in terms of both the input and the output process. A classic example of this is shown in an Institute of Grocery Distribution study. This research rated the factors which retailers and manufacturers believe linked

Table 2.2 Criteria Necessary if Brands are to Succeed		
Criteria	**Retailers (%)**	**Manufacturers (%)**
Understanding retailer objectives	20	1
Product innovation	13	16
Consumer understanding	13	13
Long-term development strategy	11	2
Strong brand image	10	8
Quality	9	–
Category management	6	10
Price	6	13
Brand distinctiveness	2	9
Supply chain linkages	1	10
Open relations	1	9

(Source: Institute of Grocery Distribution 1994)

the success of brands. Table 2.2 lists the criteria retailers use to stock new brands and manufacturers use to produce them. Major mismatches on criteria ratings, such as 'understanding retailer objectives' and 'price' explain one of the reasons for the lack of brand success.

Drawing on the points discussed so far, we can better clarify the term 'brand' through our definition:

> *A brand is a cluster of functional and emotional values that enables organizations to make a promise about a unique and welcomed experience.*

Our definition recognises that brands are about building on their value to create value and promising a unique and welcomed experience for the customer. Brands deliver a variety of benefits, which for ease can be classified as satisfying buyers' rational and emotional needs. They do this through their functional and emotional values, which enable a welcomed and unique promise to be made. For example, a teenager uses Bebo not only to share messages and photographs with friends, but also to bond with their peer group and highlight their popularity. The extent to which the brand satisfies relational needs will be assessed by the consumer trying different brands, examining the packaging, looking at functionality, considering price, etc. Besides these rational needs, they will also be seeking to satisfy emotional needs such as prestige, reassurance, style or distinctiveness.

The extent to which brands satisfy emotional needs will be evaluated by customers recalling promotions, or considering what context the brand is used in, or due to their prior experience of the brand.

CAESAR

Caesar appeals to the emotional needs of the consumer who wishes to provide the best for their pet

Brands succeed because they add to customers' experiences. In brand advertising, consider how Coca-Cola describes itself as 'Open Happiness' and how Persil emphasises childhood fun by claiming 'Dirt is Good'. Brands can create a phenomenon when their messages capture the public's imagination. The rise in user generated content on brand websites such as Nokia's collaborations with Spike Lee, or the use of YouTube for viral marketing campaigns such as Cadburys' gorilla have also enhanced customer experiences of brands. Another example of online campaigning is the Burger King 'Whopper Sacrifice' promotion. When Burger King offered a free Whopper to anyone who 'sacrificed' 10 of their Facebook friends, over 230,000 online friendships were sacrificed, showing customers' allegiance to the brand. Consider also how the design of Apple's new flagship stores or the wording on the label of an Innocent smoothie adds to the customer's impression of the brand experience.

In our definition, companies make a promise about their brands. Brands succeed because companies meet the promises made to their customers. For example, poorly trained service staff may take away from the strong brand messages in an expensive advertising campaign. The promise should be authentic. Customers have become more cynical about the brand messages they receive and the digital world means that there are fewer barriers to providing feedback. Customers have a choice of weapons for negative advocacy, including blogs, forums, Facebook and Twitter. For example, the top most mentioned brand on Twitter was recently named as Starbucks, with over 120,000 followers. There is clearly vast potential for negative word of mouth online. Companies need to make sure that their brands deliver on their promises, to encourage customers to act as positive advocates.

If we extend our view of brand as experience, brands can also be used as a force for good, to contribute for the benefit of society. Shields provides examples such as Dove's 'Campaign for Real Beauty', which challenges consumers to think beyond beauty stereotypes. Through the use of online media, companies can speedily deliver their message to a wide audience and capture the public's imagination. Consider Dove's 'Evolution' video on YouTube, which has achieved over 9 million views. Brands that capture the social consciousness can eventually become badges of consumer ideals. For example, if someone drives a Toyota Prius or wears a Live Strong wristband, those brand choices can make statements about the type of people they are.

Some may question whether the rational dimension dominates business-to-business branding and therefore whether there is any need to consider emotional aspects at all. Our work has shown that emotion plays an important role in the industrial brand selection process. For example, some managers do not just consider the rational aspects of an IT brand when they intend to install a new software system. They also seek emotional reassurance that the correct brand decision might reaffirm their continual career development or that they have not lost credibility among colleagues through the wrong brand choice.

CHARACTERISTICS OF BRANDS

Our definition of a brand adheres to a model that shows the extent to which a product or service can be augmented to provide added value to increasing levels of sophistication. This model, which is expanded on in Chapter 10, views a brand as consisting of four levels:

- generic;
- expected;
- augmented;
- potential.

The **generic** level is the commodity form that meets the buyer's or user's basic needs, for example, the car satisfies a transportation need. This is the easiest aspect for competitors to copy and consequently successful brands have added values over and above this at the **expected** level.

Within the **expected** level, the commodity is value engineered to satisfy a specific target's minimum purchase conditions, such as functional capabilities, availability, pricing, etc. As more buyers enter the market and as repeat buying occurs, the brand would evolve through a better matching of resources to meet customers' needs (e.g. enhanced customer service).

With increased experience, buyers and users become more sophisticated, so the brand would need to be **augmented** in more refined ways, with added values satisfying non-functional (e.g. emotional) as well as functional needs. For example, promotions might be directed to the user's peer group to reinforce his or her social standing through ownership of the brand.

With even more experience of the brand, and therefore with a greater tendency to be more critical, it is only creativity that limits the extent to which the brand can mature to the **potential** level. For example, grocery retail buyers once regarded the Nestlé range of confectionery brands as having reached the zenith of the **augmented** stage. To counter the threat of their brands slipping back to the **expected** brand level, and therefore having to fight on price, Nestlé shifted their brands to the **potential** level by developing software for retailers to manage confectionery shelf space to maximise profitability.

Experienced consumers recognise that competing items are often similar in terms of product formulation. As such, brand owners are no longer focusing only on rational functional issues, but are addressing the **potential** level of brands.

The evolution of brands is well exemplified by computers. For example, 20 years ago computer brands were competing predominantly at a functional level, and purchased mainly by business-to-business customers. Today, consumers are being attracted through different branding approaches, such as design-based emotional appeals. Likewise, additional functionality such as gaming and entertainment packages ensure computer brands become part of consumers' lifestyles.

To succeed in the long run, a brand must offer added values over and above the basic product characteristics, if for no other reason than that functional characteristics are so easy for competitors to copy. In the services sector, when all other factors are equal, this could be as simple as being addressed by name when getting foreign exchange at a bank. In the business-to-business market, it could be conveyed by the astute sales engineer presenting the brand as a no-risk purchase (due to the thoroughness of testing, the credibility of the organisation, compliance with industry standards, case histories of other users, etc.). It is most important to realise that the added values must be relevant to the customer and not just to the manufacturer or distributor. Car manufacturers who announced that their brands had the added value of electronic circuits emitting 'computer speak' when seat belts were not worn didn't take long to discover that this so-called benefit was intensely disliked by customers. A hotel promoting a quality service will win loyal customers by offering a special late night food menu rather than simply placing promotional literature around guest rooms.

Buyers perceive added value in a brand because they recognise certain clues that give signals about the offer. In business-to-business markets, for example, buyers evaluate brands on a wide variety of attributes, rather than just on price. As a consequence, price is rarely the most important variable influencing the purchase decision. So it is not unusual for a buyer to remain loyal to a supplier during a period of price rises. However, if the price of a brand rises and one of the signaling clues is weak (say, poor customer service) compared with the other signaling clues (say, product quality), the buyer may perceive that the brand's value has diminished and will therefore be more likely to consider competitive brands. One of the best examples of this is 'Marlborough Friday'. The cigarette war instigated by Liggett made the price of Marlborough extremely high by comparison, so they reduced their price whilst maintaining a premium. In the short term, the share price fell, but it wasn't long before they recovered and Marlborough is once again an outstanding example of a successful global brand.

If brands are to thrive, their marketing support will have been geared towards providing the user with the maximum satisfaction in a particular context. Buyers often use brands as non-verbal clues to communicate with their peer groups. In other words, it is recognised that people do not use brands only for their functional capabilities, but also for their badge or symbolic value. A bride may take pride in her choice of Jimmy Choo wedding shoes, not just because of their design quality, but because they add to her sense of glamour and exclusivity on her special day. People take care over their selection of clothes as, according to the situation, their brand of clothes is being used to signal messages of propriety, status or even seduction. Buyers choose brands with which they feel both physical and psychological comfort in specific situations. They are concerned about selecting brands that reinforce their own concept of themselves in specific situations. A very self-conscious young man may drink a particular brand of lager with his peer group because he believes it conveys an aspect of his lifestyle, whilst at home alone his brand consumption behaviour may be different, since he is less concerned about the situational context. A further example is Converse, with consumers owning different colours or styles according to particular trends in their peer groups. Where marketers have grounds for believing their brands are being used by consumers as value-expressive devices, they need to be attuned to the interaction of the marketing mix with the user's environment and provide the appropriate support. In some instances this may involve targeting promotional activity to the users' peer group, to ensure that they recognise the symbolic messages being portrayed by the brand.

Whilst this issue of appreciating the buyer's or user's environment relates to consumer markets, it is also evident in business-to-business markets. One researcher found that in a laboratory with a high proportion of well-educated scientists, there was a marked preference for a piece of scientific equipment that had a 'designer label' cabinet, over the same equipment presented in a more utilitarian manner. In a highly rational environment, scientists were partly influenced by a desire to select a brand of equipment which they felt better expressed their own concept of themselves.

Finally, our definition of a brand adopts a strategic perspective. Unless the values and experience received by the customer are unique and sustainable against competitive activity, the lifetime will be very short. Without such a strategic perspective, then it is questionable whether it is viable to follow a branding route. Chapter 9 reviews the way that the concept of the value chain can be used to identify where in the value adding process the firm's brand has a unique advantage over competition and also considers whether any of these strengths can be rapidly copied. For example, in the chocolate market Green and Black have a competitive advantage for their brand through their use of naturally sourced premium organic cocoa. Dyson's competitive advantage is in excellence of product design and Federal Express in their distribution planning and monitoring systems. It is our contention that unless brand instigators have a sustainable differential advantage, they should seriously consider the economics of following a manufacturer's brand route and consider becoming a supplier of a distributor's brand. In such situations it is more probable that the firm will follow a more profitable route by becoming a distributor brand supplier (i.e., a supplier of own label products) as discussed in Chapter 7.

BRANDS AS RELATIONSHIP BUILDERS

Relationship marketing goes beyond traditional marketing, and focuses more on creating a pool of committed, profitable customers. This is done by identifying a company's individual customers and creating mutually beneficial relationships that go beyond simple transactions.

Relationship building is neither easy nor straightforward. For example, business travelers on one flight became horrified when seeing, in mid-flight, the pilot and co-pilot locked outside their cockpit using an axe to force open the cockpit door. Adding extra miles on their loyalty cards no doubt was seen by some as a rather insensitive compensation for the experience.

Brands can develop different relationships with customers. For example, Hovis plays upon a nostalgic relationship, while Nintendo Wii has a relationship based on interactive fun with friends and family. A successful brand aims to develop a high-quality relationship in which customers feel a sense of commitment and belonging, even to the point almost of passion, as with the Manchester United brand, where couples have had part of their wedding ceremonies on the turf.

Relationship marketing aims to develop long-term loyal customers. The retention can be managed on three levels: through financial incentives, through social/financial bonds and through structural bonds. Financial incentives, such as frequent flyer schemes and the Boots Advantage Card, are very popular and can quickly be copied by competitors. Social and financial bonds see customers as clients, not numbers. They are quite common among professional services, such as lawyers taking clients to the opera and dentists making brief notes about their clients' social lives, enabling them to take an interest in subsequent dental visits. One Irish Supermarket achieves high sales through social bonds. Managers get to know customers by name, and when there is a minor customer service issue, their policy is to phone customers back within the same day to resolve it. One store manager spoke of a customer who requested Evian water. As it was not on shelf, he went into the store, located a case of Evian, and carried it to her car with his compliments. Structural bonds stretch social and financial ones to make clients more productive. For example Federal Express and UPS provide free PCs with logistics software for labeling and tracking. Baxter Healthcare tailors pallets for each hospital to fit individual storage and dispensing needs. Many major retailers link their computer systems into their suppliers' for fast and error-free ordering and invoicing.

SWIMMINGSUMMER

This summer it's all about colour. Look around, the beach is the canvas and you are the paint. So choose amongst our selection of beachwear, combine it with our suggestion of accessories and get ready to make the beach look prettier!

Colours, flowers, patterns or all kinds are all over this summer's fashion. When you pack for the holidays, keep that in mind. We also suggest some beachwear accessories for packing. As a rule, your beach bag or your perfect sarong should be a rectangular shape, around six feet long and three feet wide. You can choose from cotton or silk, but, mind you, it's the beach, so cotton is probably the best choice and easier on your pink account. Your beach bag can be anything, as long as it fits your bag and you can apply large amounts of sunscreen to protect your skin and feet. Make sure you take a good pair of flip flops. We love flip flops, and you can find similar fashion in any high street retailer.

You'll be mine in 2 weeks!

Create your free personalised plan online and see if you can get slimmer for summer.

Kellogg's Special K

SPECIAL K

Special K cereal uses a simple visual association of the swimsuit to position it as a healthy option breakfast cereal for the consumer. (Reproduced by kind permission of Special K)

HISTORICAL EVOLUTION OF BRANDS

Having clarified the concept of the brand, it is worth appreciating how brands evolved. This historical review shows how different aspects of branding were emphasised.

There were examples of brands being used in Greek and Roman times. With a high level of illiteracy, shop keepers hung pictures above their shops indicating the types of goods they sold. Symbols were developed to provide an indication of the retailer's specialty and thus the brand logo as a **shorthand device** signaling the brand's capability was born. Use is still made of this aspect of branding, as in the case, for example, of the Ralph Lauren polo logo which features the polo player, and indicates the exclusivity of the brand.

In the Middle Ages, craftsmen with specialist skills began to stamp their marks on their goods and trademarks. **Differentiating** between suppliers became more common. In these early days, branding gradually became a **guarantee** of the source of the product and ultimately its use as a form of **legal protection** against copying grew. Today, trademarks include words (e.g. Ray-Ban), symbols (e.g. the McDonalds logo) or a unique pack shape (e.g. the Pringles tall tub), which have been registered and which purchasers recognise as being unique to a particular brand. However, as the well-travelled reader is no doubt aware, trademark infringement is a source of concern to owners of well-respected brands such as Prada or Rolex.

The next landmark in the evolution of brands was associated with the growth of cattle farming in the New World of North America. Cattle owners wanted to make it clear to other potentially interested parties which animals they owned. By using a red hot iron, with a uniquely shaped end, they left a clear imprint on the skin of each of their animals. This process appears to have been taken by many as the basis for the meaning of the term brand, defined by the *Oxford English Dictionary* as 'to mark indelibly as proof of ownership, as a sign of quality, or for any other purpose'. This view of the purpose of brands as being identifying (**differentiating**) devices is important. However, in an enlightened era aware of the much broader strategic interpretation of brands, many managers still adhere to the brand primarily as a differentiating device. Towards the end of the nineteenth century, such a view was justified, as the next few paragraphs clarify. However, as the opening sections of this chapter have explained, to regard brands as little more than differentiating devices is to run the risk of the rapid demise of the product or service in question.

To appreciate why so much emphasis was placed on brands as differentiating devices, it is necessary to consider the evolving retailing environment, particularly that relating to groceries, where classical brand management developed.

In the first half of the nineteenth century, people bought their goods through four channels:

1. retailers;
2. from those who grew and sold their own produce;
3. from markets where farmers displayed produce;
4. from travelling salesmen.

Household groceries were normally produced by small manufacturers supplying a locally confined market. Consequently the quality of similar products varied according to retailer, who in many instances blended several suppliers' produce. With the advent of the Industrial Revolution, several factors influenced the manufacturer–retailer relationship, i.e.:

- the rapid rise of urban growth, reducing manufacturer–consumer contact;
- the widening of markets through improved transportation;
- the increasing number of retail outlets;
- the wider range of products held by retailers;
- increasing demand.

A consequence of this was that manufacturers' production increased, but with their increasing separation from consumers, they came to rely more on wholesalers. Likewise, retailers' dependence on wholesalers increased, from whom they expected greater services. Until the end of the nineteenth century, the situation was one of wholesaler dominance. Manufacturers produced according to wholesalers' stipulations, who, in turn, were able to dictate terms and strongly influence the product range of the retailer. As an indication of the importance of wholesalers, it is estimated that by 1900 wholesalers were the main suppliers of the independent retailers, who accounted for about 90 percent of all retail sales.

During this stage, most manufacturers were:

- selling unbranded goods;
- having to meet wholesalers' demands for low prices;
- spending minimal amounts on advertising;
- selling direct to wholesalers, while having little contact with retailers.

In this situation of competitive tender, the manufacturer's profit depended mostly on sheer production efficiency. It was virtually commodity marketing, with little scope for increasing margins by developing and launching new products.

The growing levels of consumer demand and the increasing rate of techno-logical development were regarded by manufacturers as attractive opportuni-ties for profitable growth through investing in large-scale production facilities.

Such action, though, would lead to the production of goods in *anticipation* of, rather than as a *response* to, demand. With large investments in new production facilities, manufacturers were concerned about their reliance on wholesalers. To protect their investment, patents were registered and brand names affixed by the owners. The power of the wholesalers was also bypassed by advertising brands direct to consumers. The role of advertising in this era was to stabilise demand, ensuring predictable large-scale production was protected from the whims of wholesalers. In such a situation, the advertising tended to focus on promoting awareness of reliability and guaranteeing that goods with brand names were of a consistent quality. The third way that manufacturers invested in protecting the growth of their brands was through appointing their own salesmen to deal directly with retailers.

By the second half of the nineteenth century, many major manufacturers had embarked on branding, advertising and using a sales force to reduce the dominance of wholesalers. In fact, by 1900 the balance of power had swung to the manufacturer, with whom it remained until the 1960s. With branding and national marketing, manufacturers strove to increase the consistency and quality of their brands, making them more recognisable through attractive packaging that no longer served the sole purpose of protection. Increased advertising was used to promote the growth of brands and with manufacturers exercising legally backed control over prices, more and more manufacturers turned to marketing branded goods.

This changing of the balance of power from wholesaler to manufacturer by the end of the last century marked another milestone in the evolutionary period of brands. Brand owners were concerned with using their brands as legal registrations of their unique characteristics. Besides this, they directed their efforts towards consumers to make them aware that their brand was different in some way from those of competitors. Furthermore, they wanted their brand names to encourage belief in a consistent quality level that most were prepared to guarantee. Thus, whilst the **differentiating** aspect of the concept was initially regarded as the key issue, this soon also encompassed **legal protection** and **functional communication**.

Manufacturers' interest in branding increased and with more sophisticated buyers and marketers, brands also acquired an **emotional dimension** that reflected buyers' moods, personalities and the messages they wished to convey to others. However, with greater choice through the availability of more competing products, the level of information being directed at buyers far exceeded their ability to be attentive to the many competing messages. Because of their limited cognitive capabilities, buyers began to use brand names as **shorthand devices** to recall either their brand experiences or marketing claims and thus saved themselves the effort of having continuously to seek information. Chapter 3

SINK INTO FAIRY CLEAN & CARE

FAIRY LIQUID

Fairy liquid create a glamorous associate with dishwashing to reinforce their message of 'kind to hands'.
(Reproduced by kind permission of Fairy)

provides more information about the role of the brand as a shorthand device to facilitate buyers' decision processes.

The internet is swinging the balance of power into the hands of consumers, as Chapter 8 explores. Consumers are becoming more empowered to decide what degree of information search they wish to undertake and are co-creating brands with suppliers.

BRAND EVOLUTION: DISTRIBUTORS' BRANDS AND GENERICS

Another major landmark in the growth of branding was the metamorphosis from manufacturers' brand to distributors' brand that began to occur around the 1900s. To appreciate this, one must again consider the changing nature of the retailing environment. Around the 1870s, multiple retailers — those owning 10 or more outlets —emerged, each developing their own range of brands for which they controlled the production and packaging. These distributor brands (usually referred to as own labels or private labels) became common in emergent chains such as Home & Colonial, Lipton and International Stores. The early versions of distributor brands tended to be basic grocery items. Not only did the chains undertake their own production, but they also managed the wholesaling function, with branding being almost an incidental part of the total process.

The reason for the advent of distributor brands was that, due to resale price maintenance, retailers were unable to compete with each other on the price of manufacturers' brands and relied upon service as the main competitive edge to increase store traffic. The multiples circumvented this problem by developing their own distributor brands (own label). The degree of retailer production was limited by the complexity of the items and the significant costs of production facilities. Thus, it became increasingly common for multiple retailers to commission established manufacturers to produce their distributor brands which were packaged to the retailer's specifications. Before World War II, distributor brands accounted for 10—15 percent of multiples' total sales; but with multiple retailers accounting for only 17 percent of food sales, the overall importance of distributor brands was far exceeded by manufacturer brands. During World War II, distributor brands were withdrawn due to shortages and were not reintroduced until the 1950s.

One of the consequences of the increasing growth of the multiples was the decline of independent retailers (those owning no more than nine shops). As a means of protecting themselves, some independent retailers joined together during the 1950s and agreed to place all their purchases through specific wholesalers. The formation of this new category of retailers referred to as symbol/voluntary groups (e.g. Spar), enabled the store owners to maintain some degree of individual control. With a significant element of their purchasing channelled through a central wholesaler, they were able to achieve more favourable terms from manufacturers. A further consequence of this allegiance was the introduction of a new category of brands, i.e., symbol/voluntary brands, designed to compete against the multiples' brands. These brands carry the symbol/voluntary groups' names and are priced cheaper than the equivalent manufacturers' brands. Some grocery retailers offer tiers of own label brands to compete across varied market segments. For example, Tesco's 'Finest' brand range competes at a premium price level, while the Tesco 'Value' brand attracts a more price-conscious consumer.

In the industrial sector it is less common to see distributor brands, due to the considerable investment in production, the need to understand the technology and the greater reliance upon direct delivery, with less reliance on distributors. Chapter 7 provides more detail about distributor brands.

In the packaged grocery sector, where the first alternative tier to manufacturer brands appeared, innovative marketing in the late 1970s also led to further alternative generics. In fact, the term 'generics' may be a misnomer, since it implies a return to the days when retailers sold commodities rather than brands. This trend was originally started by Carrefour in 1976, when they launched 50 'produits libres' in France, promoted as brand-free products. Some UK grocery retailers noted the initial success of these lines and thought the time was right to follow in the UK. At the time, there was growing consumer skepticism about the price premium being paid for branding and with consumers becoming more confident about selecting what in many cases were better quality distributor brands, it was thought that in a harsh economic environment, generics would be a popular alternative to manufacturer brands, further increasing distributors' control of their product mix.

The thrust behind generics was that of cutting out any superfluous frills. They were distinguishable by their plain packaging, with the marketing emphasis placed on the content, rather than on the promotional or pack features. On average, generics were priced 40 percent lower than the brand leader and approximately 20 percent lower than the equivalent distributor brands. Whilst the quality level varied by retailer, they were nonetheless generally inferior to manufacturers' brands.

Retailers in the UK who stocked a generic range developed a policy regarding the product, pricing, packaging and merchandising that only too clearly enabled consumers to associate a particular generic range with a specific store. But the withdrawal of generics was not surprising, since consumers perceived generics as similar to distributor brands. They were not perceived as a unique tier and they weakened the image and hence the sales, of the distributor brands of those retailers stocking generics. Furthermore, as they were perceived to be similar to distributor brands, more switching occurred with these, rather than with the less profitable manufacturers' brands.

However, with the entrance of value supermarkets, consumer consumption has shifted—with brands such as Aldi and Lidl finding favour with price sensitive shoppers. Therefore, the role of the distributor's own brand has become strategically significant, as it often offers a cost saving to the customer and prevents customer defection. A recent study found that 39 percent of UK customers were 'frequent buyers' of private label products, with many considering private labels to be on par with market leading brands.

Chapter 7 considers in more detail the marketing issues associated with generics. However, it is worth emphasising that any organisation operating in the consumer, services or industrial markets never offers a commodity and is always able to differentiate their offering. The marketing of generics trims some of the marginal costs away, but leaves the organisation competing on product dimensions that can be easily copied and which have little impact compared with other attributes (e.g. service, availability, imagery, etc.). Business-to-business manufacturers who believe they are marketing generic products and therefore have to offer the lowest prices are deluding themselves. For example, purchasers are not just buying tanker loads of commonly available chemicals for their production processes. They are also buying a reliable delivery service, a well-administered reordering process, advice from the supplier about the operating characteristics of the chemical, etc. By just considering issues such as these, it is easier to appreciate the fallacy of marketing generics.

BRAND CATEGORISATION

An advertising perspective

This brief historical review has shown how brands evolved and emphasised different aspects of branding. One of the weaknesses with the current views on branding is that the term is used to encompass a very broad range of issues, encouraging the possibility of confusion.

One of the early attempts to categorise brands was that of Langmaid and Gordon (1988). Based on a consideration of advertisements, they classified brands into nine categories, each representing a role in advertising, varying from simple through to complex branding. For example, at the simple end of the scale there are those brands which operate through straightforward association—e.g. Duracell batteries with their copper coloured top. By contrast, at the most complex end of the spectrum, they identify structural branding, in which for example, objects (a fairytale castle, mouse ears, etc.) are shown in order to ensure a link with Disney products. Fig. 2.1 shows the interpretations of brand types drawn up by these researchers. However, whilst

Simple

- Simple association (verbal), e.g. Schweppes
- Simple association (aural), e.g. Meteor melody
- Simple association (visual), e.g. iPod earphones
- Branding devices, e.g. Starbucks mug
- Branding symbols, e.g. Kellogg's Snap, Crackle and Pop
- Branding analogies, e.g. Generation X
- Branding metaphors, e.g. 'Australianess'
- Branding tone of voice, e.g. Virgin airline's cheeky tone
- Structural branding, e.g. Disney

Complex

FIGURE 2.1

Brand typology *(after Langmaid and Gordon 1988)*

their typology is of value to advertisers, its overt advertising bias restricts its value as an aid in evaluating how to employ the other elements of the marketing mix.

An output process

Our research and work with various organisations have shown that there are other interpretations of the role played by brands, which we will now make explicit, all of which will be addressed in more detail in subsequent chapters. However, a key problem with many of these interpretations is that they place considerable emphasis on branding as something that is done *to* consumers, rather than branding as something consumers do things *with*. It is wrong, in other words, to focus on branding as an **input** process. Clearly we need to consider carefully how marketing resources are being used to support brands, but it is crucial to understand the **output** process as well, since, as mentioned earlier, the final evaluation of the brand is in the buyer's or user's mind. Consumers are not just passive recipients of marketing activity. They consume marketing activity, twisting messages to reinforce prior expectations.

Several highly regarded branding advisers stress the importance of looking at brands as perceptions in consumers' minds, which reinforces the importance of what consumers take out of the process rather than what marketers put into it. Whilst it is clear that marketers design the firm's offer, the ultimate judge about the nature of the brand is the consumer. When buying a new brand, consumers seek clues about the brand's capabilities. They try to assess the brand through

DIOR

Dior's fragrance captures the imagery of Paris to create a customer experience. (Reproduced by kind permission of Dior)

a variety of perceptual evaluations, such as its reliability, whether it is the sort of brand they feel right with or whether it's better than another brand; so that a brand becomes not the producer's but the consumer's idea of the product. The result of good branding is a perception of the values of a product or service, interpreted and believed so clearly by the consumer that the brand adopts a personality. This is often so well developed that products with few functional differences become regarded as different, purely because of the brand personality—for example, Mini and Google.

Thus, recognising the inherent flaw when marketers focus upon branding as an input process, we have highlighted eight different ways that managers emphasise different elements of brands.

An eight-category typology

1 Brand as a sign of ownership

An early theme, given much prominence in marketing circles, was the distinction between brands on the basis of whether the brand was a manufacturers' brand or a distributors' brand ('own label', 'private label'). Branding was seen as being a basis of showing who instigated the marketing for that particular offering and whether the primary activity of the instigator was production (manufacturers' brand) or distribution (distributors' brand). However, this drew a rather artificial distinction, since nowadays consumers place a far greater reliance on distributor brands − particularly when brands such as Marks & Spencer and Harrods are perceived as superior brands in their own right. In fact, some would argue that with the much greater marketing role played by major retailers and their concentrated buying power, the concept of USP (Unique Selling Proposition) should now be interpreted as 'Universal Supermarket Patronage'!

With greater recognition that a brand has to appeal to stakeholders rather than just consumers, that corporate endorsement engenders greater trust and provides an appreciation of unifying core values, there is a move to the corporation as a brand, rather than stressing branding at the individual product line level. This provides focus when developing a brand identity. There is a danger, however, that consumers do not recognise the values that the corporate brand stands for and how these run through all the product line brands. A further danger is that as a corporation widens its portfolio, its core values become diluted.

2 Brand as a differentiating device

The historical review earlier in this chapter indicated that, at the turn of the century, a much stronger emphasis was placed on brands purely as differentiating devices between similar offerings. This perspective is seen today in many markets. Yet with more sophisticated marketing and more experienced consumers, brands succeed not only by conveying differentiation but also by being associated with added values. For example, the brand Persil not only differentiates it from other washing powder lines, but is a successful brand since it has been backed by a coherent use of resources that deliver the added value of a high-quality offering with a well-defined image. By contrast the one-man operation, 'Tom's Taxi Service', is based upon branding as a differentiating device, with little thought to communicating added values.

Small firms seem to be prone to the belief that putting a name on their product or service is all that is needed to set them apart from competitors. They erroneously believe that branding is about having a prominent name, often based around the owner's name. Yet there is ample evidence that brands fail if organisations concentrate primarily on developing a symbol or a name as a differentiating device. Chapter 4 gives examples showing the danger of adopting brands solely as differentiating devices. Brands will succeed if they offer unique benefits, satisfying real consumer needs.

Brand distinctiveness allows customers to identify products and services. This occurs not only from the brand itself, i.e., through the packaging, advertising or naming, as shown in Fig. 2.2. There are further sources to distinguish the brand. First, consumers perceive the brand in their own way. As explained before, value is in the eye of the beholder and each person can draw very different conclusions.

Secondly, people interacting with the consumers affect their perception of the brand. Especially with consumer goods such as clothing, consumers focus their attention on certain brands as a result of conversations with peer groups.

3 Brand as a functional device

Some marketers overtly emphasise a brand's functional capability. This stemmed from the early days of manufacturers' brands when firms wished to protect

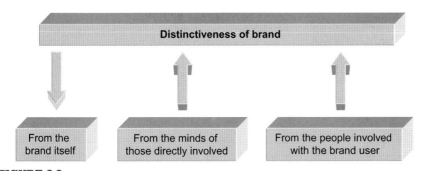

FIGURE 2.2

How a brand can be distinctive

their large production investments by using their brands to guarantee consistent quality.

As consumers began to take for granted the fact that brands represented consistent quality, marketers strove to establish their brands as being associated with specific, unique, functional benefits — for example, not just marketing a credit card, but a credit card augmented with a credit card protection policy and possibly also a concierge service.

Functional capabilities should always be focused on consumers, rather than on internal considerations. For instance, providing televisions in waiting rooms to make queues less annoying is less likely to be appreciated than a redesigned system to eliminate queues altogether.

Firms emphasising brands as functional communicators have run the risk of an excessive reliance on the functional (rational) element of the consumer choice, as all products and services also have some degree of emotional content in the buying process. For example, www.Amazon.co.uk personalises its customer's website with recommendations based on customers' past purchase history.

4 Brand as a symbolic device

In certain product fields (e.g. perfume and clothing) buyers perceive significant badge value in the brands, since brands enable them to communicate something about themselves (e.g. emotion, status, etc.). In other words, brands are used as symbolic devices, because of their ability to help users express something about themselves to their peer groups, with users taking for granted functional capabilities.

Consumers personify brands and when looking at the symbol values of brands, they seek brands which have very clear personalities and select brands that best match their actual or desired self-concept. For example, in the beer market there are only marginal product differences between brands. Comparative consumer trials of competing beer brands without brand names present, showed no significant preferences or differences. Yet, when consumers repeated the test with brand names present, significant brand preferences emerged. On the first comparative trial, consumers focussed on functional (rational) aspects of the beers and were unable to notice much difference. On repeating the trials with brand names present, consumers used the brand names to recall distinct brand personalities and the symbolic (emotional) aspect of the brands influenced preference.

Through being members of social groups, people learn the symbolic meanings of brands. As they interpret the actions of their peer group, they then respond, using brands as non-verbal communication devices (e.g. feelings, status).

To capitalise on symbolic brands therefore, marketers can use promotional activity to communicate the brand's personality and signal how consumers can use it in their daily relationships. Nonetheless, whilst there are many product fields where this perspective is useful, it must also be realised that consumers rarely consider just the symbolic aspect of brands. Research across a wide variety of product and service sectors showed that consumers often evaluated brands in terms of both symbolic (emotional) and functional (rational) dimensions. Marketers should, therefore, be wary of subscribing to the belief that a brand acts *solely* as a symbolic device.

A successful example of combining both is given by Billabong clothing which offers the functional advantages of well-made surfing clothing with the appeal of showing the user's sense of adventure and love of outdoor sports.

5 Brand as a risk reducer

Many marketers believe that buying should be regarded as a process whereby buyers attempt to reduce the risk of a purchase decision, rather than maximising their gain. When a person is faced with competing brands in a new product field, they feel risk, for example, uncertainty about whether the brand will work, whether they will be wasting money, whether their peer group will disagree with their choice, whether they will feel comfortable with the purchase, etc. Successful brand marketing should therefore be concerned with understanding buyers' perceptions of risk, followed by developing and pre-senting the brand in such a way that buyers feel minimal risk. An example of an industry appreciating perceived risk is the pharmaceutical industry. A company marketing a new drug suitable for children needs to consider the parent's concerns about risk. One company launching such a product worked with a series of pain specialists in pediatric hospitals providing clinical data and product trials, so that specialist endorsement of the brand would reassure parents and help to minimise their perception of risk.

To make buying more acceptable, buyers seek methods of reducing risk by, for example, always buying the same brand, searching for more information, only buying the smallest size, etc. Research has shown that one of the more popular methods employed by buyers to reduce risk is reliance upon reputation. Some marketers, particularly those selling to organisations rather than to final consumers, succeed with their brands because they find out what dimensions of risk the buyer is most concerned about and then develop a solution through their brand presentation, which emphasises the brand's capabilities along the risk dimension considered most important by the buyer. This interpretation of branding has the virtue of being output driven. Marketers, however, must not lose sight of the need to segment customers by similar risk perception and achieve sufficient numbers of buyers to make risk reduction branding viable.

6 Brand as a shorthand device

Glancing through advertisements, one becomes aware of brands whose promotional platform appears to be based on bombarding consumers with considerable quantities of information. To overcome the problem of sifting through large amounts of information, brands are used as shorthand devices by consumers to recall from memory sufficient brand information to make a decision. There is merit in this approach, as people generally have limited memory capabilities. To overcome this, they bundle small bits of information into larger chunks in their memory and use brand names as handles to recall these larger information chunks. By continuing to increase the size of these few chunks in memory, buyers in consumer, business-to-business and service sectors can process information more effectively. At the point of purchase, they are able to recall numerous attributes by interrogating their memory.

There is a danger of concentrating too heavily on the quantity, rather than on the quality of information directed at purchasers. It also ignores the perceptual process which is used by buyers to twist information until it becomes consistent with their prior beliefs.

7 Brand as a legal device

With the appearance of manufacturers' brands at the turn of this century, consumers began to appreciate their value and started to ask for them by name. Producers of inferior goods realised that to survive they would have to change. A minority, however, changed by illegally packaging their inferior products in packs that were virtually identical to the original brand. To protect themselves against counterfeiting, firms turned to trademark registration as a legal protection. Some regarded the prime benefit of brands as being that of legal protection. They put notable efforts towards effective trademark registration along with consumer education programmes about the danger of buying poor grade brand copies. For example, recent advertising for Kellogg emphasised the fact that the company doesn't make cereals for anyone else and that the red Kellogg signature tells consumers they are buying a genuine article.

The success of retailers' own labels intensified the need for manufacturers to protect their brands. Consumers can become confused by similar-looking competing brands. A 2004 survey found that up to 25 percent of consumers consider own-label products to be the same as branded products. Of even more concern, another study funded by the British Brand Group (BBG) in 2009 found that one in three grocery shoppers bought the wrong product because of its similar packaging. Such 'parasitic packaging' is not a counterfeit, as it does not infringe trademark or design rights. However, it hijacks distinctive features of a brand's packaging and the BBG study found that the packaging induces customers to recall the original product—which can boost sales by up to 50 percent. Over the last few years there have been famous cases of brand warfare: Coca-Cola vs. Sainsbury's Classic Cola; Van den Bergh Foods vs. Tesco's 'Unbelievable' Spread; Kellogg vs. Tesco Corn Flakes; United Biscuit vs. Asda's Puffin Bar. The last case is significant since this was the first time a manufacturer took a retailer to court and won the case, with the judge requiring an Asda pack change. This causes a dilemma for manufacturers who have to find a balance between defending their brands against seemingly unfair competition and fighting against their own customers.

Because of their intangible nature, service brands have even greater difficulties coping with counterfeit brands. Financial services have shown a way to circumvent this problem by adapting a particular house style when interacting with consumers.

Finally, more enlightened marketers are adopting the view to which the authors subscribe, that brands should be treated as strategic devices. The assets constituting the brand need to be audited, the forces affecting the future of the brand evaluated and, by appreciating how the brand achieved its added value, a positioning for the brand needs to be identified so that the brand can be successfully protected and achieve the desired return on investment. To take full advantage of brands as strategic devices, marketing analysis and brand planning is required, yet many firms are too embroiled in tactical issues and do not gain the best possible returns from their brands. Some of the key strategic issues associated with capitalising on strategic branding are covered in this book.

A good example of successful branding through majoring upon a differential advantage and ensuring the sustainability of such an advantage is the online search engine Google. Its innovations in technology made it the number one search engine. In addition, it has since developed such a successful AdWords methodology that 99 percent of its revenue is accounted for by online advertising.

This section has described the way that different characteristics of brands can be emphasised and has also highlighted the inherent weaknesses of each approach. A successful brand blends together these characteristics to present a unified, holistic promise which organisations can successfully deliver.

BRAND CONSUMERISATION SPECTRUM

As Mary Goodyear illustrates (1996), market evolution can be better understood in terms of a continuum of consumerization, which characterises the degree of dialogue between marketers and consumers. Drawing on this concept, she identifies a chronological schema for categorising brands, as shown in Fig. 2.3. From a commodity market, as a dialogue begins, so the role of the brand progresses to being a reference for quality. Seeking to compete against more intense competitors, marketers begin to understand consumers' emotional needs, enabling the brand to grow into its role of projecting a particular personality.

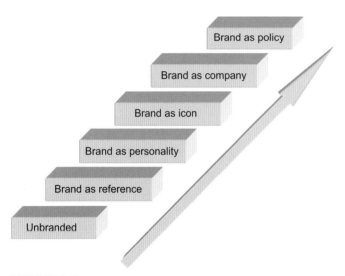

FIGURE 2.3
Brand consumerization spectrum *(after Goodyear 1996)*

A closer relationship with consumers, and a highly symbolic language characterise the development of a brand into an icon, as Coca Cola became during World War II. Douglas Holt has identified brands such as Apple, Nike, Harley Davidson, Volkswagen, Budweiser and Coca Cola as icons, because they embody ideals admired by consumers and give consumers a means to express themselves. Icons need to be continually refreshed or they fail to stay on their admired pedestal. Holt explains that these icons need to tread a fine line between milking their icon status and constantly striving to be ahead and refreshed.

To continue the brand's growth trajectory, organisations recognise that stakeholders interact with their corporation through different points of contact. To ensure a consistent brand experience they therefore concentrate on the way the whole *company* interacts with its stakeholders. In the most developed role, brands become a synonym of the company's *policy* to take larger responsibilities regarding economic values, social commitment, cultural awareness and political issues. Firms can take advantage of this role, as Body Shop and Virgin did, or become entangled in environmental problems, as was once the case for Shell.

THE VALUE OF BRANDS TO MANUFACTURERS, DISTRIBUTORS AND CONSUMERS

Manufacturers invest effort in branding for a variety of reasons. If the trademark has been effectively registered, the manufacturer has a legally protected right to an exclusive brand name, enabling it to establish a unique identity, reinforced through its advertising and increasing the opportunity of attracting a large group of repeat purchasers.

With the high costs of developing new brands, many organisations are attracted by the opportunities from line extensions. Virgin provides a good example of a brand being stretched. From its origin in the record business it has stretched into the airline, wine retailing and financial services sectors, amongst others. Caterpillar has successfully extended from heavy machinery into shoes, clothing and handbags.

Good brands have strong identities reducing the cost of new line additions through trust and other positive associations. However, marketers need to be cautious about overstretching the brand's core values.

Brand extension is such a popular choice because it offers an apparently easy and low risk way of leveraging the brand equity. It is essential, however, to realise that there is a cost to it. First, if a brand loses its credibility in one sector, the whole umbrella range could be affected. Secondly, successful brand extensions may dilute the brand values of the core product. Managers should then carefully consider whether it is worth running the risk of tarnishing the brand image and reducing the core brand equity. Consumer research can be invaluable in helping to decide whether to follow a brand extension strategy or to develop a new brand.

Manufacturers with a history of strong brands are likely to find distributors more receptive to presentations of brand extensions or even of new brands. Those manufacturers with strong brands maintain greater control over the balance of power between the manufacturer and distributor and, indeed, some argue that this is one of the key benefits of strong brands.

It is also possible for a manufacturer with strong brand names to market different brands in the same product field that appeal to different segments. This is seen in the washing detergents and the soap market, where Unilever and Procter & Gamble market different brands with minimal cannibalisation between brands from the same manufacturer. Furthermore, by developing sufficiently differentiated manufacturer brands that consumers desire, higher prices can be charged, as consumers pay less attention to price comparisons between different products because of brand distinctiveness. This clearly enhances profitability. Indeed, Table 2.3 shows the profit impact of powerful brand leaders in the grocery market.

Retailers see strong manufacturer brands as being important, since through manufacturers' marketing activity (e.g. advertising, point of sale material, etc.), a fast turnover of stock results. Also, with more sophisticated marketers

Table 2.3 Market Share and Average Net Margins for UK Grocery Brands

Rank	Net margin (%)
1	17.9
2	2.8
3	−0.9
4	−5.9

Source: Doyle (2008)

recognising the importance of long term relationships with their customers, many manufacturers and distributors have cause to recognise that their future success depends on each other and therefore strong manufacturer brands are seen as representing profit opportunities both for distributors and manufacturers. Some retailers are interested in stocking strong brands, since research indicates that the positive image of a brand can enhance their own image

Recalling the discussion in the previous section about brand names acting as a means of short-circuiting the search for information, consumers appreciate manufacturers' brands since they make shopping a less time-consuming experience. As already noted, manufacturers' brands are recognised as providing a consistent guide to quality and consistency. They reduce perceived risk and make consumers more confident. In some product fields (e.g. clothing, cars), brands also satisfy strong status needs.

Why, then, do so many manufacturers also supply distributors' brands? First, it is important to understand why distributors are so keen on introducing their own brands. Research has shown that they are particularly keen on distributor brands because they enable them to have more control over their product mix. With a strong distributor brand range, retailers have rationalised their product range to take advantage of the resulting cost savings; and many stock a manufacturer's brand leader, their own distributor's brand and possibly a second manufacturer's brand. Trade interviews also indicate that distributor brands can offer better margins than the equivalent manufacturer's brand, with estimates indicating the extra profit margin to be about 5 percent more than the equivalent manufacturer's brand. Some of the reasons why manufacturers become suppliers of distributors' brands are:

- economies of scale through raw material purchasing, distribution and production;
- any excess capacity can be utilised;
- it can provide a base for expansion;
- substantial sales may accrue with minimal promotional or selling costs;
- it may be the only way of dealing with some important distributors (e.g. Marks & Spencer).

Consumers benefit from distributors' brands through the lower prices being charged.

THE IMPORTANCE OF BRAND PLANNING

As previous sections of this chapter have shown, brands play a variety of roles and for a number of reasons satisfy many different needs. They are the end result of much effort and by implication represent a considerable investment by the organisation. With the awareness of the financial value of brands, companies question whether these financially valuable assets are being effectively managed to achieve high returns on investment. To gain the best return from their brands, firms must have a well-conceived vision for their brands and not just focus in isolation on tactical issues of design and promotion. Instead, they need to audit the capabilities of their firm, evaluate the external issues influencing their brand (briefly overviewed in the next section) and then develop a brand plan that specifies realistic brand objectives and the strategy to achieve them (covered in more detail in Chapters 9 and 11).

Brand planning is an important but time-consuming activity, which, if undertaken in a thorough manner involving company-wide discussion, will result in a well-grounded consensus about how resources can be best employed to sustain the brand's differential advantage. Without well-structured brand plans there is the danger of what we call brand 'vandalism'. Junior brand managers are given 'training' by making them responsible for specific brands. Their planning horizons tend to be in terms of a couple of years (i.e., the period before they move on) and their focus tends to be on the tactical issues of advertising, pack design and tailor-made brand promotions for the trade. At best, this results in 'fire fighting' and a defensive, rather than offensive, brand plan. The core values of the brand are in danger of being diluted through excessive brand extensions. By not preparing well-documented strategic brand plans, firms are creating their own obstacles to success. Some of the characteristics that internally hinder any chance of brand success are:

- Brand planning is based on little more than extrapolations from the previous few years.
- When it doesn't look as if the annual budget is going to be reached, Quarter Four sees brand investment being cut (i.e., advertising, market research, etc.).
- The marketing manager is unable to delegate responsibility and is too involved in tactical issues.
- Brand managers see their current positions as good training grounds for no more than two years.
- Strategic thinking consists of a retreat once a year with the advertising agency and sales managers.
- A profitability analysis for each major customer is rarely undertaken.
- New product activity consists of different pack sizes and rapidly developing 'me-too' offers.
- The promotions budget is strongly biased towards below-the-line promotional activity, supplemented only occasionally with advertising.
- Marketing documentation is available to external agencies on a 'need to know' basis only.

Brand strategy development must involve all levels of marketing management and stands a better chance of success when all the other relevant internal departments and external agencies are actively involved. It must progress on the basis of all parties being kept aware of progress.

British Airways exemplify the notion of brand development as an integrating process, having used this to achieve a greater customer focus. For example, the simple operation of taking a few seats out of an aircraft can be done with confidence, as engineering are consulted about safety implications, finance work out the long-term revenue implication, scheduling explore capacity implications and the cabin crew adjust their in-flight service routines.

THE ISSUES INFLUENCING BRAND POTENTIAL

When auditing the factors affecting the future of brands, it is useful to consider these in terms of the five forces shown in Fig. 2.4. The brand strategist can evaluate the intensity and impact of the following brand-impeding issues.

Corporation

It is not unusual for an organisation to be under utilising its brand assets through an inability to recognise what is occurring inside the organisation. Have realistic, quantified objectives been set for each of the brands and have they been widely disseminated? Aims such as 'to be the brand leader' give some indication of the threshold target but do little in terms of stretching the use of resources to achieve their full potential. Brand leadership may result before the end of the planning horizon, but this may be because of factors that the organisation did not incorporate into their marketing audit. But luck also has a habit of working against the player as much as working for the player!

Firms such as Ritz-Carlton and GE have shown how brand and corporate culture are closely interlinked and how they affect each other. Since the culture of an organisation strongly influences its brands, mergers and acquisitions can alter brand performance dramatically.

Marketers should audit how well brand and culture match each other.

Has the organisation made full use of its internal auditing to identify what its *distinctive* brand competences are and to what extent these match the

FIGURE 2.4

The five forces influencing brand potential *(Source: Interbrand)*

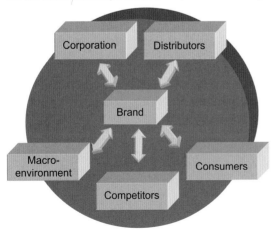

factors that are critical for brand success? For example, Apple's values of simplicity and innovation are evident in their products such as iPhone and iPod.

Is the organisation plagued by a continual desire to cut costs, without fully appreciating why it is following this route? Has the market reached the maturity stage, with the organisation's brand having to compete against competitors' brands on the basis of matching performance but at a reduced price? If this is so, *all* aspects of the organisation's value chain should be geared towards cost minimisation (e.g. eliminating production inefficiencies, avoiding marginal customer accounts, having a narrow product mix, working with long production runs, etc.). Alternatively, is the firm's brand unique in some way that competitors find difficult to emulate and for which the firm can charge a price premium (e.g. unique source of high quality raw materials, innovative production process, unparalleled customer service training, acclaimed advertising, etc.)? Where consumers demand a brand which has clear benefits, the manufacturer should ensure all departments work towards maintaining these benefits and signal this to the market (e.g. by the cleanliness of the lorries, the politeness of the telephone operators, promptness in answering a customer enquiry, etc.). In some instances, particularly in services, the brand planning document can overlook a link in the value chain, resulting in some inherent added value being diminished (e.g. a financial adviser operating from a shabby office).

Significant cultural differences between different departments of a company can affect brand success. The firm should not only audit the process to deliver the branded product or service, but also the values and attitudes of staff, to assess whether different departments' cultures are in harmony with the desired corporate brand identity and whether the firm has the appropriate culture to meet the brand's vision.

Extending the role of brands within the firm, Aaron Shields has recently advocated that all brands can be used to contribute to society. Thus, rather than assuming that this role applies to only those firms who explicitly promote CSR, such as Body Shop, it is more beneficial to consider how all brands can be used to unify views, to shape behaviour and to contribute to a caring society. Shields suggests that such brands offer a transparent view into the culture and values of the firm. He gives the examples of Johnson & Johnson's credo, which has been in place for over 60 years, or Sony's work to elevate the image of post-war Japan. To achieve this type of brand, Shields recommends working with the employees of the firm to choose the right brand values. He then advocates unifying the internal culture under these values, by recognising the dominant culture and using progressive narratives to bring the culture and brand values together. Those brand messages must then be externally promoted in a progressive way. For example, Shields uses the example of Nike, a company which promotes winning, but whose communications also raise issues around race and gender inequality. Lastly, he advocates continually raising the game, by adding additional meaning to keep the brand relevant over time. An example of this would be the Co-operative Bank, who were named Best Financial Services Provider in the 2009 Which? Service awards. The bank attributed its win to the trust associated with the brand, its ethical policy and its financial management during the credit crunch. Clearly, brands which evolve must do so within the wider business environment in order to remain relevant for consumers.

Distributors

Except for those dealing directly with customers, brand strategy cannot be formulated without regard for distributors. Both parties rely on each other for their success and even in an era of increasing retailer concentration, notwithstanding all the trade press hype, there is still a recognition amongst manufacturers and distributors that long-term brand profitability evolves through mutual support.

Suppliers need to identify retailers' objectives and align their brands with those retailers whose aims most closely match their own. Furthermore, they should be aware of the strengths and weaknesses of each distributor.

Brand suppliers who have not fully considered the implications of distributors' longer-term objectives and their strategies to achieve them, are deluding themselves about the long-term viability of their brands.

In the UK there are numerous instances of growing retailer power, with a few major operators controlling a significant proportion of retail sales (e.g. groceries, DIY, jewellery, footwear). The danger of increasing retailer power is that weaker brand manufacturers acquiesce to demands for bigger discounts, without fully appreciating that the long-term well-being of their brands is being undermined. It is crucial for brand manufacturers to analyse regularly what proportion of their brand sales goes through each distributor and then to assess for each individual distributor how important each brand is in its category. For example, Table 2.4 shows a hypothetical analysis for a confectionery manufacturer.

If this hypothetical example were for a Nestlé brand, it is clear that the particular Nestlé brand is more reliant upon Tesco than Tesco is on the particular Nestlé brand. Such an analysis better enables manufacturers to appreciate which retailers are more able to exert pressure on their brand. It indicates that if the brand manufacturer wants to escape from a position of being in the retailer's power, they need to consider ways of growing business for their brands in those sectors other than at Tesco, at *a faster rate* than is envisaged with this distributor.

When working with a distributor, the brand manufacturer should take into account whether the distributor is striving to offer a good value proposition to

Table 2.4 Power Analysis

Hypothetical Nestlé Brand Sales to Distributors		Hypothetical Market Share of Confectionery Brands Through Tesco	
Distributor	*%*	*Brand*	*%*
Tesco	25	Cadbury	35
Sainsbury	24	Mars	30
Asda	19	Nestlé	20
Kwik Save	17	Other	15
Co-op	10		100
Independent	5		
	100		

the consumer (e.g. Asda, Aldi, Primark or Matalan) or a value-added proposition (e.g. high-quality names at Harrods). In view of the loss of control once the manufacturer's brand is in the distributor's domain, the brand manufacturer must annually evaluate the degree of synergy through each particular route and be prepared to consider changes.

Does the manufacturer have an offensive distribution strategy or is it by default that its brands go through certain channels? What are the ideal characteristics for distributors of its brands and how well do the actual distributors used match these criteria? How do distributors plan to use brands to meet their objectives? How do manufacturers' and their competitors' brands help distributors achieve their objectives? Which other forms of support (e.g. discounts, merchandising) are used?

The brand manufacturer must have a clear idea of the importance of specific distributors for each brand and in Chapter 7 a matrix is presented, which enables the manufacturer to rank the appropriateness of distributors for each of their brands.

Finally, manufacturers must recognise that when developing new brands, distributors have a finite shelf space and market research must not solely address consumer issues, but must also take into account the reaction of the trade. One company developed a pyramid pack design researched well amongst consumers but on trying to sell this into the trade it failed — due to what the trade saw as its ineffective use of shelf space!

Consumers

To consumers, buying is a process of problem solving. They become aware of a problem (e.g. not yet arranged summer holidays), seek information (e.g. go to travel agent and skim brochures), evaluate the information and then make a decision (e.g. select three possible holidays, then try to book one through the travel agent). The extent of this buying process varies according to purchasers' characteristics, experience and the products being bought. Nonetheless, consumers 'work' to make a brand selection. The brand 'selection' and brand 'usage' are not necessarily performed by the same person. Therefore, marketers need to identify all individuals and identify to whom the brand appeals. In business-to-business markets several groups are involved in the purchase decision. Marketers need to formulate brand strategy that communicates the benefits of the brand in a way which is relevant to each group. Chapter 3 looks at consumer behaviour involved in brand selection in more detail, while Chapter 5 concentrates on business-to-business markets.

Brands offer consumers a means of minimising information search and evaluation. Through seeing a brand name which has been supported by continual marketing activity, consumers can use this as a rapid means of interrogating memory and if sufficient relevant information can be recalled, only minimal effort is needed to make a purchase decision. As a consequence of this, brand strategists should question whether they are presenting consumers with a few high-quality pieces of information or whether they are bombarding consumers with large quantities of information and ironically causing confusion. Likewise, in business-to-business markets, it is important to consider how firms make brand selections. This is covered in Chapter 5.

Not only should strategists look at the stages consumers go through in the process of choosing brands but they also need to consider the roles that brands play. For example, a business person going to an important business presentation may feel social risk in the type of clothes he or she wears and select a respected brand mainly as a risk reducer. By contrast, in a different situation he or she may decide to wear a Gucci watch, because of a need to use the brand as a device to communicate a message (e.g. success, lifestyle) to their peer group. Likewise, one purchasing manager may buy a particular brand, since experience has taught him or her that delivery is reliable, even though there is a price premium to pay. By contrast, another purchasing manager may be more concerned about rapid career advancement and may choose to order a different brand on the basis that he is or she is rewarded for minimising unnecessary expenditure on raw materials. Success depends on understanding the way purchasers interact with brands and employing the company's resources to match these needs.

Competitors

Brands are rarely chosen without being compared against others. Although several brand owners benchmark themselves against competition, it often appears that managers misjudge their key competitors. Managers should undertake interviews with current and potential consumers to identify those brands that are considered similar. Once marketers have selected the critical competitors, they need to assess the objectives and strategies of these companies as well as fully understand their brand positioning and personalities. It is also essential not to be restricted to a retrospective, defensive position but to gather enough information to anticipate competitive response and be able to continuously update the strategy for brand protection.

Research has shown that return on investment is related to a product's share of the market. In other words, products with a bigger market share yield better returns than those with a smaller market share. Organisations with strong brands fare better in gaining market share than those without strong brands. Thus, firms who are brand leaders will become particularly aggressive if they see their position being eroded by other brands. Furthermore, as larger firms are likely to have a range of brands backed by large resources, it is always possible for them to use one of their brands as a loss leader to underprice the smaller competitor; once the smaller brand falls out of the market, the brand leader can then increase prices. Using a low cost strategy, Ryanair has revolutionised the airline industry and has challenged other leading airlines to reduce fares in order to compete.

Brand strategists need to have given some thought to anticipating likely competitor response, as Chapter 9 argues in more detail. When the UK high street heralded the arrival of European stores such as Zara and H&M, existing brands such as River Island responded with new flagship stores. Following the UK arrival of American brands such as Abercrombie and Fitch and Banana Republic, Topshop's bold response included the launch of Topshop and Topman stores in New York.

Macro-environment

Brand strategists need to scan their macro-environments continually to identify future opportunities and threats. In essence, the challenge is to understand how the political, environmental (green), social, economic and technological environments could affect the brand. To draw an analogy with military thinking, good surveillance helps achieve success.

CONCLUSIONS

This chapter has provided an overview of the key issues involved in planning for the future of brands. It has shown that brands succeed when marketers regard them as the result of a well-integrated marketing process. To view branding as naming, design or advertising, is too myopic and such a perspective will shorten the brand's life expectancy. Branding is about a cluster of functional and emotional values, which allow an organisation to make a promise about a unique and welcomed experience.

The historical evolution of brands has shown that brands initially served the roles of differentiating between competing items, representing consistency of quality and providing legal protection from copying. However, because of the elements characterising brands, managers emphasise these in different ways to blend their individual brands. The elements are: a sign of ownership in terms of the extent to which the corporation's name is evident; a differentiating device that provides a meaningful name; a functional device; a symbolic device that enables buyers to express something about themselves; a risk-reducing device; a shorthand communication device; a legal device and a strategic device. To capitalise upon the asset represented by their brand, firms need to recognise where they are on Mary Goodyear's brand consumerisation spectrum and adopt strategic brand planning as a way of life. Finally, a model showing the five main factors that influence brand potential was reviewed.

MARKETING ACTION CHECKLIST

It is recommended that after reading this chapter, marketers undertake the exercises which follow to help clarify the direction of future brand marketing programmes.

1. Either by looking through previous market research reports, or by putting yourself in the position of a buyer, write down the four main reasons, in order of importance, why one of your company's brands is being bought. Then show an advertisement for this brand (or a catalogue page describing it) to one of your buyers and ask them to tell you the four key points they took from the message. If the results from the first and second part of this exercise are the same, your brand is correctly majoring upon relevant buyer choice criteria. However, any discrepancy is indicative of inappropriate brand marketing.
2. Write down the core values of one of your brands and ask the other members of your organisation to do the same. Compile a summary of the replies (without participants' names) and circulate the findings, asking for comments about: (a) reasons for such varied replies, and (b) which three core values are the key issues your company is trying to major on. Repeat the exercise until a consensus view has been reached.
3. Having identified your team's views about the core values of your brand, ask your team to write down: (a) what your brand communicates about your company's relationship with the purchaser, and (b) how the different resources supporting the brand are satisfying this relationship objective. Collate the replies from all parties and consider how well your team appreciates your brand.
4. Write down what you understand by the term 'brand' and compare your views with those of your colleagues. Where there are a large number of

comments relating to 'differentiation', 'logo', or 'unique design', your firm may not be fully capitalising its brand asset.

5. After making explicit what the added values are of one of your brands, estimate how long it would take a major organisation to buy-in resources which would help it copy each of these added values. When considering the results of this exercise, also ask yourself: (a) how relevant to the buyer are the added values that this major competitor would find easiest to emulate, and (b) what added values the buyer does not yet get from any brand and how difficult it would be to develop these added values.

6. For each of the added value benefits that your brand represents, write down how each element of the marketing mix helps to achieve it. If there are any instances where different elements of the mix are not operating in the same direction, consider why this is so and identify any changes necessary.

7. For each stage of the life cycle of one of your brands (introduction, growth, maturity, saturation, decline) identify how the added values may have to change to adapt to buyer sophistication.

8. Identify the clues that buyers use to evaluate the brand's added values and consider how much emphasis is placed on these clues in your marketing activity.

9. Make explicit the main situations within which one of your brands is: (a) bought, and (b) consumed. Evaluate the appropriateness of your current brand strategy for each of these situations.

10. What type of relationship does your brand suggest to your customers (e.g. personal commitment, intimacy, dependence, intimidating superiority)? What strategies have you established to reinforce the bond between the brand and your customers (e.g. financial incentives, social bonds, structural bonds)?

11. If you were to remove the name of your brand from its packaging, is there anything else that would signify the identity of the brand to buyers? How well is this protected against copying?

12. To what extent are you emphasising the elements of your brand, i.e.: a sign of ownership; a differentiating device; a functional communicator; a symbolic device; a risk-reducing device; a shorthand information device; a legal device; a strategic device? In view of the promise you are making to your consumers with this brand, are you over emphasising some elements at the expense of others?

13. What strategies have you established to protect your brands from look-alikes? Are you monitoring potential counterfeits? How long does it take for one of your brands to be launched? Are you aware of the risk of information leakage during that period?

14. Rank the importance of the different reasons why people interested in your brand value it. What are you doing to protect the brand on each of these valued attributes?

15. Where on Mary Goodyear's brand consumerization spectrum (Fig. 2.3) does your brand fall? Which other brands are at the same level as your brand and what are you doing to move beyond these?

16. Does your firm annually prepare a brand plan which audits the forces influencing the brand, have quantified brand objectives and consider strategies that are able to satisfy these objectives?

17. Using the audit in the section 'The importance of brand planning', what is your view about whether your firm is helping or hindering brand development?

18. How well positioned is each of your brands in relation to the five forces affecting brand potential? (outlined in the section 'The importance of brand planning'.)

STUDENT BASED ENQUIRY

■ Using a brand you are familiar with, examine how it creates a consumer experience through traditional and online media.

■ Select a brand of your choice and identify its generic, expected, augmented and potential components. From this analysis, suggest three new ways to build the brand's potential.

■ Choose either: i) the brand typology or ii) the eight-category typology presented in this chapter. Identify brands which fit at each element of your chosen typology and explain why you have chosen each brand.

■ Select a brand from the Interbrand Top 10 Brands list. Assess how it positions itself in relation to each of the five forces that affect brand potential.

References

Doyle, P. (2008). *Value-Based Marketing: Marketing Strategies for Corporate Growth and Shareholder Value*. London: John Wiley and Sons.

Goodyear, M. (1996). Divided by a common language. *Journal of the Market Research Society, 38*(2), 110−122.

Langmaid, R. & Gordon, W. (1988). 'A great ad − pity they can't remember the brand − true or false'. In *31st MRS Conference Proceedings*, 15−46. London: MRS.

Further Reading

Allan, J. (1981). Why fine fare believes in private label. Paper presented at *Oyez Seminar: Is the Brand Under Pressure Again?* London, September 1981

Allison, R. I., & Uhl, K. (1964). Influence of beer brand identification on taste perception. *Journal of Marketing Research, 1*(3), 36−9.

Assael, H. (1987). *Consumer Behavior and Marketing Action*. Boston: Kent Publishing.

Barwise, P., Higson, C., Likierman, A., & Marsh, P. (1989). *Accounting for Brands*. London: London Business School and the Institute of Chartered Accountants.

Berwin, L. (2009). *Value Supermarkets continue to shrink Tesco's market share*. Available at: www. retail-week.com. Accessed 15 June 2009.

British Market Research Bureau. (2009). *A Study into the Impact of Similar Packaging on Consumer Behaviour. A Report for the British Brands Group*. Berkshire: Don Edwards & Associates Limited.

Buck, S. (1997). The continuing grocery revolution. *Journal of Brand Management, 4*(4), 227−238.

Co-operative Bank named Best Financial Services Provider, Fair Investment.co.uk (online). Available at http://www.fairinvestment.co.uk/News/banking-news-Co-operative-Bank-named-Best-Financial-Services-Provider-3459.html. Accessed 6 July 2009.

Copeland, M. (April 1923). Relation of consumers' buying habits to marketing methods. *Harvard Business Review, 1*, 282−9.

de Chernatony, L. (1987). Consumers' perceptions of the competitive tiers in six grocery markets. Unpublished PhD thesis. London: City University Business School.

de Chernatony, L. (1988). Products as arrays of cues: how do consumers evaluate company brands? In T. Robinson, & C. Clarke-Hill, (Eds.), *Marketing Education Group Proceedings*, 1988. Huddersfield: MEG.

de Chernatony, L. & McWilliam, G. (1990). Appreciating brands as assets through using a two-dimensional model. *International Journal of Advertising, 9*(2), 111−19.

de Chernatony, L. (2001). *From Brand Vision to Brand Evaluation*. Oxford: Butterworth Heinemann.

de Ruyter, K., & Wetzels, M. (2000). The role of corporate image and extension similarity in service brand extensions. *Journal of Economic Psychology, 21*, 639−659.

Dowling, G. (2000). *Creating Corporate Reputations*. Oxford: Oxford University Press.

Economist Intelligence Unit. (Oct. 1968). Own brand marketing. *Retail Business, 128*, 12−19.

Euromonitor. (1989). *UK Own Brands 1989*. London: Euromonitor.

Furness, V. (Oct 2002). Straying beyond the realms of own label. *Marketing Week, 24*, 23−24.

Gordon, W., & Corr, D. (1990). The space between words. *Journal of the Market Research Society, 32*(3), 409−434.

Hawes, J. (1982). *Retailing Strategies for Generic Grocery Products*. Ann Arbor: UMI Research Press.

Holt, Douglas B. (2004). *How Brands Become Icons: the Principles of Cultural Branding*. Harvard Business School: Harvard Business Press.

Jacoby, J., & Mazursky, D. (1984). Linking brand and retailer images − do the potential risks outweigh the potential benefits? *Journal of Retailing, 60*(2), 105−122.

Jacoby, J., Speller, D., & Berning, C. (June 1974). Brand choice behavior as a function of information load: replication and extension. *Journal of Consumer Research, 1*, 33−42.

Jarrett, C. (Sept. 1981). The cereal market and private label. Paper presented at *Oyez Seminar: Is the Brand Under Pressure Again?* London.

Jefferys, J. (1954). *Retail Trading in Britain 1850−1950*. Cambridge: Cambridge University Press.

Jones, J. P. (1986). *What's in a Name?* Lexington: Lexington Books.

King, S. (1970). *What is a Brand?* London: J. Walter Thompson.

King, S. (1984). *Developing New Brands*. London: J. Walter Thompson.

King, S. (Oct.1985). Another turning point for brands? *ADMAP, 21*, 480−484, 519.

Kotler, P. (1988). *Marketing Management*. Englewood Cliffs: Prentice Hall International.

Lamb, D. (Jan. 1979). The ethos of the brand. *ADMAP, 15*, 19−24.

Larkin, J. (2003). *Strategic Reputation Risk Management*. Basingstoke: Palgrave.

Levitt, T. (July−Aug. 1970). The morality of advertising. *Harvard Business Review* 84−92.

Levitt, T. (Jan.–Feb. 1980). Marketing success through differentiation of anything. *Harvard Business Review* 83–91.

Marketing (6 March 1997). 17% of shoppers take own-label in error. 1.

McDonald, M. (1995). *Marketing Plans*. Oxford: Butterworth-Heinemann.

McGowan, P., & Thygesen, F. (July 2002). Defying labels. *Research* 21–24.

Meadows, R. (July–Aug. 1983). They consume advertising too. *ADMAP, 19*, 408–13. *Understanding the branding process* 63.

Morrin, M. (Nov 1999). The impact of brand extensions on parent brand memory structures and retrieval processes. *Journal of Marketing Research, 36*, 517–523.

Murphy, J. (April 1990). Brand valuation – not just an accounting issue. *ADMAP, 26*, 36–41.

Outlaw, J. (2009) *Turning Customers into Fans – the case for brand advocacy*. fhios insights (online). Available at: http://www.fhiosinsights.com/?p=265. Accessed 16 June 2009.

Patti, C., & Fisk, R. (1982). National advertising, brands and channel control: an historical perspective with contemporary options. *Journal of the Academy of Marketing Science, 10*(1), 90–108.

Pitcher, A. (1985). The role of branding in international advertising. *International Journal of Advertising, 4*(3), 241–246.

Pitta, D. A., & Katsanis, L. P. (1995). Understanding brand equity for successful brand extension. *Journal of Consumer Marketing, 12*(4), 51–64.

Porter, M. (1985). *Competitive Advantage*. New York: the Free Press.

Room, A. (1987). History of branding. In J. Murphy (Ed.), *Branding: A Key Marketing Tool*. Basingstoke: Macmillan.

Ross, I. (1971). Self concept and brand preference. *Journal of Business, 44*(1), 38–50.

Schultz, M., & de Chernatony, L. (2002). The challenge of corporate branding. *Corporate Reputation Review, 5*(2/3), 105–112.

Schutte, T. (1969). The semantics of branding. *Journal of Marketing, 33*(2), 5–11.

Shields, A. (2009). *A technique for producing a more civil society through brands* (online). Available at http://civilbranding.com/paper/. Accessed: 17 June 2009.

Simmons, M. & Meredith, B. (1983). Own label profile and purpose. *Paper Presented at Institute of Grocery Distribution Conference*. Radlett: IGD.

Sinclair, S., & Seward, K. (1988). Effectiveness of branding a commodity product. *Industrial Marketing Management, 17*, 23–33.

Staveley, N. (Jan. 1987). Advertising, marketing and brands. *ADMAP, 23*, 31–35.

Steenkamp, J., Geyskens, I., Gielens, K., & Koll, O. (2004). A global study into drivers of private label success. *Report Commissioned by AIM, the European Brands Association*.

The Secret to Google's Success, Available at http://www.businessweek.com/magazine/content/06_10/b3974071.htm. Accessed: 15 June 2009.

Thermistocli & Associates. (1984). *The Secret of the Own Brand*. London: Thermistocli &Associates.

Watkins, T. (1986). The Economics of the Brand. London: McGraw Hill.

www.brandrepublic.com/Digital/News/904325/top-100-mentioned-brands-Twitter/ (2009). Accessed: 16 June 2009

http://www.campaignforrealbeauty.com/ (2009). (Accessed: 17 June 2009).

http://www.livestrong.org/site/c.khLXK1PxHmF/b.2660611/k.BCED/Home.htm (2009). (Accessed: 17 June 2009).

http://www.whoppersacrifice.com/ (2009). (Accessed: 17 June 2009).

PART

2

Brand Management in Different Sectors

CHAPTER 3

How Consumers Choose Brands

OBJECTIVES

After reading this chapter, you will be able to:

- Explain the stages of consumer decision-making for extended, routine and limited problem-solving, as they apply to brand purchases.
- Understand the concept of involvement and recognise the relationship between brand positioning and consumer involvement.
- Discuss the cues used by customers to evaluate brands and understand the challenge to branding of consumer perception.
- Recognise the role of perceived risk in consumer brand choice.

SUMMARY

The purpose of this chapter is to show how an understanding of consumers' buying processes can help in developing successful brands. It opens by looking at how consumers process information and shows that, depending on the extent to which consumers perceive competing brands to differ and on their involvement in the brand purchase, four buying processes can be identified. The implications of the predominance of low consumer involvement with brands is addressed. The chapter considers how customers choose brands according to their clusters of needs, thereby resulting in their having brand repertoires in each product category. Through recognising that different consumers offer different profit opportunities in product categories, the benefits of differential brand marketing are considered. The chapter shows how consumers search for information about brands, explains why this search is limited, goes through the arguments for giving consumers only a few pieces of high-quality brand information and illustrates how consumers evaluate brands as arrays of clues, with the brand name emerging as a very high-quality clue. The influence of perception on branding is addressed. Building on earlier concepts in the chapter, brand naming issues are reviewed, along with a consideration of the way that brands can be presented as risk-reducing devices.

Creating Powerful Brands. DOI: 10.1016/B978-1-85617-849-5.10003-0

BRANDS AND THE CONSUMER'S BUYING PROCESS

There are many theories about the way consumers buy brands and debate still continues about their respective strengths and weaknesses. For example, some argue that brand choice can be explained by what is known as 'the expectancy-value model'. In this model, it is argued that consumers intuitively assign scores to two variables, one being the degree to which they expect a pleasurable outcome, the other being the value they ascribe to a favourable outcome. When faced with competing brands, this model postulates that consumers assign scores to these expectancy-value parameters and, following an informal mental calculation, make a selection based on the highest overall score.

We find this hard to accept, since people have limited mental processing capabilities and many brands, particularly regularly purchased brands, are bought without much rational consideration.

In reality, consumers face a complex world. They are limited both by economic resources and by their ability to seek, store and process brand information. For this reason, we are also sceptical of the economist's view of consumer behaviour. This hypothesises that consumers seek information until the marginal value gained is equal to, or less than, the cost of securing that knowledge. Many researchers have shown that consumers do not acquire perfect information – in fact even when presented with the economist's view of 'perfect' information, they are unable to comprehend it! Just think of the technical information that accompanies mobile handsets, cars, audio systems and PCs.

It is not our aim to become embroiled in a review of the merits of the different consumer behaviour models that could explain the brand selection process. The interested reader can appreciate this by consulting any of the numerous texts on this subject or by consulting the references at the end of the chapter.

Instead, we subscribe to a more well-accepted model of consumer behaviour. This shows the consumer decision process occurring as a result of consumers' seeking and evaluating small amounts of information to make a brand purchase. Consumers rely upon a few pieces of selective information which, they feel confident, will help them decide how the brand might perform. For example, why does someone flying from London to Amsterdam choose BA rather than KLM? Both airlines offer good service, a high degree of reliability and many convenient departure slots. They may then choose BA, when there is so little difference, only because the name is more familiar, it reflects their national pride or because they belong to a "loyalty scheme".

The stages in the buying process, when consumers seek information about brands and the extent of the information search, are influenced by an array of factors such as time pressure, previous experience, their situation, advice from

AFTER GOING THROUGH HELL FOOD DESERVES HELLMANN'S

THE ORIGINAL MAYONNAISE

HELLMAN'S

Hellman's strives to remain top of mind by claiming to be 'the original mayonnaise'. This advertisement represents the customer's situational involvement, as Hellman's creates an association between the product and the usage situation of a barbecue

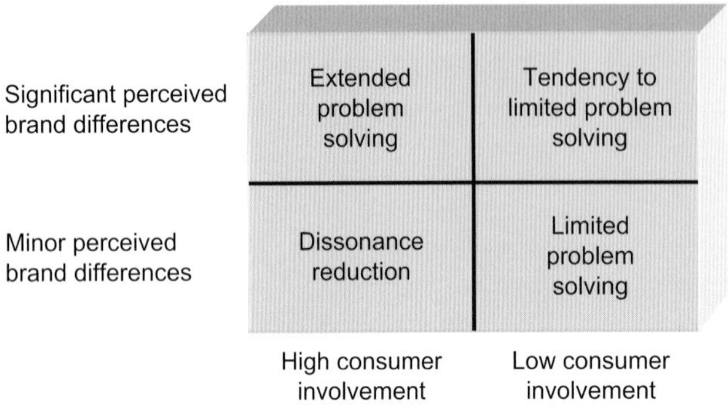

	High consumer involvement	Low consumer involvement
Significant perceived brand differences	Extended problem solving	Tendency to limited problem solving
Minor perceived brand differences	Dissonance reduction	Limited problem solving

FIGURE 3.1

Typology of consumer decision processes *(adapted from Assael 1987)*

friends and so on. However, two factors are particularly useful in explaining how consumers decide. One is the extent of their involvement in the brand purchase, the other is their perception of any differences between competing brands. For example, a housewife may become *very involved* when buying a washing machine, because with her large family it is important that she replaces it quickly. As such, she will show an *active* interest in evaluating different washing machine brands, all of which she can probably evaluate because of her experience. She will be able to evaluate the few brands that broadly appeal and will buy the brand which comes closest to satisfying her needs on one or a number of the key attributes important to her. By contrast, the same housewife is likely to show *limited involvement* when buying a brand of baked beans, as they are of little personal importance and evoke little interest in her regular grocery shopping. She may perceive minimal difference between competing brands and because of the low importance of this purchase, does not wish to waste time considering different brands. As such, she is likely to make a rapid decision, based predominantly on previous experience.

With an appreciation of the extent of consumers' *involvement* in a purchase decision and their perception of the degree of *differentiation* between brands, it is possible to categorise the different decision processes using the matrix shown in Fig. 3.1.

The strength of this matrix, as will now be shown, is that it categorises and illustrates simply, the stages through which the consumer is likely to pass when making different types of brand purchases. Each of the quadrants is now considered in more detail.

EXTENDED PROBLEM SOLVING

Extended problem solving occurs when consumers are *involved* in the purchase and where they perceive *significant differences* between competing brands in the same product field. This type of decision process is likely for high-priced brands that are generally perceived as a risky purchases due to their complexity (e.g. washing machines, cars, TVs, home computers) or brands that reflect the buyer's self image (e.g. clothing, cosmetics, jewellery). It is characterised by consumers *actively* searching for information to evaluate alternative brands.

DULUX

Dulux uses perceptual organisation, as the consumer sees a smile in this advert, which promises a positive emotional experience with a change of paint colour

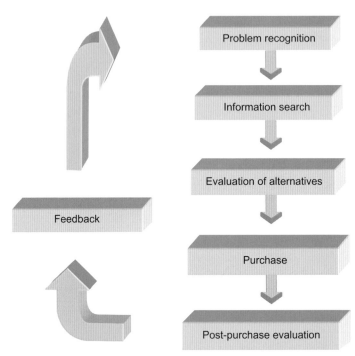

FIGURE 3.2
Stages in complex decision making

When making a complex purchase decision, consumers pass through the five stages shown in Fig. 3.2.

The decision process starts when the consumer becomes aware of a problem. For example, a young person may have tried their friend's new mobile phone and become aware of how inferior their own phone is. This recognition would trigger a need to resolve the problem and, if they feel particularly strongly, they will embark on a course to replace their phone. Depending upon their urgency to act and their situation (e.g. time availability, financial situation, confidence, etc.) they might take action quickly or, more likely, will become more attentive to information about mobile phones and buy a brand some time later.

The search for information would start first in their own memory and, if they feel confident that they have sufficient information already, they will be able to evaluate the available brands. Often, though, consumers do not feel sufficiently confident to rely on memory alone (particularly for infrequently bought brands), so they will begin to scan the external environment (e.g. visit shops, browse online, become attentive to certain advertisements, talk to friends). As they get more information, the highly involved consumers will start to learn how to interpret the information in their evaluation of competing brands.

Even so, consumers do not single-mindedly search for information about one particular purchase. The New York Times recently reported findings from a market research study by Yankelovich, which suggest that a person living in a city 30 years ago saw up to 2,000 advertisement a day, compared with up to 5,000 adverts in 2009. The newspaper cites extreme examples of advertising's ubiquity, including chicken eggs stamped with names of TV shows, software adverts on airline seat-backs, and advertising for children's brands on the liners of pediatrician's tables.

Of these messages, people are attentive to less than 2%. Consumers' perceptual processes protect them from information overload and helps them search and interpret new information. The issue of what these perceptual processes are is dealt with later in this chapter. Should something interest them, their attention will be directed to this new source. Even here, however, of the few advertise-ments that they take notice of, they are likely to ignore the points that do not conform to their prior expectations and interpret some of the other points within their own frame of reference.

Thus, the brand marketer has to overcome, amongst other issues, three main problems when communicating a brand proposition. First, they have to fight through the considerable 'noise' in the market to get their brand message noticed. If they can achieve this, the next challenge is to develop the content of the message in such a way that there is harmony between what the marketer puts into the message and what the consumer takes out of the message. Having overcome these two hurdles, the next challenge is to make the message powerful enough to be able to reinforce the other marketing activities designed to persuade the consumer to buy the brand.

As the consumer in our example mentally processes messages about competing brands, they would evaluate them against those criteria deemed to be most important. Brand beliefs are then formed (e.g. 'the Nokia model is easy to use', 'I could get apps with the iPhone', etc.). In turn, these beliefs begin to mould an attitude and if a sufficiently positive attitude evolves, so there is a greater likelihood of a positive intention to buy that brand.

Having decided which mobile phone brand to buy, the consumer would then make the purchase — assuming a distributor can be found for that particular brand and that the brand is in stock. Once the consumer has the phone, the consumer would discover its capabilities and assess how well their expectations were met by the brand. As can be seen from the model shown in Fig. 3.2, they would be undertaking post-purchase evaluation. Satisfaction with different aspects of the brand will strengthen positive beliefs and attitudes towards the brand. Were this to be so, the consumer would be proud of their purchase and praise its attributes to their peer group. With a high level of satisfaction, the consumer would look favourably at this company's brands in any future purchase.

Should the consumer be dissatisfied though, they would seek further information after the purchase to provide reassurance that the correct choice was made. For example, they may go back to the retailer, where the brand was bought, and check that the SIM card is correctly inserted and that the device is not faulty. If they find sufficiently reassuring information confirming a wise brand choice, they will become more satisfied. Without such positive support, they will become disenchanted with the brand and over time will become more dissatisfied. They are likely to talk to others about their experience, not only vowing never to buy that brand again, but also convincing others that the brand should not be bought.

In the event that the consumer is satisfied with the brand purchase and repeats it in a relatively short period of time (e.g. buys a mobile phone for their son's or daughter's birthday), they are unlikely to undergo such a detailed search and evaluation process. Instead, the person is likely to follow what has now become a more routine problem-solving process. Problem recognition would be followed by memory search which, with prior satisfaction, would reveal clear intentions, leading to a purchase. Brand loyalty would ensue, which would be reinforced by continued satisfaction (should quality be maintained). This process is shown in Fig. 3.3.

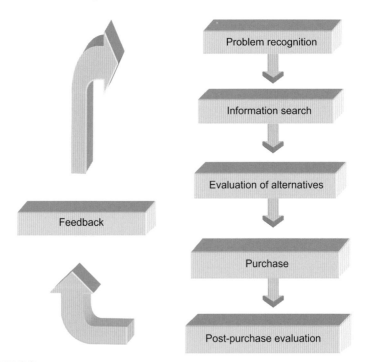

FIGURE 3.3
Routine problem-solving behaviour

When consumers are deeply involved with the brand purchase and when they perceive large differences between brands, they are more likely to seek information actively in order to make a decision about which brand to buy. As such, brand advertising may succeed by presenting relatively detailed information explaining the benefits of the brand, as well as reinforcing its unique differential positioning. It is important for the brand marketer to identify those attributes consumers perceive to be important and focus on communicating them as powerfully as possible. In circumstances such as those just described, as the consumer is likely to actively seek information from several different sources, the brand marketer should use a consistent multimedia promotional approach. Also, it is important to ensure that all retail assistants likely to come in contact with our inquisitive consumer are well versed in the capabilities of the brand.

DISSONANCE REDUCTION

This type of brand buying behaviour is seen when there is a high level of consumer involvement with the purchase but the consumer perceives only *minor differences* between competing brands. Such consumers may be confused by the lack of clear brand differences. Without any firm beliefs about the advantages of any particular brand, a choice will most probably be made based on other reasons such as, for example, a friend's opinion or advice given by a shop assistant.

Following the purchase, the consumer may feel unsure, particularly if they receive information that seems to conflict with their reasons for buying. The consumer would experience mental discomfort or what is known as 'post-purchase dissonance' and would attempt to reduce this state of mental uncertainty. This would be done either by ignoring the dissonant information, for example, by refusing to discuss it with the person giving conflicting views or by selectively seeking those messages that confirm prior beliefs.

In this type of brand purchase decision, the consumer makes a choice without firm brand beliefs, then changes attitude *after* the purchase – often on the basis of experience with the chosen brand. Finally, learning occurs on a selective basis to support the original brand choice by the consumer being attentive to positive information and ignoring negative information. This brand buying process is shown in Fig. 3.4.

Dissonance reduction may be illustrated through the following example. A young businesswoman is very proud of having just bought her first flat. To reflect her successful life, she wants to furnish the place with modern and stylish comforts. Although she is very busy and financial considerations are hardly crucial, she thinks she should consider carefully which carpet would be the

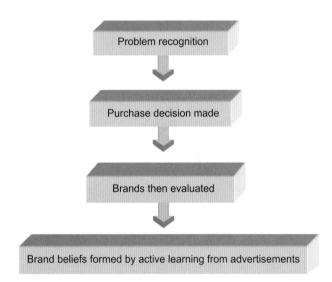

FIGURE 3.4

Brand purchasing under dissonance reduction

most appropriate for her dining room, in which she hosts many leisure and business parties. So she goes to different shops, but finds the experience very confusing. She is overwhelmed by the flood of information about carpet attributes she has never heard of before: type of pile, proportion of the wool, whether it is Scotchguarded, etc. Unable to fully evaluate the advantages and disadvantages of the different carpets, she makes a rapid decision based on her perceptions and the reassuring explanations of a particularly helpful salesperson. At her first dinner party after the new carpet has been fitted one of her guests spills a glass of red wine, which she rapidly cleans up as recommended by the salesperson. The following morning she checks whether there is any stain left. When she sees that the carpet is spotless, she realises that the Scotch-guarding has protected it against this kind of accident. When her guest calls her to apologise for the damage, she is pleased to reassure them that everything is perfectly fine. As a consequence of this experience, she starts reading leaflets to understand the different carpet types. It is only then, with hindsight, that she compliments herself on her wise choice.

When consumers are involved in a brand purchase, but perceive little brand differentiation or lack the ability to judge between competing brands, the advertising should reduce post-purchase dissonance through providing reassurance *after* the purchase. For example, in the software market, consumers may not have the technical expertise to distinguish between software options (for example, functionality, simplicity, likelihood of problems, etc.). However, they are involved in the purchase due to the risk they perceive in running the software. By featuring 'Frequently Asked Questions' sections on websites and by using online weblogs featuring comments from software experts, software manufacturers can reassure customers after purchase to encourage loyalty and recommendation.

Also, close to the point of purchase, as consumers are unsure about which brand to select, promotional material is particularly important in increasing the likelihood of a particular brand being selected. For example, information about traceability may influence consumers' choice of meat products. Likewise, any packaging should try to stress a point of difference from competitors and sales staff should be trained to be 'brand reassurers', rather than 'brand pushers'.

PERSIL

Persil Small and Mighty seeks to reduce dissonance by offering a money back guarantee. (Reproduced by kind permission of Persil)

An example of this is in retail pharmacy, where pharmacy assistants are ethically obliged to recommend the brand that meets the best interests of the customer.

LIMITED PROBLEM SOLVING

When consumers do not regard the buying of certain products as important issues and when they perceive only minor differences between competing brands in these product fields (e.g. packaged groceries, household cleaning materials), then their buying behaviour can be described by the 'limited problem solving process'. The stages that the consumer passes through are shown in Fig. 3.5.

Problem recognition is likely to be straightforward. For example, an item in the household may be running low. As the consumer is not particularly interested personally in the purchase, they are not motivated actively to seek information from different sources. Whatever information they have will probably have been passively received, say, via a television commercial that the consumer wasn't paying particular attention to.

Alternative evaluation, if any, takes place *after* the purchase. In effect, fully formed beliefs, attitudes and intentions are the *outcomes* of purchase and not the cause. The consumer is likely to regard the cost of information search and evaluation as outweighing the benefits.

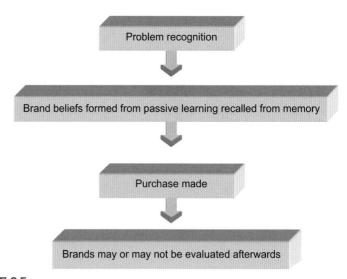

FIGURE 3.5
Limited problem solving brand purchase

Promotions providing information, however, still do have a role to play in low involvement brand purchasing. But whilst they have a positive role, it is different from that in high involvement buying. Consumers passively receive information and process it in such a way that it is stored in their memories without making much of an impact on their existing mental structure. Having stored the message, no behavioural change occurs until the consumer comes across a purchase trigger (e.g. an in-store display of the brand) at the point at which they need to purchase the product in question. After trying the brand, the consumer can then decide how satisfactory it is. If the brand is satisfactory, there will be belief in the particular brand, albeit a fairly 'weak' one, which will lead to the likelihood of repeat brand buying.

When *regularly* purchasing these kinds of brands, consumers would establish buying strategies that reduce the effort in decision making (i.e. routine problem solving). Following a similar flow chart to that in Fig. 3.3, any further purchase decisions about the brand would be based on a memory scan which, if holding details about a satisfactory experience, would result in brand loyalty. Thus, in extended problem solving situations, the consumer considers it important to buy the 'right brand' and it is difficult to induce brand switching. By contrast, in low involvement buying, the 'right brand' is less central to the consumer's lifestyle and brand switching may be more easily achieved through coupons, free trial incentives, etc.

In low involvement purchasing, consumers occasionally show variety-seeking behaviour. There is little involvement with the product and therefore the consumer feels little risk in switching between brands. Over time, consumers feel bored buying the same brand and occasionally seek variety by switching.

Consumers pay minimal attention to advertisements for these kinds of brands. Consequently, the message content should be kept simple and the advertisements should be shown frequently. A single (or a low number) of benefits should be presented in a creative manner that associates a few features with the brand. In low involvement brand buying, consumers are seeking acceptable, rather than optimal purchases, i.e. they seek to minimise problems rather than maximise benefits. Consequently, it may be more appropriate to position low involvement brands as functional problem solvers (e.g. a brand of washing up liquid positioned as an effective cleaner of greasy dishes) rather than as less tangible benefit deliverers (e.g. a brand of washing up liquid positioned as being kind to hands).

Trial is an important method by which consumers form favourable attitudes after consumption, so devices such as money-off coupons, in-store trial and free sachets are particularly effective.

As consumers are not motivated to search out low involvement brands, manufacturers should ensure widespread availability. Any out-of-stock

RADOX
Radox appeals to the self-indulgent need state with their series of 'be selfish' advertisements and a supporting brand

situations would probably result in consumers switching to an alternative brand, rather than visiting another store to find the brand. Once inside a store, little evaluation will be made of competing brands, so locating the brand at eye level or very close to the check-out counter is an important facilitator of brand selection. Packaging should be eye-catching and simple.

TENDENCY TO LIMITED PROBLEM SOLVING

While the 'limited problem solving' aspect of the matrix describes low involvement purchasing with *minimal* differences between competing brands, this can also be used to describe low involvement brand purchasing when the consumer perceives *significant* brand differences. When a consumer feels minimal involvement, they are unlikely to be sufficiently motivated to undertake an extensive search for information. So even though there may be notable differences between brands, (e.g. Pringles' tub packaging), because of the consumer's low involvement they are less likely to be concerned about any such differences. Brand trial would take place and, in an almost passive manner, the consumer would develop brand loyalty. The brand selection process is very similar to that described in 'limited problem solving' and similar marketing issues need to be addressed. Chapter 2 earlier identified the emphasis marketers place on brands as legal devices. It is now more evident why brand owners feel concern about competitors developing similar looking brands – 'look-alikes'. In packaged groceries, retailers' own labels often employ similar packages and designs to leading brands. Where these look-alikes compete in low involvement categories, consumers are unlikely to invest their time searching for information. Therefore, some of the consumers who superficially glance along the shelves may pick up a pack because it appears to be the leading brand. An indication of the scale of this error is considered in Chapter 2 in 'Brand as a legal device'.

CONSUMERS' LOW INVOLVEMENT WITH BRANDS

For many product and service categories, consumers exhibit a low level of involvement in changing between brands. Furthermore, consumers often show low levels of attention, low interest and limited absorption of brand communications. Decisions are often affected by data sensed and stored in memory some time ago, rather than by undertaking detailed search activities. Heath (2001) suggests that consumers now expect reputable brands to be similar, so they do not regard brand learning as very important. Therefore, brand decisions are made intuitively instead of rationally. Low involvement processing is used, which requires little working memory and thus messages are stored as simple associations with the brand. He cites the example of the blue colour used with Pepsi and the red colour used with Coca-Cola as signals we pick up about those brands. When associations are repeatedly stored via low involvement processing, this forms strong links to the brand in long-term memory. In turn, this affects intuitive decision making, yet the consumer finds it difficult to explain the reasons for their choice and does not realise that advertising had an impact on behavoiur.

Heath (2001) provides an example of low involvement in practice, where the customer is buying luxury toilet paper. One advertisement uses a puppy to suggest softness, the other claims that their product is quilted. Heath (2001) explains that over time, the customers' involvement level with the advertisement will be low, as the customer is not working at involvement. Therefore, the customer stores 'puppy' for one brand, and 'quilted' for the other. When the customer stands at the shelf in a hurry and has to choose between the two brands, they consider the markers they have in memory and choose the brand with the best associations. For example, Heath (2001) explains that the 'puppy' association may also have associations of 'loving' and 'family', which could influence the customers' brand choice.

Romaniuk and Sharp (2004) emphasise the importance of brand salience as distinct from brand attitude. Salience, they explain, is 'the propensity of the brand to be thought of by buyers in buying situations' (Romaniuk and Sharp, 2004, p.328). They explain that salience might arise from a number of cues. For example, we might never have seen an advertisement for McDonalds. If the first advertisement we see for McDonalds features salads, we will associate salads with McDonalds. If we had already seen an advertisement for McDonalds, our association for McDonalds is strengthened, as we are adding new links to the information we already have about the brand. We might also add cues together. For example, we might consider usage situations when buying a soft drink (I am going to a barbecue/I am buying this for use at home); or we might consider benefits (this is refreshing, it is low fat); or we might think of functional qualities (it has a sports lid, it is plastic so it will not break). Romaniuk and Sharp (2004, p.335) explain that the more cues that are associated with a brand, the more likely it is that the brand will come to mind when the consumer is in a purchase situation and the brand will be more likely to be thought of as an option to buy.

Another factor besides level of usage that influences the level of image response is the extent to which a brand's attribute is a defining characteristic of the category. For example, security and interest rates are two of several characteristics that link brands to the category of banks. The more attributes a brand has of this category, the more likely consumers make this association. Furthermore, these two attributes are regarded as being more prototypical of brands belonging to the category of banks then, say, the attribute "is Swiss". Romaniuk and Sharp (2000) draw on this to consider how image data can be used to refine a brand's positioning. Respondents are interviewed and for a specified list of brands, they are asked whether they perceive each brand as having a particular attribute. The data is then tabulated for the total sample, such that the attributes are presented in descending order of frequency of mention going down the page and the brands are presented across the page in terms of descending order of users.

As there is a usage effect shown by the number of responses a brand gets across attributes and a proto-typical effect in the frequency of responses to attributes, these can be used to calculate the expected number of responses for each brand-attribute cell. A statistically based formula, the chi-squared formula, is employed, i.e.:

$$\text{Expected score for a cell} = \frac{(\text{Row total}) \times (\text{Column total})}{\text{Total for all cells}}.$$

By then subtracting the actual brand-attribute scores for each cell from the expected scores, a new grid can be produced showing how each brand deviates from the expected score on each attribute. Column by column, going down each brand's deviation scores, one can see on which attributes a brand is particularly outstanding [high positive deviations] and particularly weak [notable negative deviations]. Checking these few attributes against the desired positioning attributes in the brand plan, the discrepancies indicate areas for further development work, i.e. reinforcing some attributes and undertaking remedial work on others. Furthermore, those attributes that are not associated with any of the brands need considering in terms of whether they offer attractive potential positioning options.

This section has shown that, given an appreciation of the degree of involvement consumers have with the brand purchase and their perception of the degree of differentiation between brands, it is possible to identify their buying processes. With an appreciation of the appropriate buying process, the marketer is then able to identify how marketing resources can best be employed. As the next section explains, a further benefit of appreciating consumers' buying processes is that brands can be developed and presented in such a way that consumers perceive them as having added values over and above the basic commodity represented by the brand.

CONSUMERS' NEED-STATES

Brands satisfy different needs. One way of appreciating this is through the three categories of needs identified by Park and his colleagues (1986). Functional brands focus on technical features and mainly solve externally-generated

Table 3.1 Need-States and Yoghurt Brands (*after Gordon 1994*)	
Need-state	**Brand**
Sophisticated me	Real Greek Yoghurt
Mummy me	Munch Bunch
Healthy me	Shape
Budget me	Tesco own label
Indulgent me	Marks & Spencer

consumption needs, e.g. Microsoft Office solves the need for efficient office communication. Symbolic brands, such as Chanel or Rolex, stress intangible benefits and fulfill internally generated needs for self enhancement, role position, group membership or ego identification. Experiential brands, e.g. Center Parc, Build-a-Bear Workshop and Rainforest Cafe satisfy desires for sensory pleasure, variety and cognitive stimulation.

Wendy Gordon's work (1994) shows that brands are bought to reflect consumers' needs in a particular context and she introduces the term 'need-state' to describe this. She has demonstrated that there can be more differences between the same consumer purchasing on two different occasions than between two different consumers choosing a brand on the same occasion. Consumers use brands to meet specific needs at a particular time in a specific situation. Consider, for example, the changing need-states of a housewife buying yoghurt during a week, as shown in Table 3.1. On a Friday, she prepares for an important dinner party. One of the recipes needs 'Real Greek Yoghurt', which in her opinion is proof of being a sophisticated hostess. To ease her feelings of guilt at the children having to stay upstairs watching television during this dinner party, she buys a pack of Munch Bunch yoghurts. The day after the party she weighs herself and is concerned at her increased weight as a result of the creamy dessert she has eaten and sees the solution in a diet yoghurt. On Tuesday her weekly shopping brings back the reality of buying within a budget and she chooses the cheaper, own label yoghurts. Two days later her husband takes the children bowling and she stays at home to watch her favorite television show. Although she is still aware of her diet plans, she wants to indulge herself and chooses a Marks & Spencer brand.

This example shows the need to reconsider the concept of brand loyalty. It would be more realistic to consider it in terms of consumers having a repertoire of brands in a particular product field and switching between these brands. Consumers using two brands in a short period of time should not be considered 'disloyal'. Instead this is more an expression of active discernment, choosing brands to meet specific need-states. Most consumers switch between brands.

DIFFERENTIAL BRAND MARKETING

Gordon's work provides some enlightenment about the reasons why consumers show a greater propensity to use a particular brand, yet will also, from time to time, use other brands. Ehrenberg's empirical work (1993) provides considerable insight about consumers' 'loyalty' behaviour, in particular:

- heavy category users have a repertoire of brands;
- big brands are bought within the repertoire more frequently than smaller brands;
- a brand's most loyal consumers are the least valuable, as they buy the category infrequently;
- a brand's best consumers are mostly competitors' consumers who occasionally buy it.

One of the implications from these findings is that marketers should not regard each consumer in a target segment as being equally attractive and assume that the same brand marketing strategy is equally appropriate across all the target segment. More recently, Mitchell (2005) identified two UK studies undertaken across numerous product categories that add further insights to loyalty behaviour. He cites a research study by Reinartz that found that loyal customers become more demanding, therefore they cost more to serve. In addition, those who were classified 'loyal' and were showered with retention marketing programmes, had already 'moved on' and were unresponsive and ultimately not profitable. In addition, the more comfortable customers were with a company's systems, the better they were about 'squeezing' better value from them. Mitchell (2005) also cited a study by East, who found that new customers were the best advocates for the brand. Therefore, better information about customer segments will provide better opportunities to target marketing programmes to harness customer loyalty.

In fact, Hallberg's research (1995) shows the need for different strategies for the same brand across the target segment as a result of different buyers having differing degrees of attractiveness. He argues that consumers should be categorised by their 'profit differentials', particularly since 10–15% of buyers in a category produce the majority of a brand's profits. As such, differential brand marketing is needed, whereby the more valuable consumers are, the greater the special treatment they deserve.

The way to develop a differential brand marketing strategy is, as Hallberg shows, to segment consumers into four profit opportunity groups – i.e. high, medium, low and no profit segments. To do this requires good market research data, good management information systems showing detailed costs and a database of the brand's consumers, ideally over a number of years. A case study in the yoghurt category showed that a yoghurt brand was only bought by

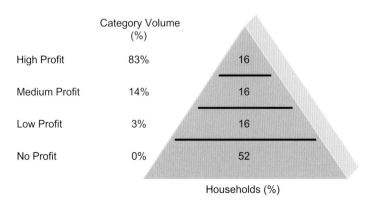

Category Volume
(%)

High Profit	83%
Medium Profit	14%
Low Profit	3%
No Profit	0%

16

16

16

52

Households (%)

FIGURE 3.6
Yoghurt category profit matrix *(after Hallberg 1995)*

48% of households in a year. The 52% of yoghurt-consuming households not buying the brand in that year represent the "no profit opportunity" group and, particularly if a national television campaign had been used, reflected an unproductive use of resources. The firm knew how many yoghurts each household bought in that year and arranging households in descending order according to the numbers of yoghurts bought, then divided them into three equally-sized groups, each accounting for 16% of households, as shown in the profit matrix pyramid of Fig. 3.6. Analysis then revealed that the high profit segment, the top 16% of households, accounted for 83% of the brand's profits. With analysis such as this showing just how profitable a small group of consumers are, firms can develop much more focused brand strategies. Consumers can be categorised by their profit attractiveness and, particularly since their addresses are known from the firm's database, more productive use can be made of resources through tailoring different strategies to different profit opportunity consumers.

Using this analysis, marketers can benefit from tracking lost consumers and, particularly for high profit potential consumers, through understanding their reasons for defecting, make appropriate brand changes. Lost consumers can be categorised according to whether they leave the brand due to changing need-states, for example, a baby growing out of nappies, or through dissatisfaction. Again, good market research should help amend the brand's strategies to reduce the defection rate.

In an online context, Koças and Bohlmann (2008) highlighted the relevance of customer segmentation for Internet retailers, even for undifferentiated homogenised goods such as books or CDs. In their study, they found that firm-specific loyalty alone was not sufficient to explain differentiated pricing. Instead, pricing is driven by the ratio of segments within which a firm competes to its loyal segment size. Therefore as an example, smaller online retailers can limit price switchers to soften price competition.

CONSUMERS' PERCEPTIONS OF ADDED VALUES

Brands are able to sustain a price premium over their commodity form since consumers perceive relevant added values. For brands with which consumers become personally involved, there may be a complex cluster of added values over and above the brand's basic functional purpose, such as the ability of a clothing brand to signify membership of a particular social group whose distinctiveness is the result of their ability always to be at the forefront of chic fashion. By contrast, for low involvement brands, the added value could be as simple as the friendly smile consistently apparent from a newsagent, encouraging the busy commuter buying his daily newspaper to remain a loyal consumer.

The concept of added values is an extremely important aspect of brands, being their *raison d'être*. As there are many different ways in which added values interact with commodity forms in the branding process, we do not have just one chapter dedicated to this. Instead the topic is addressed at appropriate points, in particular in Chapter 10. Here we briefly overview some of the aspects of added value and consider its role in relation to consumer behaviour.

The brand's added values are those that are relevant and appreciated by consumers and are over and above the basic functional role of the product. For example, well-travelled international sales executives may recognise the prime functional benefit of Hilton as being clean, comfortable establishments to sleep in, but they also appreciate the brand's 'no surprises strategy' as an added value. This hotel group has a policy of consistency of standards throughout the world. For brands with which the consumer feels low involvement, the added values may often be other functional benefits. For example, while the prime benefit sought from a bleach may be its ability to kill germs, Domestos not only meets this requirement but also has the added value of a directional nozzle to ensure that difficult-to-reach corners are reached. The marginal cost from the directional nozzle is an added value contributing to the overall value of the brand, enabling a price premium to be charged. Likewise, while consumers may buy a brand of salt for its taste, they may pay a premium for a particular brand which contains a lower sodium level. A further example of added value can be seen in Green & Black's chocolate. According to the company's website, the product is about being 'consciously respectful of our earth while also indulging in the best things life has to offer'. To emphasise this, the 'green' element of the brand is built into its brand name.

For the marketer, the challenge is to appreciate how all of the marketing resources supporting a brand interact to produce the added values which consumers perceive as being unique to a specific brand. The physical (or service) component is combined with symbols and images communicated through the internet, advertising, PR, packaging, pricing and distribution to create meanings. These meanings not only differentiate the brand but also give it added values. Consumers interpret the meaning of the marketing activity behind a brand and project values onto the brand, endowing brands with a personality. Many researchers in the 1960s and 1970s demonstrated the added value of strong brand personalities, showing that consumers tend to choose brands with the same care as they choose friends. Aaker (1997) categorised personality into five dimensions, which were Sincerity, Excitement, Competence, Sophistication and Ruggedness. These traits can be expanded in meaning; for example, Sophistication means 'upper class' or 'charming'. We can see this personality trait apply to brands such as Chanel or Rolex. On the other hand, ruggedness means 'outdoorsy' or 'tough', which would apply to brands such as Land Rover or Marlboro. Consumers can project an image of their own personality by selecting a brand that reflects that personality. By interpreting the personality of brands, consumers feel more comfortable buying particular brands. This can be for a variety of reasons; for example, dilutable Ribena has a personality that is warm, caring and maternal, while the ready to drink format of the same brand is seen by children as fun and lively.

Chapter 4 discusses the added value of brand personality in more depth, but it is worth stressing here that a strong brand personality evolves because of a consumer-focused marketing investment programme. It takes time and resources to build the brand personality and a lack of commitment to brand investment will weaken the brand's personality. For example, counterfeit versions of luxury brands may negatively impact on their perceived sophistication.

A useful framework to help us understand the diverse types of brand added values was introduced by Jones (1986). It consists of four types:

1. *Added values from experience*. With repeated trial, consumers gain confidence in the brand and, through its consistent reliability, perceive minimal risk. Particularly in grocery shopping, where a consumer is typically faced with over 35,000 lines in a grocery superstore, this added value is particularly appreciated, enabling consumers to complete their shopping rapidly by choosing known names.
2. *Added values from reference group effects*. The way that advertising uses personalities to endorse a brand is perceived by many target consumers as relating it to a certain lifestyle to which they may well aspire. An example of this is the promotion of Gillette Champions – Thierry Henry, Tiger Woods

Table 3.2 The Impact of Branding on Taste Tests

	Blind %	Branded %
Prefer Diet Pepsi	51	23
Prefer Diet Coke	44	65
Equal/don't know	**5**	**12**
	100	100

and Roger Federer, whose sporting success communicates the Gillette slogan 'the best a man can get'.

3. For example, for branded painkillers, it has been shown that approximately a quarter of the pain relief was attributable to branding. In the suits market, one often hears of men remarking that for a certain event they 'feel' more comfortable with a certain type of suit. Further evidence of this added value is seen when consumers taste brands without their names, then with their names. In Table 3.2 it can be seen that functionally Diet Pepsi is preferred to Diet Coke in blind testing but, through the strong image recalled by the name, the overall preference in the branded product test is for Diet Coke.

4. *Added values from the appearance of the brand.* Consumers form impressions of brands from their shapes and packaging and develop brand preferences based on their attraction to the pack design. This is particularly so in the premium biscuit market, where designers use metal tins to enhance the premium positioning of their brand and in the confectionery market where Toblerone offers a triangular product as a point of difference.

Thus, successful brands have consumer-relevant added values which buyers recognise and value sufficiently to pay a price premium. Accepting this, the task for the marketer is to communicate a brand's added values to consumers in such a way that the message penetrates consumers' perceptual defences and is not subject to perceptual distortion. This is not a small problem since, as the next section shows, consumers are attentive to only a small amount of marketing information.

The extent to which consumers search for brand information

The way consumers gain information is shown in Fig. 3.7, i.e. first from memory but if insufficient information is held, then from external sources.

Information may be stored in the memory as a result of an earlier active search process, as in the case for example, of the assessment of a newly-bought brand immediately after purchase. Alternatively, information may be stored in the

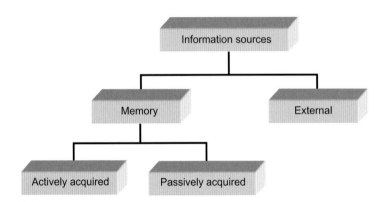

FIGURE 3.7
The way consumers get brand information

memory as a result of a passive acquisition process — an advertisement might-have caught the reader's eye whilst a newspaper was being casually skimmed.

Two relevant factors that influence the degree to which consumers search their memory are the amount of stored information and its suitability for the particular problem. In a study amongst car purchasers, those repeatedly buying the same brand of car over time undertook less external search than those who had built up a similar history in terms of the number of cars bought but who had switched between brands of cars. Repeatedly buying the same brand of car increased the quantity of *suitable* information in the memory and limited the need for external search.

If there is insufficient information already in a consumer's memory and if the purchase is thought to warrant it, external search is undertaken. Research into consumer behaviour, however, shows that external search is a relatively limited activity, although there are variations between different groups of consumers. In one of the early studies on consumers' search for information, recent purchasers of sports shirts or major household goods (e.g. TVs, fridges, washing machines, etc.) were asked about their pre-purchase information search. Only 5% of electrical appliance buyers showed evidence of a very active information search process, whilst one-third claimed to seek virtually no pre-purchase

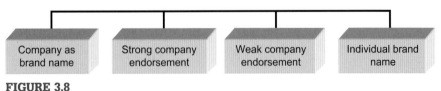

FIGURE 3.8
Brand name spectrum

information. Just under one-half (47%) of appliance purchasers visited only one store and only 35% considered another attribute in addition to brand name and price. Even less evidence of information search was found amongst purchasers of sports shirts, the conclusion being 'that many purchases were made in a state of ignorance, or at least of indifference'. As was pointed out, however, the apparent lack of deliberation does not indicate irrational decision behaviour. Some purchasers may have found it difficult to evaluate all the features of a product and instead have relied upon a limited number of attributes that they felt more comfortable with.

A further detailed study of consumers buying cars and major household appliances again showed evidence of limited external search. Less than one-half of the purchasers interviewed (44%) used no more than one information source, 49% experienced a deliberation time of less than two weeks and 49% visited only one retail outlet when making these major brand purchases. Numerous other instances have been reported of consumers undertaking limited external search for expensive brand purchases in such product fields as financial services, housing, furniture and clothing.

Not surprisingly, for low cost, low risk items (e.g. groceries), external search activity is also restricted. No doubt due to the low level of involvement that these brands engender, far more reliance is placed on memory. For example, when shopping for washing powders, consumers simplify purchase by considering only one or two brands and by using only three to five brand attributes. Amongst consumers of breakfast cereals, only 2% of the available information was used to make a decision. When using in-store observations of grocery shoppers, 25% made a purchase decision without any time for deliberation and 56% spent less than eight seconds examining and deciding which brand to buy.

With the increase in e-commerce, consumers have taken their information searches online. Su (2008) explains that benefits of online shopping include greater product-related information and reduced search costs. Searching for an appropriate seller is a cross-site search, while searching for a product or price is an in-site search. He found that respondents went directly to their preferred websites, as cross-site searches reduced satisfaction. In addition, in-site searching reduced customers' liking for online shopping, 39% of respondents could not navigate the site, 56% of attempts to search for information failed, and 62% of online shoppers had given up. Therefore, Su (2008) calls for increased access convenience through search engine optimisation, and increased search convenience through clearer menu categories. Successful online brands such as Amazon achieve this, for example, through simple one-click-to-purchase approaches.

Reasons for limited search for external information

Several reasons exist for the apparently limited external search. Consumers have finite mental capacities, which are protected from information overload by perceptual selectivity. This focuses consumers' attention on those attributes considered important. For example, one study reported that because of perceptual selectivity, only 35% of magazine readers exposed to a brand advertisement noticed the brand being advertised. Information is continually bombarding consumers and information acquisition is a continuous process.

The search for external information represents a cost (the time and effort) and some consumers do not consider that the benefits outweigh these costs. This is particularly so for low involvement brands.

In research studies into consumer behaviour, a lot of emphasis has been placed on measuring the *number* of sources consumers use rather than considering the *quality* of each informational source.

The prevailing circumstances of consumers also have an impact on the level of external search. Consumers may feel time pressure (e.g. newly married couples seeking a home when there is a lack of rented accommodation) or they may not find the information easy to understand (e.g. food labels). The search for information is also affected by the consumer's emotional state. For example, one study reviewing the way funeral services are marketed noted that due to their emotional state, consumers pay little attention to information during this traumatic period.

BRAND INFORMATION: QUALITY OR QUANTITY?

The preceding sections have shown the relatively superficial external search for information undertaken in selecting brands. The question marketers need to consider, therefore, is whether increasing levels of information help (or hinder) consumer brand decisions. In one of many studies to assess the decision-making process, consumers were presented with varying levels of information about brands of washing powders and asked to make brand selections. Prior to the experiment, they were asked to describe their preferred washing powder brand. The researchers found that accuracy (in the sense that consumers selected the brand in the survey that matched their earlier, stated brand preference) was *inversely* related to the number of brands available. Initially, accuracy of brand choice improved as small amounts of brand information were made available but a point was reached at which further information reduced brand selection accuracy.

In another study, housewives were given varying levels of information about different brands of rice and pre-prepared dinners and were asked to choose the brand they liked best. Again, prior to the experiment, they were asked about their preferred brand. Confirming the earlier survey on washing powders, increasing information availability from low levels helped decision making but continuing provision of information reduced purchasing accuracy and resulted in longer decision periods.

The conclusion is that marketers need to recognise that increasing the quantity of information will not necessarily increase brand decision effectiveness, even though it may make consumers more confident.

It is becoming apparent that consumers follow two broad patterns when searching for information. Some people make a choice by examining one brand at a time, i.e. for the first brand they select information on several attributes, then for the second brand they seek the same attribute information and so on. This strategy is known as **choice by processing brands**.

An alternative strategy is seen when consumers have a particularly important attribute against which they assess all the brands, followed by the next most important attribute and so on. This is known as **choice by processing attributes**.

It has been shown that consumers with limited knowledge of a product or service tend to process information by attributes, while more experienced consumers process the information by brands. Furthermore, choice by processing attributes tends to be the route followed when there are few alternatives, when differences are easy to compute and when the task is easier in general.

CLUES TO EVALUATE BRANDS

Rather than engaging in a detailed search for information when deciding between competing brands, consumers look for a few clues that they believe will give an indication of brand performances. For example, when consumers buy a new car, they talk about them as being 'tinny' or 'solid' based on the sound heard when slamming the door. Thus, some consumers use the sound as being indicative of the car's likely performance.

There are many examples of consumers using surrogate attributes to evaluate brands, e.g. the sound of a lawn mower engine as indicative of power; the feel of a bread pack as indicative of the freshness of bread; the clothing style of banking staff as indicative of their understanding of financial services; and high prices as being indicative of good quality. It is now widely recognised that consumers conceive brands as arrays of clues (e.g. price, colour, taste, feel, etc.). Consumers assign information values to the available clues, using only those few clues which have a high information value. A clue's information value is a function of its predictive value (how accurately it predicts the attribute being evaluated) and its confidence value (how confident the consumer is about the predictive value assigned to the clue). This concept of brands as arrays of clues also helps explain why consumers undertake only a limited search for information. If, through experience, consumers recognise a few clues offering high predictive and high confidence values, these will be selected. More often than not, the most sought-after clue, as explained later in this chapter, is the presence of a brand name, which rapidly enables recall from memory of previous experience. However, when the consumer has limited brand experience, the brand name will have low predictive and confidence values and thus more clues will be sought, usually price, followed by other clues. Learning through brand usage enables the consumer to adjust predictive and confidence values internally and these stabilise over time.

There are numerous studies showing that when faced with a brand decision, consumers place considerable importance on the presence or absence of brand names. Not only do brand names have a high predictive value but consumers are also very confident, particularly from experience with this one. Of all the marketing variables, it is the brand name which receives the most attention by consumers and it is a key influencer of their perceptions of quality.

BRAND NAMES AS INFORMATIONAL CHUNKS

The previous section has explained that brand names are perceived by consumers as important information clues, which reduce the need to engage in a detailed search for information. An explanation for this can be found in Miller's (1956) work, which investigated the way the mind encodes information. If we compare the mind with the way computers work, it can be seen that we can evaluate the quantity of information facing a consumer in terms of the number of 'bits'. All the information on the packaging of a branded grocery item would represent in excess of a hundred bits of information. Researchers have shown that at most the mind can simultaneously process seven bits of information. Clearly, to cope with the information deluge from everyday life, our memories have had to develop methods for processing such large quantities of information.

This is done by a process of aggregating bits of information into larger groups or 'chunks', which contain more information (Buschke, 1976). A further analogy may be useful. The novice yachtsman learning the Morse code initially hears 'dit' and 'dot' as information bits. With experience, they organise these bits of information into chunks (letters), then mentally build these chunks into larger chunks (words). In a similar manner, when first exposed to a new brand of convenience food, the first scanning of the label would reveal an array of wholesome ingredients with few additives. These would be grouped into a chunk interpreted as 'natural ingredients'. Further scanning may show a high price printed on a highly attractive, multicolour label. This would be grouped with the earlier 'natural ingredients' chunk to form a larger chunk, interpreted as 'certainly a high quality offering'. This aggregation of increasingly large chunks would continue until final eye scanning would reveal an unknown brand name but, on seeing that it came from a well-known organisation (e.g. Nestlé, Heinz, etc.), the consumer would then aggregate this with the earlier chunks to infer that this was a premium brand − quality contents in a well-presented container, selling at a high price through a reputable retailer, from a respected manufacturer. Were the consumer not to purchase this new brand of convenience food but later that day see an advertisement for the brand, they would be able to recall the brand's attributes rapidly since the brand name would enable fast accessing of a highly informative chunk in the memory.

The task facing the marketer is to facilitate the way consumers process information about brands, such that ever larger chunks can be built in the memory which, when fully formed, can then be rapidly accessed through associations from brand names. This relates to the brand characteristic described in Chapter 2 of a shorthand device. Frequent exposure to advertisements containing a few claims about the brand should help the chunking process through either passive or active information acquisition. What is really important, however, is to reinforce attributes with the brand name rather than continually repeating the brand name without at the same time associating the appropriate attributes with it.

THE CHALLENGE TO BRANDING FROM PERCEPTION

To overcome the problems of being bombarded by vast quantities of information and having finite mental capacities to process it all, consumers not only adopt efficient processing rules (for example, they only use high information value clues when choosing between brands and aggregate small pieces of information into larger chunks) but they also rely upon their perceptual processes. These help brand decision-making by filtering information (perceptual selectivity) and help them to categorise competing brands (perceptual organisation).

Amongst others, Bruner (1957) made a major contribution in helping lay the foundations for a better understanding of the way consumers' perceptual processes operate. He showed that consumers cannot be aware of all the events occurring around them and with a limited attention span, they acquire information selectively. With this reduced data set, they then construct a set of mental categories which allows them to sort competing brands more rapidly. By allocating competing brands to specific mental categories, they are then able to interpret and give more meaning to brands. A consequence of this perceptual process is that consumers interpret brands in a different way from that intended by the marketer. The classic example of this was the cigarette brand Strand. Advertisements portrayed a man alone on a London bridge, on a misty evening, smoking Strand. The advertising slogan was 'You're never alone with a Strand'. Sales were poor, since consumers' perceptual processes accepted only a small part of the information given and interpreted it as, 'if you are a loner and nobody wants to know you, console yourself by smoking Strand'. It is clearly very important that brand marketers appreciate the role of perceptual processes when developing brand communication strategies and the two key aspects of perception are reviewed next.

Perceptual selectivity

Marketers invest considerable money and effort communicating with consumers, yet only a small fraction of the information is accepted and processed by consumers. First of all, their brand communication must overcome the barrier of **selective exposure**. If a new advertisement is being shown on television, even though the consumer has been attentive to the programme during which the advertisement appears, when the commercial breaks are on, the consumer may, out of preference, choose to engage in some other activity rather than watching the advertisements.

The second barrier is that of **selective attention**. The consumer may not feel inclined to do anything else while the television commercials are on during their favourite programme and might watch the advertisements for entertainment, taking an interest in the creative aspects of the commercial. At this stage, selective attention filters information from advertisements, so building support for existing beliefs about a brand ('Oh, it's that Toyota advert. They are good reliable cars. Let's see if they drive the car over very rough ground in this advert') and avoiding contradicting claims ('I didn't realise this firm produces mobile phone handsets, besides the PC I bought from them. In view of the problems I had with my PC, I just don't want to know about their products any more').

The third challenge facing a brand is **selective comprehension**. The consumer would start to interpret the message and would find that some of the information does not fit well with their earlier beliefs and attitudes. They would then 'twist' the message until it became more closely aligned with their views. For example, after a confusing evaluation of different companies' life insurance policies, a young man may mention to his brother that he is seriously thinking of selecting an AXA policy. When told by his brother that he knew of a different brand that had shown a better return last year, he may discount this fact, arguing to himself that his brother as a software engineer, probably knew less about money matters than he himself did as a sales manager.

With the passage of time, memory becomes hazy about brand claims. Even at this stage, after brand advertising, a further challenge is faced by the brand. Some aspects of brand advertising are **selectively retained** in the memory, normally those claims which support existing beliefs and attitudes.

From the consumer's point of view, the purpose of selective perception is to ensure that they have sufficient, relevant information to make a brand purchase decision. This is known as **perceptual vigilance**. Its purpose is also to maintain their prior beliefs and attitudes. This is known as **perceptual defence**. There is considerable evidence to show that information that does not concur with consumers' prior beliefs is distorted and that supportive information is more readily accepted. One of the classic examples of this is a study that recorded *different* descriptions from opposing team supporters who all saw the *same* football match. Many surveys show that selectivity is a positive process, in that consumers actively decide which information clues they will be attentive to and which ones they will reject.

Thus, as a consequence of perceptual selectivity, consumers are unlikely to be attentive to all of the information transmitted by manufacturers or distributors. Furthermore, in instances where consumers are considering two competing brands, the degree of dissimilarity may be very apparent to the marketer but if the difference, say, in price, quality or pack size is below a critical threshold, this difference will not register with the consumer. This is an example of Weber's Law — the size of the least detectable change to the consumer is a function of the initial stimulus they encountered. Thus, to have an impact upon consumers' awareness, a jewellery retailer would have to make a significantly larger reduction on a €5000 watch than on a €500 watch.

Perceptual organisation

'Perceptual organisation' allows consumers to decide between competing brands on the basis of their similarities within mental categories conceived earlier. Consumers group a large number of competing brands into a few categories since this reduces the complexity of interpretation. For example, rather than evaluating each marque in the fragrance market, consumers would have mental categories such as Chanel No. 5 as a 'sophisticated' scent and DKNY Night as a 'mysterious, evening' scent. By assessing which category the new brand is most similar to, consumers can rapidly group brands and are able to draw inferences without detailed search. If a consumer places a brand such as Tesco's own label washing powder into a category they had previously identified as 'own label', then the brand will achieve its meaning from the class it is assigned to by the consumer. In this case, even if the consumer has little experience of the newly categorised brand, they are able to use this perceptual process to predict certain characteristics of the new brand. For example, the consumer may reason that stores' own labels are inexpensive, thus this own label should be inexpensive and should also be quite good.

However, in order to be able to form effective mental categories in which competing brands can be placed and that lead to confidence in predicting brand performance, relevant product experience is necessary. The novice to a new product field has less well-formed brand categories than more experienced users. When new to a product field, the trial consumer has a view (based upon perceptions) about some attributes indicative of brand performance. This schema of key attributes forms the initial basis for brand categorisation, drives their search for information and influences brand selection. With experience, the schema is modified, the search for information is redirected and brand categorisation is adjusted, eventually stabilising over time with increasing brand experience.

An interesting study undertaken amongst beer drinkers is a useful example of how learning moulds brand categorisation. Without any labels shown, the beer drinkers were generally unable to identify the brand they drank most often and expressed no significant difference between brands. In this instance, the schema of attributes to categorise brands was based solely on the physical characteristics of the brands (palate, smell and the visual evaluation).When the study was later repeated amongst the same drinkers, but this time with the brands labelled, respondents immediately identified their most often drunk brand and commented about significant taste differences between brands! With the labels shown, respondents placed more emphasis on using the brand names to recall brand images as well as their views about how the brands tasted. As a consequence of consumers using this new schema of attributes to evaluate the brands, a different categorisation of the brands resulted.

Whilst it might be thought that the simplest way for consumers to form mental groups is to rely solely on one attribute and to categorise competing brands according to the extent to which they possess this attribute, evidence from various studies shows that this is often not the case. Instead, consumers use several attributes to form brand categories. Furthermore, it appears that they weight the attributes according to the degree of importance of each attribute. Thus, marketers need to find the few key attributes that are used by consumers to formulate different brand categorisations and concentrate upon the relevant attributes to ensure that their brand is perceived in the desired manner.

Gestalt psychologists provide further support for the notion of consumers interpreting brands through 'perceptual organisation'. This school of thought argues that people see objects as 'integrated wholes' rather than as a sum of individual parts. The analogy being drawn is that people recognise a tune rather than listen to an individual collection of notes.

To form a holistic view of a brand, consumers have to fill in gaps of information not shown in the advertisement. This concept, referred to as 'closure', is used by brand advertisers to get consumers more involved with the brand. For example, Kellogg's once advertised on billboards with the last 'g' cut off and it was argued that consumers' desire to round off the advertisement generated more attention. Likewise, not presenting an obvious punch line in a pun when advertising a brand, can again generate involvement through closure. This was one of the reasons for the successful 'Absolut' advertising (e.g. Absolut Spring, Absolut Design, etc.), where some advertising did not even feature the product but rather an object which resembled the shape of the unique bottle. When a brand has built up a respected relationship with consumers, the concept of closure can be successfully employed in brand advertising. For example, Nike is such a well-regarded brand that the brand name does not have to appear in its advertising campaigns. When consumers see advertising featuring the swoosh logo and the words 'just do it' they automatically think of the brand.

NAMING BRANDS: INDIVIDUAL OR COMPANY NAME?

From the previous sections, it should now be clear that consumers seek to reduce the complexity of buying situations by cutting through the vast amount of information to focus on a few key pieces of information. A brand name is, from the consumer's perspective, a very important piece of information and is often the key piece. It is, therefore, essential that an appropriate brand name is chosen that will reinforce the brand's desired positioning by associating it with the relevant attributes that influence buying behaviour.

A brief consideration of some very well-known brand names shows that rather unusual reasons formed the basis for name selection. However, in today's more competitive environment, far more care is necessary in naming a brand. For example, Dell was named after its founder, Michael Dell; Yves Saint Laurent Paris perfume is a tribute to the Parisians who were dressed by the designer; Mercedes was named after a friend's daughter; and ASICS was derived from an acronym of 'Anima Sana In Corpore Sano', meaning 'a sound mind in a sound body'. Today, however, because of the increasing need to define markets on a global basis, idiosyncratic approaches to naming brands can lead to failure. For example, General Motor's Nova failed in Spain because the name means 'doesn't go'; while Roll Royce's plans to wrap a new model in mysterious fog because of the name 'Silver Mist' were fortunately halted when it was noted that 'mist' in German means 'dung', which obviously would have elicited different images.

By contrast, a financial services corporation embraced the challenge of global marketing and incorporated this in its thinking about brand naming. AXA was chosen as the brand name since it is the same regardless of whether spelt from left to right or right to left. As it starts with "A", it will appear early in any directory listings. Furthermore, as an invented name, it has no limiting meanings.

When examining brand names, it is possible to categorise them broadly along a spectrum, with a company name at one end (e.g. Sony, Cadbury, HSBC), all the way through to individual brand names which do not have a link with the manufacturer (e.g. Ariel, Gillette, Pantene and Olay emanating from Procter & Gamble).

There are varying degrees of company associations with the brand name — for example, brand names with strong company endorsement would include Cadbury's Crunchie, Heinz Baked Beans and Kellogg's Frosties and there are brand names with weak company endorsement, such as KitKat from Nestlé.

There are many advantages to be gained from tying the brand name in with the firm's name. With the goodwill that has been built up over the years from continuous advertising and a commitment to consistency, new brand additions can gain instant acceptance by being linked with the heritage. Consumers feel more confident trying a brand that draws upon the name of a well-established firm. Building upon high awareness and strong associations of the brand with healthy eating and children's tastes, Weetabix was augmented by brand extensions including Weetabix Bitesize, Westabix Minis and the new Oatibix range. The new brands were able to benefit from similar associations built up over the years by the Weetabix name. In this example, however, the brand name was extended to a sector not dissimilar from that where the original brand's strengths were built. If this is not the case, the company's image could be diluted by following a corporate endorsement naming policy.

In the financial services sector, it is very common to see brands being strongly tied to the corporation (e.g. MBNA Platinum Card, First Direct Regular Saver and NatWest Savings Accelerator Credit Card). Part of the reason for this is that consumers and financial advisers have traditionally evaluated policies by considering the parent corporation's historical performances, so company-linked brand names in this sector are common. With these organisations recognising that they can differentiate themselves in a more sustainable manner through their organisational culture, this is a further reason for the popularity of using the corporate name as the key brand identifier.

Nonetheless, whilst a brand can gain from an umbrella of benefits by being linked with a company name, the specific values of each brand still need to be conveyed. For example, whilst organisations such as Co-operative Bank have been promoting the ethical benefits associated with their corporate name, there is a danger of not adequately promoting the benefits of the individual brands (Privilege Account, Privilege Premier Account, Smart Saver Account and

Guaranteed Fund) leading to the possibility of consumer confusion. Interestingly, when a corporation has developed a particularly novel concept with values that differ from the corporation, then the brand is launched without such a strong corporate name tie. First Direct is a very different approach to financial services and the brand was launched very much as a stand-alone brand using the black and white logo to communicate the no-nonsense approach, albeit with a small logo identifying it as a division of HSBC.

There are advantages in all aspects of communication to be gained from economies of scale when an organisation ties a brand name in with its corporate name. This advantage is sometimes given an undue importance and weighting by firms thinking of extending their brands into new markets. This question of brand extension is a complex issue that involves more than just the name and is dealt with in Chapter 9. However, it is worth mentioning here that more products and services are likely to be marketed under the same corporate-endorsed brand name. Nonetheless, to help the brand fight through the competing noise in the market, it is still essential to know what the brand means to the consumer, how the brand's values compare against competitive brands and how marketing resources are affecting brand values.

There are also very good reasons why in certain circumstances it is advisable to follow the individual brand name route. As the earlier Procter & Gamble example showed, this allows the marketer to develop formulations and positioning to appeal to different segments in different markets. However, the economics of this need to be carefully considered since firms may, on closer analysis, find that by trying to appeal to different small segments through different brand offerings they are encountering higher marketing costs resulting in reduced brand profitability.

When striving to have coverage in different segments of the market as, for example, Blackberry does with the Storm, Bold, Curve and Pearl handsets, it is important that individual brand names sufficiently reinforce their different brand positioning. Some firms try to differentiate their brands in the same market through the use of numbers. When this route is followed, however, the numbers should be indicative of relative brand performance – in the electric shower market a '4000' model would be expected to have approximately double the power of the '2000' model. In some markets, firms do not appear to have capitalised on naming issues. For example, in the digital camera market, Nikon brands five of its models as D3X, D700, D90, D5000 and D40. Consumers cannot infer much about relative differences from these brand nomenclatures.

Another advantage of using individual brand names is that if the new line should fail, the firm would experience less damage to its image than if the new brand had been tied to the corporation. Keller and Sood (2003) cite a study in which the acceleration problems with the Audi 5000 automobile had a greater

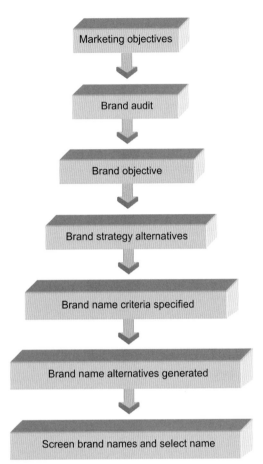

FIGURE 3.9

Schema to identify most appropriate brand name

spillover to the Audi 4000 model than the Audi Quatro and this was attributed to the different marketing and branding of the Quatro.

A STRATEGIC APPROACH TO NAMING BRANDS

Adopting a strategic approach to brand naming is important, as over the long term the brand name becomes linked with numerous associations that may constrain the brand. For example, The Carphone Warehouse, a key independent mobile phone retailer, devised its name to reflect its origins. However, it no longer sells out of warehouses and does not sell "carphones". A view is needed also about the territories in which the brand will be marketed. For example, Vodafone acquired the Irish brand Eircell to enter the Irish mobile phone market. The re-branding exercise took place in 2002. The Eircell logo was purple, so a key aspect of the re-branding was a change to Vodafone's red logo, with the slogan 'Red is the new purple: Vodafone is the new name for Eircell'. When there is no pre-existing name, a suggested way of developing the brand's name would be to follow the more strategic process outlined in the flow chart in Fig. 3.9.

Let us now consider each of these steps in turn.

Marketing objectives

The marketer needs to be certain about the marketing objectives that the brand must achieve. Clearly stated, quantified and timed targets must be available for each segment showing the level of sales expected from each of the product groups comprising the company's portfolio. The marketing objectives will give an indication as to whether emphasis is being placed on gaining sales from existing products to existing customers or whether new horizons are envisaged (e.g. through either product extensions or new customer groups). By having clearly defined marketing objectives, brand managers are then able to consider how each of their brands needs to contribute towards satisfying the overall marketing objectives.

The brand audit

The internal and external forces that influence the brand need to be identified, such as company resources, competitive intensity, supplier power, threats from substitutes, buyer concentration, economic conditions and so on. This audit should help identify a few of the criteria that the name must satisfy. For example, if the brand audit showed that the firm has a superior phone screen that consumers valued because of the reproduced colour being so true to life, then one issue the name would have to satisfy would be its reinforcement of the critical success factor of 'colourful realism'.

Brand objectives

In the brand planning document, clear statements about individual brand objectives should be made, again helping clarify the criteria that the brand name must satisfy. Statements about anticipated levels of sales through different distributors, to specified customers, will help the marketer to identify criteria for the name to meet. For example, if the primary market for a new brand of rechargeable batteries is 8–10 year old boys who are radio-controlled car racing enthusiasts, and if the secondary market is fathers who help their sons—the primary target's need may be for long intercharge periods, whilst the secondary market may be more concerned about purchase costs. The primary need for the brand name would be to communicate power delivery, with an undertone about cost.

Other brand objective statements about positioning and brand personality would further clarify some of the criteria that the name must satisfy.

Brand strategy alternatives

The marketer must be clear about what broad strategies are envisaged for the brand in order to satisfy the brand objectives. Issues here would include:

- manufacturer's brand or distributor's brand?
- specialist or niche brand?
- value-added or low-price positioning?

Again, these would clarify issues that the name must satisfy.

Brand name criteria specified

From the previous sections, the marketer should be able to list the criteria that the brand name must satisfy. They might also wish to learn from other companies' experiences what appears to work best with brand names. This issue will be addressed in the next section of this chapter.

Brand name alternatives generated

With a clear brief about the challenges that the new brand name must overcome, the marketer can now work with others to stimulate ideas for possible brand names. It is unlikely that the brand managers would work on this alone.

Instead, they would be joined by others from the marketing department, by advertising agency staff, by specialist name-generating agencies where appropriate and by other company employees. Also, at this stage an agency may be commissioned to undertake some qualitative research to help generate names. Some of the methods that might be used to generate names could include:

- brainstorming;
- group discussions;
- management inspiration;
- word association;
- in-company competitions amongst employees;
- computer-generated names.

It is important to stress that, during the name generation stage, any intentions to judge the names must be suppressed. If names are evaluated as they are generated, this impedes the mind's creative mode and results in a much lower number of names.

Screen and select the brand name

By scoring each name against the criteria for brand name effectiveness, an objective method for judging each option can be employed. Each name can be scored in terms of how well it matches each of the criteria and by aggregating each name's score, a value order will result. The more sophisticated marketer can weight each of the criteria in terms of importance and arrive at a rank order on the basis of the highest aggregated weighted score. Whatever numerical assessment procedure is employed, it should be developed only on the basis of an agreed internal consensus and after discussions with key decision makers. Not only does this enhance commitment to the finally selected brand name but it also draws on the relevant experience of many executives.

By following this schema, the marketer is able to select a name that should satisfy the company's ambitions for long-term, profitable brand growth. This process should also result in a name that is well able not only to defend a sustainable position against competitive forces but also to communicate added values to consumers effectively — for example, the spread, I Can't Believe it's Not Butter!, from Unilever, and Vanish, the stain remover from Reckitt Benckiser.

ISSUES ASSOCIATED WITH EFFECTIVE BRAND NAMES

When considering criteria for brand name effectiveness, some of the guidelines include:

- *The brand name should be simple.* The aim should be to have short names that are easy to read and understand. Consumers have finite mental capacities and find it easier to encode short words in memory. This is the reason why names with four syllables or more are usually contracted. Listening to consumers talk clarifies the way that long brand names are simplified (e.g. Pepsi rather than Pepsi Cola, Marks' rather than Marks & Spencer's). When consumers get emotionally closer to brands, they are more likely to contract the name; for example, Mercedes becomes Merc, Budweiser becomes Bud and an Apple iMac becomes a Mac.

- *The brand name should be distinctive.* Brand names such as Kodak and Adidas create a presence through the distinctive sound of the letters and the novelty of the words. This creates attention and the resulting curiosity motivates potential consumers to be more attentive to brand attributes.

- *The brand name should be meaningful.* Names that communicate consumer benefits facilitate consumers' interpretations of brands. For example, Kwik Fit, Ray Ban and WeightWatchers leave little doubt about the benefits to be gained from these brands. Creativity should be encouraged at the expense of being too correct. A battery branded 'Reliable' would communicate its capability but would not attract as much attention as the more interesting 'Die Hard'.

- *The brand name should be compatible with the product.* The appropriateness of the name Timex with watches is more than apparent, reinforcing the meaningfulness of the brand name. However, marketers should beware of becoming too focused on specific benefits of the product, especially in a mature market. Orange offered a dramatic and refreshing alternative in a sector where the tradition of brand naming was built on the suffixes 'tel' and 'com'.

- *Emotion helps for certain products.* For those product fields where consumers seek brands primarily because they say something about the purchaser, as for example in the perfume market, emotional names can succeed. Examples here include Clinique Happy and Ralph Lauren's Romance.

- *The brand name should be legally protectable.* To help protect the brand against imitators, a search should be undertaken to identify whether the brand name is available and, if so, whether it is capable of being legally registered.

- *Beware of creating new words.* Marketers developing new words for their brand have to anticipate significant promotional budgets to clarify what their invented word means. For example, the successfully invented names of Kodak, Nokia, Esso and Häagen-Daz succeeded because of significant communication resources.

- *Extend any stored-up equity.* When firms audit their portfolio of current and historical brands, they may find there is still considerable goodwill in the marketplace associated with brands they no longer produce. There may be instances when it is worth extending a historical name to a new line (e.g. Maltesers to ice cream), or even re-launching historical lines (e.g. Cadbury's re-launch of the Wispa bar in response to customer demand).

- *Avoid excessive use of initials.* Over time some brand names have been shortened, either as a deliberate policy by the firm or through consumer terminology (e.g. International Business Machines to IBM, General Electric to GE, Music Television to MTV and British Midland to bmi). It takes time for the initials to become associated with brand attributes and firms generally should not launch new brands as arrays of initials. The hypothetical brand North London Tool and Die Company certainly fails the criteria of being short, but at least, unlike the initials NLT&D, it does succeed in communicating its capabilities.
- *Develop names that allow flexibility.* The hope of any marketer is for brand success and eventually a widening portfolio of supporting brands to better satisfy the target market. Over time, more experienced consumers seek a widening array of benefits so, if possible, the name should allow the brand to adapt to changing market needs. For example, with an abstract name, Apple was able to diversify from computers into MP3 devices and mobile phone handsets, while retaining the image of the brand.
- *Develop names that are internationally valid.* It is essential to establish during the naming process what geographical coverage the brand will assume. When a name is intended for only one nation or one culture, the cultural associations linked to it are immediately evident. Whenever the brand name spans different languages and cultures, it becomes more difficult to forecast customers' responses. For instance, the Mexican food corporation Bimbo, which has no specific meaning in Spanish, has negative implications in the USA and UK.

Whilst these points should contribute to the way organisations think about the appropriateness of different brand names, we should never lose sight of the fact that it is consumers who buy brands, not the managers who manage them. For this reason it is always wise, when short-listing potential names, to undertake consumer research and evaluate consumers' responses. For example – are the words harsh sounding? Are there any negative associations with the words? Are the names appropriate for the proposed brand? Do the words 'roll off the tongue' easily? Are the words memorable? – and so on.

Once a decision has been taken about the brand name and the brand has been launched, the firm should audit the name on a regular basis. This will show whether or not the meaning of the brand name has changed over time as a result in changes in the marketplace. If the environment has changed to such an extent that the firm is missing opportunities by persisting with the original name, then consideration should be given to changing the name.

Finally, in an era of corporate acquisitions, firms need to consider how their brand naming strategy needs to be revised as a result of mergers and acquisitions. A company with a name that doesn't fit the direction of the purchaser should not be automatically dropped, in favour of the parent corporation. Rather, there should be an assessment of the inherent equity in each corporation's name, a consideration of the long-term markets for each corporation and the strengths and weaknesses of individual versus unified naming. PricewaterhouseCoopers was formed from the merger of Price Waterhouse and Coopers and Lybrand in 1998. The combination of the two brand names allowed the new brand to benefit from the heritage of its two founding firms.

THE BRAND AS A RISK REDUCER

The final issue of relevance to branding to be considered in this chapter is the concept of perceived risk. Earlier parts of this chapter showed that products and services can be conceived as arrays of clues and that the most-consulted clue when making a brand choice is the presence or absence of a brand name. This reliance upon brand name is also confirmed by the considerable volume of consumer research on the concept of perceived risk. It is clear that, when buying, consumers develop risk-reducing strategies. These are geared to either reducing the uncertainty in a purchase by buying, for example, only advertised brands or to minimising the chances of an unpleasant outcome by buying, for example, only previously tried brands. This concept also affects the way organisations buy brands and is also addressed in Chapter 5.

It must be stressed that we are talking only about consumers' **perceptions** of risk, rather than **objective** risk, since consumers react only as they perceive situations. Whilst marketers may believe they have developed a brand that is presented as a risk-free purchase, this may not necessarily be the perception consumers have. Consumers have a threshold level for perceived risk, below which they do not regard it worthwhile to undertake any risk-reduction action. However, once this threshold level is exceeded, they will seek ways of reducing perceived risk.

We can now start to view brands as being so well formulated, distributed and promoted that they provide consumers with the added value of increased confidence. For example, if the brand is available from a quality retailer, this should signal increased certainty regarding its performance. If there has been a lot of supporting advertising, this would also be read as being indicative of a low-risk product. Furthermore, should there also be favourable word-of-mouth endorsement, this too would allay concerns about the brand. Marketers can gain a competitive edge by promoting their brands as low-risk purchases.

By viewing risk as being concerned with the *uncertainty* felt by consumers about the *outcome* of a purchase, it is possible to appreciate how marketers can reduce consumers' risk in brand buying. For example, appropriate strategies to reduce consumers' worries about the *consequences* of the brand purchase would include developing highly respected warranties, offering money back guarantees for first time, trial consumers and small pack sizes during the brand's introductory period. To reduce their *uncertainty*, consumers will take a variety of actions, such as seeking out further information, staying with regularly used brands or buying only well-known brands. Marketers can reduce concerns about uncertainty by providing consumers with relevant, high-quality information, by encouraging independent parties such as specialist magazine editors to assess the brand and by ensuring that opinion leaders are well versed in the brand's potential. For example, by tracking purchasers of a newly launched microwave oven brand, home economists can call on these innovative customers and, by giving a personalised demonstration of the new brand's capabilities, ensure that they are fully conversant with the brand's advantages. This can be particularly effective, since early innovators are regarded as a credible information source.

The favoured routes to reduce risk vary by type of product or service and it is unusual for only one risk-reducing strategy to be followed. It is, however, apparent that brand loyalty and reliance on major brand image are two of the more frequently followed actions. When consumers evaluate competing brands, not only do they have an overall view about how risky the brand purchase is but they also form a judgement about *why* the brand is a risky purchase. This is done initially by evaluating which dimensions of perceived risk cause them the most concern. There are several dimensions of risk. For example:

- *financial risk*: the risk of money being lost when buying an unfamiliar brand;
- *performance risk*: the risk of something being wrong with the unfamiliar brand;
- *social risk*: the risk that the unfamiliar brand might not meet the approval of a respected peer group;
- *psychological risk*: the risk that an unfamiliar brand might not fit in well with one's self image;
- *time risk*: the risk of having to waste further time replacing the brand.

If the marketer is able to identify which dimensions of perceived risk are causing concern, they should be able to develop appropriate consumer-orientated risk-reduction strategies. The need for such strategies can be evaluated by examining consumers' perceptions of risk levels and by gauging whether this is below their threshold level. It should be realised, however, that the level of risk varies between people and also by product category. For example, cars, insurance and household appliances are generally perceived as being high-risk purchases, while toiletries and packaged groceries are low-risk purchases.

Table 3.3 shows the results of a seminal consumer study by Jacoby and Kaplan (1972) indicating how some of the dimensions of perceived risk vary by product field. From these findings, it can be seen that marketers of life

Table 3.3 Consumers' Views About the Dimensions of Risk (*after Jacoby and Kaplan 1972*)

	Life Insurance	Suit
Financial risk	7.2	6.4
Performance risk	6.7	5.8
Psychological risk	4.9	6.9
Social risk	4.8	7.3
Overall risk	7.0	5.9

1 = very low risk, 10 = very high risk

Table 3.4 Consumers' Views About Risk Types (*after Brooker 1984*)

Risk Type	Peaches (Mean)	Spaghetti (Mean)
Financial risk	3.6	3.2
Performance risk	3.6	3
Physical risk	1.8	1.7
Psychological risk	2.4	2.3
Social risk	1.6	2
Time-loss risk	2.5	2.7
Overall	3.2	2.9

insurance policies need to put more emphasis on stressing the relative cost of policies and how well-protected consumers are, compared with other competitors' policies. By contrast, the suit marketer should place more emphasis on reference group endorsement of their brand of suit by means, for example, of a photograph of an appropriate person in this suit.

The study was supported by a later work by Brooker (1984) who added time-loss risk to the dimensions presented by Jacoby and Kaplan (1972) in his study of grocery shoppers.

As the Jacoby and Kaplan (1972) study had previously reported, Brooker found that financial risk and performance risk were the greatest perceived risks and he also found that those two risks were related. For food categories, his study found that social risk was the least perceived risk. Therefore, brand managers in this sector should spend more time reassuring customers about the quality of their food produce in order to reduce their perceived risk.

CONCLUSIONS

By adopting an information processing model of consumer behaviour, this chapter has considered how a knowledge of consumers' buying processes can help to develop successful brands. Through an appreciation of the differences consumers perceive between competing brands and their involvement in the brand buying process, we identified different consumer buying processes. According to the type of buying process, an active or passive approach to brand information acquisition may be followed. If the consumer actively seeks information, it was shown that this would be for only a few key pieces of information regarded as highly indicative of the brand's capabilities.

It is often the case that consumers have a low level of involvement when choosing brands. As such, promotional campaigns should major upon key sensory associations. A further implication of their low involvement state is that tracking brand images shows only small changes over time. Assessments of brand health should consider brand association measures and brand credit. More finely tuned brand positioning can be achieved through considering the deviations resulting from the difference between the actual and expected brand attribute matrices.

Consumers choose brands to reflect their needs in a particular context. Market research can help identify, for a particular category, a consumer's differing need-states and thereby aid marketers to blend their brands more closely with a specific need-state. As a consequence, consumers have a repertoire of brands in a specific category and switching behaviour reflects, to some extent, different need-states. Differential marketing builds on the concept of brand repertoires, encouraging marketers to target small groups of highly profitable consumers with special programmes.

When evaluating competing brands, consumers are concerned with appreciating the extent to which the brands have added values over and above the commodity form of the brand. These added values may be as simple as polite service from a bank clerk, through to a complex cluster of lifestyle associations by driving a particular car marque.

One added value often overlooked by marketers is labelling which displays only a few pieces of information, facilitating brand choice. Consumers have finite mental capabilities and seek to process a few, high-quality pieces of information as quickly as possible. They often use surrogate clues, such as price, to evaluate brands and they place considerable reliance on the presence or absence of brand names.

The problem facing the marketer, however, is that consumers are selective in their search for brand information and they twist some of the information to make it fit their prior beliefs. Brand promotional activity must, therefore, be regularly audited to evaluate the extent to which consumers correctly interpret the message.

To encourage consumer appreciation of the brand's true capabilities, the right type of brand name is needed. In certain circumstances there are strengths in having a brand name tied to the corporation but in others, unique brand names are more appropriate. Brand name selection should not be based on tactical issues, but rather on a more robust, strategic basis that relates the potential name to marketing objectives and other forces influencing the brand.

Finally, marketers should appreciate that consumers perceive risk when buying brands. Through appreciating the extent of perceived risk and the factors causing consumer concern, they can develop strategies geared to reducing this perceived risk.

MARKETING ACTION CHECKLIST

To help clarify the direction of future brand marketing activity it is recommended that the following exercises are undertaken.

1. Write down how involved you believe the consumer is when buying a brand in the product field where you have a presence. From the consumer's perspective, evaluate whether competing brands in this product field are strongly or weakly differentiated, stating what the differentiating features are. Then, consulting Fig. 3.1, identify the consumer decision process reflecting brand buying behaviour. Ask other colleagues to undertake this exercise individually. If there is a lack of consensus about the level of involvement of consumers, the degree of brand differentiation or the basis for brand differentiation, your department may be basing its brand communication programme on erroneous assumptions about consumer behaviour. This can be resolved by undertaking qualitative research amongst target consumers to assess the dimensions they use to differentiate between brands and the basis for their involvement in the purchase. This should then be followed by a survey to quantify the extent of consumers' involvement and their perceptions of brand differences. On the basis of such market research, the consumer's decision process can be evaluated and, by referring to the early parts of this chapter, the appropriateness of the current brand communication strategy can be assessed.

2. When was any market research last undertaken to assess consumers' involvement and their perceptions of brand differences? If this information is more than a year old, consider whether the dynamic nature of your market necessitates a further update.

3. At what stages in the consumer's buying process is any brand information currently being directed? On the basis of the model of the buying process identified in the earlier exercises, when do consumers seek information? Are there any discrepancies?

4. Get your colleagues to write down what stages they believe consumers pass through in the buying process for one of your brands. Ask someone who does not work in the marketing or sales department (and ideally who is new to the firm) to narrate the stages they went through when buying this particular brand. In an open forum, resolve any differences between what managers' perceptions are and the reality of consumer buying activity.

5. Have you any information about the factors which encourage or discourage consumers to seek more information about your brand (e.g. role of family, previous experience, etc.)? How are you addressing each of these factors in your brand activity?

6. How much do you know about consumers' perceptions of your brands (and, if appropriate, about perceptions of your company)? Are there any differences between the brand communications objectives agreed internally (and specified to your promotions agency) and the way consumers interpret your brands?

7. How do you help consumers evaluate your brand very shortly after it has been purchased for the first time? How well-equipped are retailers to resolve consumers' doubts about the brand?

8. Has any consumer research been undertaken to evaluate the key attributes that consumers use to make a brand selection? What attributes did your last promotional campaign major on? Did these attributes match the findings from the previous research?

9. What are the added values that distinguish your brand from competitors? How relevant are these to consumers? How are *all* elements of the marketing mix being used to achieve these added values? Is there any inconsistency between the elements of the marketing mix?

10. Show several consumers a pack or brochure describing your brand, along with similar material from your competitors. Do not allow them to spend long looking at these examples. Observe how they examine the material. Remove the examples and ask what they recalled and why they recalled this. Discuss the results with your colleagues and consider which aspects of your pack/brochure are critical to the consumer and which parts clutter the central message. What informational clues were consumers using to draw inferences about the capabilities of the product or service? How much marketing attention is being focused on these informational clues?

11. List the brands against which your brand competes. Now categorise these brands into groups that show some form of similarity, starting the basis for brand groupings. Ask colleagues to do this exercise individually and collate the forms. In an open forum, discuss any differences. Commission a survey to evaluate which brands consumers see as competing in a particular product field, how they would group these brands and what their basis for categorisation was. If there is not agreement between

consumers' categorisation of brands and yours, you may well need to reconsider your brand strategy.

12. What are the strengths and weaknesses of your brand's name? How able is the name to cope with future opportunities and yet overcome any threats? State the criteria that your experience has shown are essential for a successful brand name and critically evaluate your brand name against this.

13. List the aspects of risk that consumers perceive when buying your brand and your competitors' brands. In what order of perceived risk do consumers rank these brands? What actions can you take to reduce consumers' perceptions of risk when buying your brand?

14. How do you ensure that your most profitable customers choose your brand consistently? Have you established special retention programmes tailored to their needs? What tracking is being undertaken to understand why people are defecting from your brand?

15. Consider your marketing budget. Is it spread evenly across all your customers or does it give greater emphasis to the customers that buy the brand more frequently and provide higher levels of profits?

STUDENT BASED ENQUIRY

- Select a brand of your choice and outline the stages of the decision making process a consumer might follow in deciding whether to purchase it. Pay particular attention to the sources of information the consumer might use and suggest how a brand manager might use these insights to improve their marketing communications in the future.
- Consider a brand you have purchased over the past month. Evaluate the extent to which you experienced cognitive dissonance. How could the brand managers take steps to reduce dissonance for your chosen brand?
- Select a product category of your choice. Using Wendy Gordon's work, illustrate how different brands within your chosen product category meet your different need states at different times.

References

Aaker, J. L. (1987). Dimensions of brand personality. *Journal of Marketing Research, 34*(3), 347–356.

Assael, H. (1987). *Consumer Behavior and Marketing Action*. Boston: Kent Publishing.

Brooker, G. (1984). An assessment of an expanded measure of perceived risk. *Advances in Consumer Research, 11*(1), 439–441.

Bruner, J. S. (1957). On perceptual readiness. *Psychological Review, 64*(2), 123–152.

Buschke, H. (1976). Learning is organised by chunking. *Journal of Verbal Learning and Verbal Behaviour, 15*, 313–324.

Ehrenberg, A. S. C. (Oct. 1993). If you're so strong, why aren't you bigger? *ADMAP* 13–14.

Gordon, W. (1994). Taking brand repertoires seriously. *Journal of Brand Management, 2*(1), 25–30.

Hallberg, G. (1995). *All Consumers are not Created Equal.* New York: John Wiley & Sons.

Heath, R. (2001). Low involvement processing – a new model of brand communication. *Journal of Marketing Communications, 7,* 27–33.

Jacoby, J., & Kaplan, L. (1972). The components of perceived risk. In M. Venkatesan (Ed.), *Third Annual Conference, Association for Consumer Research. Proceedings* (pp. 382–393). University of Chicago.

Jones, J. P. (1986). *What's in a Name? Advertising and the Concept of Brands.* Lexington: Lexington Books.

Keller, K. L., & Sood, S. (2003). Brand Equity Dilution. *MIT Sloan Management Review, 45*(1), 12–15.

Koças, Cenk, Bohlmann, & Johnathan, D. (2008). Segmented switchers and retailer pricing strategies. *Journal of Marketing, 72*(3), 124–142.

Miller, G. A. (1956). The magical number seven, plus or minus two: some limits on our capacity for processing information. *The Psychological Review, 63*(2), 81–97.

Mitchell, A. (2005). Back to basics as loyalty marketing comes full circle. *Marketing Week, 28*(30), 28–29.

Park, C., Jaworski, B., & MacInnis, D. (Oct. 1986). Strategic brand concept-image management. *Journal of Marketing, 50,* 135–145.

Romaniuk, J., & Sharp, B. (2004). Conceptualizing and measuring brand salience. *Marketing Theory, 4*(4), 327–342.

Romaniuk, J., & Sharp, B. (2000). Using known patterns in image data to determine brand positioning. *International Journal of Market Research, 42*(2), 219–230.

Su, B. (2008). Characteristics of consumer search on-line: how much do we search? *International Journal of Electronic Commerce, 13*(1), 109–129.

The New York Times. (2009) *Anywhere the Eye can See, Its Likely to see an Ad.* Available at: http://www.nytimes.com/2007/01/15/business/media/15everywhere.html. Accessed 6 July 2009.

Further Reading

Aaker, D. (1991). *Managing Brand Equity.* New York: The Free Press.

Adams, S. C., & Miller, A. S. (Dec 1972). How many advertising exposures per day? *Journal of Advertising Research, 12,* 3–10.

Addison, Wesley, Bettman, J. R., & Kakkar, P. (March 1977). Effects of information presentation format on consumer information acquisition strategies. *Journal of Consumer Research, 3,* 233–240.

Allison, R. I., & Uhl, K. P. (1964). Influence of beer brand identification on taste perception. *Journal of Marketing Research, 1*(3), 36–39.

Antonides, G., & van Raaij, F. (1998). *Consumer Behaviour.* Chichester: John Wiley.

Assael, H. (1998). *Consumer Behaviour and Marketing Action.* Cincinnati: South-Western College Publishing.

Barwise, P., & Ehrenberg, A. (1985). Consumer beliefs and brand usage. *Journal of the Market Research Society, 27*(2), 81–93.

Bauer, R. A. (1960). Consumer behavior as risk taking. In R. S. Hancock (Ed.), *Dynamic Marketing for a Changing World. 43rd Conference of the American Marketing Association* (pp. 389–398). Chicago: American Marketing Association.

Beales, H., Maziz, M., Salop, S., & Staelin, R. (8 June 1981). Consumer search and public policy. *Journal of Consumer Research*, 11–22.

Bennett, P., & Mandell, R. (1969). Prepurchase information seeking behaviour of new car purchasers: the learning hypothesis. *Journal of Marketing Research, 6*(4), 430–433.

Bettman, J. R. (1979). *An Information Processing Theory of Consumer Choice*. Reading, MA: Massachusetts.

Bettman, J. R., & Park, C. W. (Dec 1980). Effects of prior knowledge and experience and phase of the choice process on consumer decision processes: a protocol analysis. *Journal of Consumer Research, 7*, 234–248.

Biehal, G., & Chakravarti, D. (March 1982). Information-presentation format and learning goals determinants of consumers' memory retrieval and choice processes. *Journal of Consumer Research, 8*, 431–441.

Britt, S. H. (Feb 1975). How Weber's law can be applied to marketing. *Business Horizons, 13*, 21–29.

Bruner, J. S. (1958). Social psychology and perception. In D. T. Kollat, R. D. Blackwell, & J. F. Engel (Eds.), *1970 edition of: Research in Consumer Behavior*. New York, Holt: Rinehart and Winston.

Bucklin, L. P. (1969). Consumer search, role enactment and market efficiency. *Journal of Business, 42*, 416–438.

Capon, N., & Burke, M. (Dec. 1980). Individual, product class and task-related factors in consumer information processing. *Journal of Consumer Research, 7*, 314–326.

Chisnall, P. M. (1985). *Marketing: A Behavioural Analysis*. London: McGraw Hill.

Claxton, J. D., Fry, J. N., & Portis, B. (Dec. 1974). A taxonomy of prepurchase information gathering patterns. *Journal of Consumer Research, 1*, 35–42.

Cox, D. F. (1967). The sorting rule model of the consumer product evaluation process. In D. F. Cox (Ed.), *Risk Taking and Information Handling in Consumer Behavior*. Boston: Harvard University.

Derbaix, C. (March 1983). Perceived risk and risk relievers: an empirical investigation. *Journal of Economic Psychology, 3*, 19–38.

East, R. (1997). *Consumer Behaviour*. London: Prentice Hall.

Ehrenberg, A. S. C. (1988). *Repeat Buying: Facts, Theories and Application*. London: Griffin.

Ehrenberg, A. S. C., Goodhardt, G. J., & Barwise, T. P. (1990). Double jeopardy revisited. *Journal of Marketing, 54*, 82–91.

Gardner, D. M. (1971). Is there a generalised price-quality relationship? *Journal of Marketing Research, 8*(2), 241–243.

Gemunden, H. G. (1985). Perceived risk and information search. A systematic meta-analysis of the empirical evidence. *International Journal of Research in Marketing, 2*(2), 79–100.

Haines, G. H. (1974). Process models of consumer decision making. In G. D. Hughes, & M. L. Ray (Eds.), *Buyer/Consumer Information Processing*. Chapel Hill: University of North Carolina Press.

Hansen, F. (1972). *Consumer Choice Behavior. A Cognitive Theory*. New York: The Free Press.

Hastorf, A. H., & Cantril, H. (1954). They saw a game: a case history. *Journal of Abnormal and Social Psychology, 49*, 129–134.

Jacoby, J., Chestnut, R. W., & Fisher, W. A. (1978). A behavioural process approach to information acquisition in nondurable purchasing. *Journal of Marketing Research, 15*(3), 532–544.

Jacoby, J., Speller, D. E., & Kohn, C. A. (1974). Brand choice behavior as a function of information load. *Journal of Marketing Research, 11*(1), 63–69.

Jacoby, J., Szybillo, G. J., & Busato-Schach, J. (March 1977). Information acquisition behavior in brand choice situations. *Journal of Consumer Research, 3*, 209–216.

Kapferer, J.-N., & Laurent, G. (1986). Consumer involvement profiles: a new practical approach to consumer involvement. *Journal of Advertising Research, 25*(6), 48–56.

Katona, G., & Mueller, E. (1955). A study of purchasing decisions. In L. H. Clark (Ed.), *Consumer Behavior. The Dynamics of Consumer Reaction*. New York: New York University Press.

Kendall, K. W., & Fenwick, I. (1979). What do you learn standing in a supermarket aisle? In W. L. Wilkie (Ed.), *Advances in Consumer Research, Vol. 6* (pp. 153–160) Ann Arbor: Association for Consumer Research.

Kiel, G. C., & Layton, R. A. (1981). Dimensions of consumer information seeking behavior. *Journal of Marketing Research, 18*(2), 233–239.

Krugman, H. E. (March–April 1975). What makes advertising effective? *Harvard Business Review, 53*, 96–103.

Krugman, H. E. (1977). Memory without recall, exposure without perception. *Journal of Advertising Research, 17*(4), 7–12.

Lannon, J., & Cooper, P. (1983). Humanistic advertising: a holistic cultural perspective. *International Journal of Advertising, 2*, 195–213.

McNeal, J., & Zeren, L. (1981). Brand name selection for consumer products. *Michigan State University Business Topics, 29*(2), 35–39.

Midgley, D. F. (1983). Patterns of interpersonal information seeking for the purchase of a symbolic product. *Journal of Marketing Research, 20*(1), 74–83.

Neisser, V. (1976). *Cognition and Reality*. San Francisco: W.H. Freeman and Company.

Newman, J. W. (1977). Consumer external search: amounts and determinants. In A. G. Woodside, J. N. Sheth, & P. D. Bennett (Eds.), *Consumer and Industrial Buying Behavior*. Amsterdam: North Holland Publishing Company.

Newman, J. W., & Staelin, R. (1973). Information sources of durable goods. *Journal of Advertising Research, 13*(2), 19–29.

Olshavsky, R. W., & Granbois, D. H. (Sept. 1979). Consumer decision making – fact or fiction? *Journal of Consumer Research, 6*, 93–100.

Ray, M. L. (1973). Marketing communication and the hierarchy of effects. In P. Clarke (Ed.), *New models for Communication Research*. Beverley Hills: Sage.

Reed, S. K. (1972). Pattern recognition and categorisation. *Cognitive Psychology, 3*(3), 382–407.

Render, B., & O'Connor, T. S. (1976). The influence of price, store name and brand name on perception of product quality. *Journal of the Academy of Marketing Science, 4*(4), 722–730.

Rigaux-Bricmont, B. (1981). Influences of brand name and packaging on perceived quality. In A. Mitchell (Ed.), *Advances in Consumer Research. Vol. 9*. Chicago: Association for Consumer Research.

Robertson, K. (1989). Strategically desirable brand name characteristics. *Journal of Consumer Marketing, 6*(4), 61–71.

Rock, I. (1975). *An Introduction to Perception*. New York: Macmillan.

Roselius, T. (1971). Consumer rankings of risk reduction methods. *Journal of Marketing, 35*(1), 56–61.

Ruffell, B. (1996). The genetics of brand naming. *Journal of Brand Management, 4*(2), 108–115.

Russo, J. E., Staelin, R., Nolan, C. A., Russell, G. J., & Metcalf, B. L. (June 1986). Nutrition information in the supermarket. *Journal of Consumer Research, 13*, 48–70.

Schiffman, L., & Kanuk, L. (2000). *Consumer Behavior*. Upper Saddle River: Prentice Hall International.

Schwartz, M. L., Jolson, M. A., & Lee, R. H. (March–April 1986). The marketing of funeral services: past, present and future. *Business Horizons*, 40–45.

Sheth, J., Mittal, B., & Newman, B. (1999). *Customer Behavior*. Orlando: Dryden Press.

Shipley, D., Hooley, G., & Wallace, S. (1988). The brand name development process. *International Journal of Advertising, 7*(3), 253–266.

Solomon, M., Bamossky, G., & Askegaard, S. (1999). *Consumer Behaviour*. Upper Saddle River: Prentice Hall International.

The Irish Times Millennium 200 Edition. *Ribena: a legendary brand looks to the future,* Available at: www.business2000.ie/pdf/pdf_4/smithkline_beecham_4th_ed.pdf. Accessed 7 July 2009.

Turtle, G. (7 March 2002). Papering over the cracks. *Marketing Week,* 23–27.

Venkatesan, M. (1973). Cognitive consistency and novelty seeking. In S. Ward, & T. S. Robertson (Eds.), *Consumer Behavior: Theoretical Sources*. Englewood Cliffs, N.J: PrenticeHall.

Which?. (March 1995) *Look-alikes,* 30–31.

Zajonc, R. B. (1968). Cognitive theories in social psychology. In G. Lindzey, & E. Aronson (Eds.), *The Handbook of Social Psychology.Vol. 1*. Reading, Massachusett: Addison Wesley.

CHAPTER 4

How Consumer Brands Satisfy Social and Psychological Needs

OBJECTIVES

After reading this chapter, you will be able to:

- Evaluate the role of the brand as a symbolic device.
- Understand the inter-relationship between the brand's personality and the consumer's self-concept.
- Explain the types of relationships consumers develop with brands.
- Identify the myths associated with powerful brands.
- Discuss the role and relevance of semiotics in conveying brand meaning.

CONTENTS

SUMMARY

The purpose of this chapter is to consider the social and psychological roles played by brands. When consumers buy brands, they are not just concerned with their functional capabilities. They are also interested in the brand's personality, which they may consider appropriate for certain situations. They look to brands to enable them to communicate something about themselves and also to understand the people around them better.

The chapter focuses on the consumer rather than on organisational brands, reflecting the greater emphasis placed on brand personality and symbolism in consumer marketing. This does not necessarily mean they are inappropriate in business-to-business markets, rather, that they are not as frequently employed.

We open this chapter by considering the added values from the images surrounding brands. We then address the symbolic role played by brands, where less emphasis is placed on what brands *do* for consumers and more on what they *mean* for consumers. Different symbolic roles for brands are identified, along with a consideration of the criteria necessary for brands to be effective communication devices. We draw on self-concept theory to explain how consumers seek brands with images that match their own self-image

123

Creating Powerful Brands. DOI: 10.1016/B978-1-85617-849-5.10004-2

and avoid those brands which reflect undesirable symbolic associations. A model of the way consumers select brands is presented, which shows how consumers choose brands to project images appropriate for different situations. We then examine how personal values influence brand selection and show the importance of brand personification as a means of enabling consumers to judge brands easily. Different brand personalities give rise to different types of relationships and we explore these relationships. We also identify the role of the brand in creating stronger brand relationships through consumer communities. Finally, we review the way that semiotics, the scientific study of signs, can contribute to brand effectiveness. We illustrate how brands within a single product category can communicate very different meanings through the symbols they use in marketing communications.

ADDED VALUES BEYOND FUNCTIONALISM

Brands succeed because they represent more than just utilitarian benefits. The physical constituents of the product or service are augmented through creative marketing to give added values that satisfy social and psychological needs. Surrounding the intrinsic physical product with an aura or personality gives consumers far greater confidence in using well-known brands. Evidence of this is shown by one study that investigated the role that branding played in drugs sold in retail stores. People suffering from headaches were given an analgesic. Some were given the drug in its well-known branded form, the rest had the same drug in its generic form, lacking any branding. The branded analgesic was more effective than the generic analgesic and it was calculated that just over a quarter of the pain relief was attributed to branding. What had happened was that branding had added an image of serenity around the pharmacological ingredients and, in the consumers' minds, had made the medication more effective than the unbranded tablets.

The images surrounding brands enable consumers to form a mental vision of what and who brands stand for. Specific brands are selected when the images they convey match the needs, values and lifestyles of consumers. For example, at a physical level, drinkers recognise Guinness as a rich, creamy, dark, bitter drink. The advertising has surrounded the stout with a personality which is symbolic of nourishing value and myths of power and energy. The brand represents manliness, mature experience and wit. Consequently, when drinkers are choosing between a glass of draught Guinness or Murphy's, they are subconsciously making an assessment of the appropriateness of the personality of these two brands for the situation in which they will consume it, be it amongst colleagues at lunch or amongst friends in the evening.

Knowles (2001) starting from the position that branding is the social amplification of benefit, provides a further array of examples about how brands are selected because they enable consumers to make non-verbal statements about themselves, for example:

'I am a high achiever'	Rolls Royce, Rolex and Hermès
'I am on my way to the top'	BMW, Tag Heuer and MaxMara
'I am an individual'	Apple, Twitter and Muji
'I am a world citizen'	British Airways and Skype
'I care about the environment'	Toyota Prius and Fairtrade

Brands are an integral part of our society and each day we have endless encounters with brands. Just think of the first hour after waking up and consider how many brands you come across. From seeing the brands used, people are able to understand each other better and help clarify who they are. The young manager going to the office in the morning is proud to wear his Armani suit, as he feels it makes a statement about his status; but in the evening with his friends, he wants to make a different statement and wears his Levi jeans and Polo casual shirt.

Powerful brands make strong image statements and consumers choose them not just because of their quality, but because of the images they project. A study by the advertising agency BBDO found that consumers are more likely to find differences between competing brands where emotional appeals are used, such as for beers, than between those predominately relying on rational appeals. Functional differences between brands are narrowing due to technological advances but the emotional differences are more sustainable. Having a functional advantage, such as a faster processing speed or a greater number of pixels, may be a competitive advantage today but over time it becomes dated. By contrast, when associating a brand with particular values, such as being honest and dependable, these values have a greater chance of lasting as they are more personally meaningful and thus help ensure the longevity of the brand.

Particularly for conspicuously consumed brands, such as those in the clothing and car markets, firms can succeed by positioning their brands to satisfy consumers' emotional needs. Consumers assess the meanings of different brands and make a purchase decision according to whether the brand will say the right sort of things about them to their peer group and whether the brand reflects back into themselves the right sort of personal feeling. For example, a young man choosing between brands of suits may well consider whether the brands reflect externally that he is a trendsetter and reflect back into himself that he is confident in his distinctiveness. In other words, there is a dialogue between consumers and brands. In the main, consumers do not just base their choice on rational grounds, such as perceptions of functional capabilities,

ASICS

In this advert for ASICS running shoes, the use of the word 'weight' has two meanings, which illustrate the benefits for the consumer in both a physical and psychological sense. The message in this advertisement also reinforces ASICS brand slogan 'Sound Mind. Sound Body'. (Reproduced by kind permission of ASICS)

beliefs about value for money or availability. Instead, they recognise that to make sense of the social circles they move in and to add meaning to their own existence, they must look at what different brands symbolise. They question how well a particular brand might fit their lifestyle, whether it helps them express their personality and whether they like the brand and would feel right using it.

As Mahajan and Wind (2002) note, given the proliferation of brands, appealing to the heart rather than the mind is important. An example of this is an advertisement for BMW's Mini 'I am Mini, hear me roar', and the accompanying website (www.hearmeroar.com), which claim personality traits such as 'Attitude', 'Performance', 'Hero', 'Gutsy', and asks the consumer 'Why whisper when you can ROAR?' In high priced purchases, emotional positioning can be powerful. For example, Tiffany & Co. advertised that 'Blue is the colour of dreams', which built on the brand image created by the brand's distinctive duck egg blue packaging and the emotional experience of receiving the blue box as a gift.

Brands are part of the culture of a society and as the culture changes, they need to be updated. The vodka brand Absolut, with its creative advertising majoring on its unique bottle shape, continually reflects changing culture. For example, the brand recently provided 'Drinkspiration' — an application for the iPhone which allows consumers to select a drink to suit 'the mood, weather, color, time, location, bar vibe and more' (http://www.absolutdrinks.com/web/Absolut_Members.aspx). However it should be noted that brands don't just reflect culture, they are sometimes part of culture. For example, Harley Davidson bikes reflect the call for freedom and comradeship. In turn, by promoting the Harley Owners Group (HOG), members enact their values through regularly joining with like-minded Harley owners and enjoying weekends away, travelling around new parts.

BRANDS AND SYMBOLISM

A criticism often voiced is that many models of consumer behaviour do not pay sufficient attention to the social meanings people perceive in different products. A lot of emphasis has historically been placed on the functional utility of products and less consideration given to the way that some people buy products for good feelings, fun and, in the case of art and entertainment, even for fantasies. Today, however, consumer research and marketing activity is changing to reflect the fact that consumers are increasingly evaluating products not just in terms of what they can *do* but also in terms of what they *mean*. The subject of symbolic inter-actionism has evolved to explain the type of behaviour whereby consumers show more interest in brands for what they say about them rather than what they do for them. As consumers interact with other members of society, they learn the symbolic meanings of products and brands through the responses of other people. Their buying, giving and consuming of brands facilitates communication between people. For example, blue jeans symbolise informality and youth. Advertising and other types of marketing communication help give symbolic meanings to brands — the classic example of this being the advertising behind Levi jeans. As Elliott (1997) observes, the symbolic value of brands operates in a cyclical process. Consumers construct and make sense of the world through interpreting brands as symbols inferring meaning; then they use these meanings to surround themselves with brands and so internally they develop their self-identity.

The symbolic meaning of brands is strongly influenced by the people with whom the consumer interacts. A new member of a social group may have formed ideas about the symbolic meaning of a brand from advertisements but if such a person hears contrary views from friends about the brand, they will be notably influenced by their views. To be part of a social group, the person doesn't just need to adhere to the group's attitudes and beliefs but also to reflect these attitudes and beliefs through displaying the right sorts of brands.

To facilitate understanding of the symbolic meaning of brands, design and visual representations are important in conveying meanings, especially in the service sector where no tangible product is available. For instance, the continual visual representation of First Direct financial services through black and white advertisements communicates a no-nonsense challenger. The inclusion of groups of happy teenagers in Sketchers trainers advertising conveys a sense of group acceptance and belonging, and the imagery of mountain climbers battling Everest under extreme conditions suggests the efficacy of The North Face clothing. Visual representations also have the advantage of avoiding the logical examinations to which verbal expressions are subjected and are therefore more likely to be accepted. Customers are less likely to challenge the symbolic simplicity of Innocent smoothies than the verbal claim stating that Innocent fruit smoothies do not contain additives.

Some brands have capitalised on the added value of symbolism, i.e. meanings and values over and above the functional element of the product or service, as in the case of, for example, Häagen-Dazs. Symbolism is sought by people in all walks of life to help them better understand their environment. Different marques of cars succeed because they enable drivers to say something about who they are. We may buy different brands of handbags, such as Prada as opposed to Fendi, not just for their aesthetic design but to enhance self-esteem.

Consumers perceive brands in very personal ways and attach their own values to them. Elliott and Wattanasuwan (1998) show that even though competing firms are striving to portray unique values for their brands, the symbolic interpretation of each brand varies by type of person. Different people ascribe different meanings to the same product. This challenges the assumption amongst marketers that the brand's symbolic meaning is the same amongst all the target market.

To cope with the numerous social roles we play in life, brands are invaluable in helping set the scene for the people we are with. As such, they help individuals join new groups more easily. New members at a golf club interpret the social information inherent in the brands owned by others and then select the right brand to communicate symbolically the right sort of message about themselves. When playing golf, smart trousers may be seen to be necessary to communicate the social role but to play with a particular group of people it may be important to have the right *brand* as well. The symbolic meaning of the brand is defined by the group of people using it and varies according to different social settings.

Brands as symbols can act as efficient communication devices, enabling people to convey messages about themselves and facilitate expressive gestures. Using a BMW key ring enables the owner to say something about themselves to friends and colleagues. Using Möet & Chandon champagne for a special occasion says something about the importance of the event, while giving an Omega watch as a birthday present says something about the important relationship between the giver and receiver.

Advertising and packaging are also crucial in reinforcing the covert message that is signified by the brand. Charles Revlon of Revlon Inc. succeeded because he realised that women were not only seeking the functional aspects of cosmetics, but also the seductive charm promised by the alluring symbols with which his brands have been surrounded. The rich and exotic packaging and the lifestyle advertising supporting perfume brands are crucial in communicating their inherent messages.

Brands are also used by people as ritual devices to help celebrate a particular occasion. For example, Interflora is synonymous with the celebration of birthdays, Valentine's Day, Mothers' Day, the arrival of new babies and the emotion of love.

They are also effective devices for understanding other people better. The classic example of this was the slow market acceptance of Nescafé instant coffee in the USA. In interviews, American housewives said they disliked the brand because of its taste. Yet, blind product testing against the then widely-accepted drip ground coffee, showed no problems. To get to the heart of the matter, housewives were asked to describe the sort of person they thought would be using a particular shopping list. Two lists were given to the samples. Half saw the list of groceries including Nescafé instant coffee and the other half saw the same list, but this time with Maxwell House drip ground coffee rather than

Nescafé. The results of these interviews using the different coffee brands led respondents to infer two different personalities. The person who had the Nescafé grocery list was perceived as being lazy, while the drip ground person was often described as a 'good, thrifty housewife'. As a consequence of this research, the advertising for Nescafé was changed. The campaign featured a busy housewife who was able to devote more attention to her family because Nescafé had freed up time for her. This change in advertising helped Nescafé successfully establish their brand of instant coffee.

Brands are also effective devices for expressing something to ourselves symbolically. For example, amongst final year undergraduates there is a ritual mystique associated with choosing the right clothes for job interviews and spending longer on shaving or hair grooming than normal. These activities are undertaken not only to conform to the interview situation but also to give the person a boost in self-confidence. In these situations, the consumer is looking for brands that will make them 'feel right'. A further example of this is on the Coca-Cola website, where customers were invited to add content which reflected 'the Coke side of life'. The emphasis of brands here is to help consumers communicate something about themselves. So, consumers look to brands in highly conspicuous product fields as symbolic devices to communicate something about themselves or to better understand their peer groups. The symbolic interpretations of some brands are well accepted. For example, the UEFA Champion's League logo features eight stars which represent the final eight teams. Consumers associate the logo and the theme music from Handel's 'Zadok the Priest' with the competition, as they are aired at the same time in all countries. In addition, the trophy is absent until the final, to enhance its prestige.

Consumers also strive to understand their environment better through decoding the symbolic messages surrounding them. A client working with an architect sees things like certificates on the architect's wall, the tastefully designed office, the quality of the paper on which a report is word-processed, the binding of the report and the list of clients the architect has worked for. All of these are decoded as messages implying a successful practice.

Symbols acquire their meaning in a cultural context, so the culture of the society consuming the brands needs to be appreciated to understand the encoding and decoding process. People learn the inherent meaning of different symbols and through regular contact with each other there is a consistent interpretation of them. To take a brand into a new culture may require subtle changes to ensure that the symbol acquires the right meaning in its new cultural context. For example, the Red Cross becomes the Red Crescent in the Middle East.

If a brand is to be used as a communication device, it must meet certain criteria. It must be highly visible when being bought or being used. It must be bought by a group of people who have clearly distinguishable characteristics, which in turn facilitates recognition of a particular stereotype. For example, *The Guardian* newspaper reader has been stereotyped as a well-educated person, possibly working in education or local government. In the newspaper market, some readers select different brands as value-expressive devices. They provide a statement about who they are, where they are in life and what sort of people they are. Since brands can act as self-expressive devices, the user prefers brands which come closest to meeting their own self-image. The concept of self-image is important in consumer branding and is reviewed in the next section.

SELF-CONCEPT AND BRANDING

In consumer research, it is argued that consumers' personalities can be inferred from the brands they use, from their attitudes towards different brands and from the meanings brands have for them. Consumers have a perception of themselves and they make brand decisions on the basis of whether owning or using a particular brand, which has a particular image, is consistent with their own self-image. They consider whether the ownership of certain brands communicates the right sort of image about themselves. Brands are only bought if they enhance the conception that consumers have of themselves or if they believe the brand's image to be similar to that which they have of themselves. Just as people take care to choose friends who have a similar personality to themselves, so brands, which are symbolic of particular images, are chosen with the same concern. As brands serve as expressive devices, people therefore prefer brands whose image is closest to their own self-image.

This way of looking at personality in terms of a person's self image can be traced back to Roger's self-theory. Motivation researchers advanced the idea of the self-concept, which is the way people form perceptions of their own character. By being with different people, they experience different reactions to themselves and through these clues start to form a view about the kind of people they are. A person's self-concept is formed in childhood. From many social interactions, the person becomes aware of their **actual self concept**, i.e. an idea of who they think they are. Parker (2009) calls this 'me as I am'. However, when they look inward and assess themselves, they may wish to change their actual self-concept to what is referred to as the **ideal self-concept**, i.e. who they think they would like to be, or 'the good me' (Parker, 2009). To aspire to the ideal self concept, the person buys and owns brands that they believe support the desired self-image.

One of the purposes of buying and using particular brands is either to maintain or enhance the individual's self-image. Choosing brands with appropriate image associations helps to enhance consumers' self image and their psychological well-being (Aaker, 1997; Parker, 2009). By using brands as symbolic devices, people are communicating certain things about themselves. Most importantly, when they buy a particular brand and receive a positive response from their peer group, they feel that their self-image is enhanced and they will be likely to buy the brand again. In effect, they are communicating that they wish to be associated with the kind of people they perceive as consuming that particular brand. For example, research has shown that self-image congruence can predict perceptions of quality and merchandise purchase decisions among sports fans. This is important as such merchandise is a revenue source as well as a means of strengthening supporters' loyalty to the team (Kwak and Kang, 2009).

There has been a lot of debate about which type of self-concept (actual or ideal) is more indicative of purchase behaviour. To understand this better, a study was designed that looked at 19 different product fields – ranging from headache remedies, as privately consumed products, through to clothes, as highly conspicuously consumed products. There was a significant correlation between the purchase intention for the actual and ideal self-concept results. This indicated that both are equally good indicators of brand selection.

On the other hand, people also select brands to avoid undesired stereotypes. Psychologists have identified an 'undesired self' which often comprises experience-based negative aspects, such as memories of dreaded experiences or unwanted emotions, but which is necessary as a standard to judge one's well-being (Ogilvie, 1987). Ogilvie suggested that the distance between our undesired self and our real self can be a better reference point for satisfaction than the distance between our real self and our desired self. In other words, we may be more likely to judge our well-being in relation to how distant we are from our most negative image of ourselves. In the context of brands, recent research (Bosnjak and Brand, 2008) suggests that the 'undesired self' can be used to explain consumption-related attitudes for high involvement products such as automobiles. They recommend that brand managers should take negative brand-related associations and stereotypes into account when trying to understand why consumers choose – or do not choose – brands.

The behaviour of individuals also varies according to the situation they are in. The brand of lager bought for drinking alone at home in front of the television is not necessarily the same as that bought when out on a Saturday night with friends in a pub. Situational self-image – the image the person wants others to have of them in a particular situation – is an important indicator of brand choice. According to the situation, the individuals match their self image to the social expectations of that particular group and select their brands appropriately. The impact of situation on brand choice can be modelled, as shown in Fig. 4.1.

Consumers anticipate and then evaluate the people they are likely to meet at a particular event, such as those going to an important dinner party. They then draw on their repertoire of self-images to select the most appropriate self-image

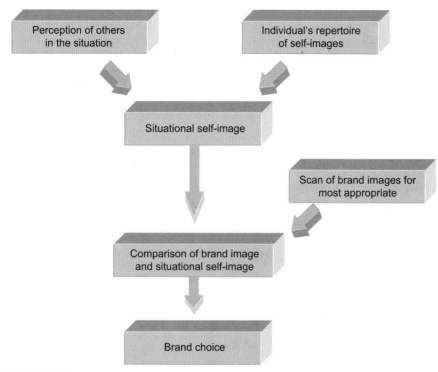

FIGURE 4.1

The impact of situation and self-image on brand choice (*after Schenk and Holman 1980*)

for the situation ('I can't let my hair down on Saturday night as there are too many of my husband's colleagues there. Better be a lot more reserved, especially as his boss is hosting this party'). If the situation requires products to express the situational self-image, such as a certain type of clothing, the consumer may decide to buy new clothes. When shopping, they will consider the images of different clothes and select the brand which comes closest to meeting the situational self-image they wish to project at, for instance, the dinner party.

Finally, it needs to be realised that there is an interaction between the symbolism of the brand being used and the individual's self-concept. Not only does the consumer's self-image influence the brands they select, but the brands have a symbolic value and this in turn influences the consumer's self-image.

BRAND VALUES AND PERSONALITY

As people become more integrated into a particular society, they learn about the values of that society and act in ways that conform to these values. Research has shown that *values* are a powerful force influencing the behaviour of people.

PERSIL

This advertisement for Persil Non-Biological powder reflects a changing male image, using a football analogy to describe the 'sensitive type' of consumer targeted by the brand. (Reproduced by kind permission of Persil)

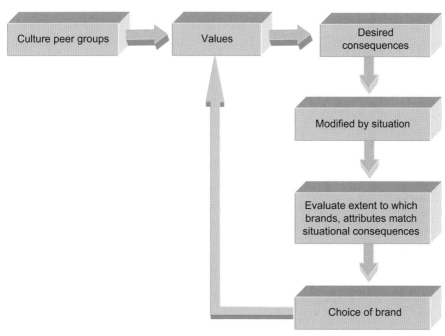

FIGURE 4.2
Consumers' values and brand selection (*adapted from Gutman 1982*)

A helpful understanding of the term 'values' is provided by Rokeach, who defines a value as an enduring belief that a particular type of behaviour (for example, being honest or courageous) or state of existence (such as happiness or security) is worth striving for. Identifying the values a consumer adheres to helps in understanding their brand selections and can be used to develop brands. Fig. 4.2 helps clarify this.

A young woman may belong to a sports club with a culture where everyone is considered to be a member of the team and trains with the team irrespective of weather conditions or whether they have been selected for particular matches. By training with the team, she begins to appreciate the contribution of her peers, particularly after successful matches. Reinforcement of the team ethos may be emphasised by the social aspect of the club but also by the informal dress code of the team members. In terms of the diagram in Fig. 4.2, her desired consequence is to enjoy playing sport but also to be seen as 'one of the team'. When the young woman is socialising with her friends at weekends, she will wear particular clothes which reflect her personal style. She may opt for high fashion brands, have her hair styled and wear her favourite perfume. However, when she socialises with the team, she may wear sports clothes or adopt clothing brands her teammates wear, to show her identification with the group.

DON'T WORRY, WE DON'T LIKE LOTS OF CHEMICALS EITHER.

Sanex Naturprotect deodorant replaces the active artificial chemical with the natural mineral Alum.
More natural, but highly effective.

Sanex. Keeps skin healthy.
www.sanexnaturprotect.com

Winner Deodorant Category. Survey of 12,026 people by T

SANEX

The reassuring tone of the Sanex advertisement suggests a membership reference group, which understands the consumer's preference to avoid chemicals, while still highlighting the efficacy of the product.

With the team, her priority is to be a team member; while with her friends, she is an individual and her values are enjoyment and self-expression.

For the consumer, to continually assess the extent to which each of the competing brands in a particular product field reflects their individual values would make brand selection a lengthy task. An easier approach is for them to personify brands, since by considering brands in more human terms they are then able to rapidly recognise the values portrayed by competing brands. There is a lot of evidence showing that for powerful brands, consumers have clear perceptions of the type of people they might be, were they to come to life. For example, two brands of credit cards have very similar performance attributes but it is in their brand personalities that consumers perceive differences. One of these brands might be described as 'Easy to get on with, has a sense of humour, knows lots of people and is middle class'. By contrast the other brand could be described as 'A little bit dull, harder to find them and quite sophisticated'. The primary source is the cluster of values marketers select to define the brand. Their intention is to then create a personality for the brand (e.g. through short lifestyle sketches or using a celebrity endorsement) to act as a powerful metaphor to communicate the brand's values.

The personality of the brand grows from many sources. One of the main influences on the brand's personality comes from the product or service itself. A brand of bleach, for example, which is highly caustic and requires safety clothing, may be perceived to be more masculine than a brand which is less concentrated. Another major influence is advertising and the use of celebrities in endorsements. The events sponsored by the brand owners also influence the brand's personality. For example, a refined and genteel persona is associated with the sponsorship of a classical music event, as opposed to a rugged or solid personality associated with sponsoring a rugby match. The challenge for the brand marketer is to manage all the points of interaction that their brand has with consumers, to ensure a holistic brand personality.

The importance of value as an influencer of brand choice behaviour has attracted research interest. A particularly insightful stream of work has resulted from Jagdish Sheth and his colleagues (1991). They argue, as Fig. 4.3 indicates, that consumer choice behaviour is influenced by five consumption values.

These five consumption values are:

- *Functional value* – reflecting the utility a consumer perceives from a brand's functional capability, for example, the perceived engineering excellence of an Audi, the cleaning capabilities of a Dyson vacuum cleaner or the easy handling of a Bugaboo pushchair.

CADBURY FUDGE

This advert extends the Fudge brand image to dessert. To retain the associations consumers may have with the original 'finger' bar, the slogan builds on the original slogan of 'A finger of fudge is just enough to give yourself a treat'. (Reproduced by kind permission of Cadbury)

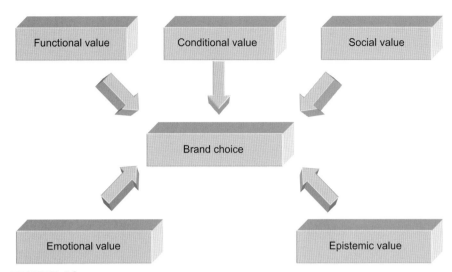

FIGURE 4.3

How values affect brand choice (*after Sheth et al. 1991*)

- *Social value* – representing the utility a consumer perceives through the brand being associated with a particular social group, such as Facebook users who are perceived as friendly and socially popular.
- *Emotional value* – is the utility a consumer perceives from the brand's ability to arouse particular feelings, such as the statement of love associated with diamonds.
- *Epistemic value* – is the utility a consumer perceives when trying a new brand mainly to satisfy their curiosity, such as switching from their habitual purchase of Pantene, on one occasion, to a newly launched shampoo brand just because they wondered what it might be like.
- *Conditional value* – reflects the perceived utility from a brand in a specific situation. For example, a t-shirt might be worn at home for comfort but wearing the same t-shirt to a football game may represent team support.

These values make different contributions in different contexts. When out to impress his new girlfriend, a young man in a restaurant may order a particular wine based on his concerns about maximising social, emotional and conditional values. By contrast, a few years later he may buy some wines to lay them down for future years, being primarily concerned about functional values. Understanding which values the consumer is particularly concerned about in a specific purchase context enables the brand marketer to give the brand the most appropriate set of values.

BRAND PERSONALITY AND RELATIONSHIP BUILDING

Appreciating consumers' values provides a basis for developing brand personalities to build stronger relationships. When a brand has a well-defined personality, consumers interact with it and develop a relationship, just as people in life do. The nature of a relationship between two people can be inferred by observing their individual attitudes and behaviours. Likewise, as Blackston (1992) argues, the nature of a relationship between consumers and brands can be determined through the attitudes and behaviours they display towards each other, in particular:

■ how consumers perceive and react to a brand;
■ how the brand behaves and reacts to the consumers.

Brand research has traditionally focused on how consumers perceive and react to a brand. By considering both aspects of the interaction, a more appropriate brand strategy can result. A hypothetical relationship between doctor and patient illustrates how the nature of the relationship can change, once both perspectives are taken into account. The patient may perceive the doctor as professional, caring and capable. This would suggest a positive relationship. However, the picture changes when we consider what the patient thinks the doctor thinks of them. The patient's view is that the doctor sees them as a boring hypochondriac. It does not matter whether the doctor really regards the patient as such, because the relationship depends on the patient's perception of the doctor's attitude. This private view makes the relationship difficult and unpleasant for the patient, though it may go unnoticed by the doctor. Brand owners may think that consumers have a positive relationship with the brand, but in order to be sure, they also need to ask consumers what they think the brand thinks of them.

The messages conveyed between a brand and its consumers build a two-way communication, in which consumers express their views on the brand and the brand 'responds' by stating a specific attitude towards consumers. As a result, marketers have to deal with two sets of attitudes in consumers' minds: first, they need to understand how consumers perceive the brand as the *object* of their attitudes; secondly, they need to discover the *subjective* brand, with its own set of attitudes. This dual perspective provides a more realistic insight into consumers' perceptions than the single-sided traditional analysis of consumers' views of the brand. Since the second perspective – the brand attitude – can often be the true brand discriminator, marketers need to undertake market research to unearth what consumers think the brand thinks of them (Blackston, 1992).

Brands can also act as relationship partners for consumers, enabling them to resolve personal issues. The nature of these relationships is varied and Fournier

(1998) has identified 15 different types of relationships between consumers and brands:

- *Committed partnership* — is a voluntary, long-term relationship, such as when someone becomes an advocate for ASICS trainers after years of marathon running.
- *Marriage of convenience* — i.e. a chance encounter leading to a long-term bond, such as when someone regularly uses a ketchup brand after trying it at a barbecue.
- *Arranged marriage* — is an imposed long-term partnership, such as when a student uses a brand of software because it was the only one available to run on their PC.
- *Casual friendship* — is one of few expectations and infrequent interactions, such as when an infrequent crisp eater rotates between brands, without preferring a particular one.
- *Close friendship* — relates to bonding through a sense of shared reward, such as when a golfer believes that using a particular brand of putter improves their game.
- *Compartmentalised friendship* — is a specialised friendship dependent on a particular situation. This would be a man choosing his beer according to the group of people he is drinking with.
- *Kinship* — is an involuntary union, such as when a student living away from home uses the same brand of tea that their mother used.
- *Rebound relationship* — is about a wish to replace a prior partner; for example, a woman who switches her perfume to avoid associations with an ex boyfriend.
- *Childhood friendship* — is about the brand evoking childhood memories, such as when a man chooses a brand of breakfast cereal because he recalls eating it in childhood.
- *Courtship* — is a testing period before commitment, such as a when woman experiments with samples of two makeup brands before committing herself to one.
- *Dependency* — relates to obsessive attraction, as in the case of a man who is upset when his favorite magazine is sold out.
- *Fling* — is about short-term engagement, such as when a woman tries another perfume for one evening, though she feels guilty about neglecting her traditional brand.
- *Adverserial relationship* — reflects bad feelings, as seen when consumers refuse to buy a brand of jeans they 'hate'.
- *Enslavement* — is an involuntary forced relationship, such as when a young couple is unhappy with their bank but they are tied into staying by a fixed rate mortgage.
- *Secret affair* — is a private, risky relationship, such as when someone indulges in shopping for expensive clothing brands, hiding the cost from their partner.

For a good brand relationship, the following attributes are necessary:

- *Love* and *passion* — Consumers must feel affection for the brand and want to have it at all costs.
- *Self-concept connection* — The brand must give consumers a sense of belonging or make them feel younger.
- *Interdependence* — The brand must become part of the consumer's everyday life.
- *Commitment* — Consumers need to be faithful to the brand through good and bad times, as in the case of Coca-Cola and Persil.
- *Intimacy* — Consumers should be very familiar with the brand and understand it well.
- *Partner quality* — Consumers seek those traits in the brand, such as trustworthiness, which they would in a friend.
- *Nostalgic attachment* — The brand should evoke pleasant memories because the consumer, or somebody close to them, used it in the past.

By focusing on these attributes, managers can identify strengths and weaknesses in the relationship between consumers and the brand. Consumers are looking for community and connection, as well as a way to express their status and lifestyles. Brands offer these benefits, as they present an identity that its users can relate to. For example, Martin (2003) cites the examples of Saturn, who host owner reunions and barbecues, as well as Harley Davidson, whose HOGS (Harley's Owners Groups) facilitate bonding through charity rides and other gatherings. Perhaps the strongest example of a relationship is that forged between Apple computers and their users. Apple has portrayed itself as pioneering and being independent, since its 1984 Superbowl television commercial which was centred around the Orwellian concept of 'big brother' and showed Macintosh computers as a means to resist oppression. Recent research has found Mac users to be 'active, avant garde and early adapters', and positions Apple's customers among the 'Sociologically Elite' (Sellers, 2008). There is clearly a symbiotic relationship between the Apple brand and its users. When consumers are intrinsically linked to a brand, it could be asked: 'Are consumers attracted to the brand because of its values, or is it the collective image people have of Apple that results in a desirable, co-created brand personality?'

A recent study by Mulyanegara et al. (2009) illustrates the role of the brand-consumer relationship in expressing personality among young fashion consumers. Using the 'Big Five' personality constructs: neuroticism, extroversion, openness, agreeableness and conscientiousness, the researchers explored the extent to which the young person's personality was expressed in the brand personalities of their fashion choices. They presented brands which were trusted, sociable, exciting or sincere. The research indicated that those who are conscientious or score high on neuroticism tend to prefer trusted brands, while extroverts prefer sociable brands. Therefore, Mulyanegara et al. (2009) recommend that brand managers should consider personality segmentation or employ advertising which emphasises the personality of their brands. This will allow consumers to select brands which are in line with their own personalities.

THE CONTRIBUTION OF SEMIOTICS TO BRANDING

People make inferences about others from the brands they own, since some brands act as cultural signs. Semiotics is the scientific study of signs. It helps clarify how consumers learn meanings associated with products and brands. If marketers are able to identify the rules of meaning that consumers have devised to encode and decode symbolic communication, they can make better use of advertising, design and packaging. For example, gold has been enshrined in our culture as a symbol of wealth and authority and can convey meanings of luxury, love, importance, warmth and eternity. But to use this as the prime colour on a box for a cheap, mass produced, plastic moulded toy car runs the risk of it being interpreted as vulgar.

Some researchers postulate that brands act as communicative sign devices at four levels. At the most basic level, the brand acts as a utilitarian sign. For example, a particular brand of washing machine may convey the meaning of reliability, effectiveness and economic performance. At the second level, a brand acts as a commercial sign conveying its value. For example, Porsche and Skoda signify extremes in value perceptions. At the third level, the brand acts as a socio-cultural sign, associating consumers with particular groups of people; for example, having certain brands 'to keep up with the Joneses', or wearing particular colours to signify allegiance to a football team. At the fourth level, the brand can be decoded as a mythical sign. For example, Nike, Interflora and Guinness all build on mythical associations.

Semiotics provides a better understanding of the cultural relationship between brands and consumers. Checking the communications briefs for brand advertising against the way consumers interpret the messages can result in the more effective use of brand resources. For example, British Airways once wished to increase the number of female executives using its airline. They developed an advertising campaign, targeted at women business travellers, which spoke about the ergonomics of seats. Semiotics Solutions, a UK research consultancy specialising in semiotics, undertook research to evaluate the new campaign. They found that it was not sufficiently sensitive to the fact that women are not as tall as men and the copy was rejected by women business travellers who felt that 'talking about 6 1/2 foot women was an insult'.

Semiotics can help in the design of brands, as was the case with a hypermarket in the Mammouth chain. Using group discussions, the different values consumers ascribed to hypermarkets were identified and designs developed to match these. Patterns of similarity were sought in terms of the way consumers associated different values with hypermarkets and four segments were tentatively identified from the qualitative research:

- *Convenience values* – characterised by 'Find the product quickly, always enough in stock, always on the same shelf'.
- *Critical values* – characterised by 'My husband isn't interested in frills and friendliness. He's only bothered about his wallet. He looks at the quality of the products and at the prices'.
- *Utopian values* – expressed by comments such as 'I like being somewhere on a human scale and not somewhere vast and overwhelming'.
- *Diversionary values* – such as 'I get the basic stuff out of the way first, and then I give myself a little treat, such as browsing in the book department'.

In the group discussions, consumers spoke about spatial issues, and it was inferred that:

- Convenience values were associated with interchanges and avenues.
- Critical values were associated with roundabouts and orientation maps.
- Utopian values related to markets and public gardens.
- Diversionary values encompassed covered arcades and flea markets.

Further analysis revealed that consumers expressing convenience and critical values wanted simple, continuous space. By contrast, consumers with utopian and diversionary values preferred complex, discontinuous space. Then, by considering customers in these two broad categories, the semiotic analysis led to the suggested design shown in Fig. 4.4. It was anticipated that as they gained more experience shopping in the hypermarket, the two consumer groups would use separate entrances. As such, a different store design was conceived for each entrance. Most consumers would seek all four values of the hypermarket but would be particularly drawn to the section that reflected their values. Decisions about where to locate the produce were aided by the group discussions. After the store accepted them, the designs had to be adjusted to cope with operational issues, such as ease of rapidly replenishing shelves, lighting, safety regulations, etc. However, overall, Mammouth found this approach helpful in conceiving a new design for their hypermarket.

Semiotics, as Alexander (1996) showed, can be a helpful tool to identify, evaluate and exploit the cultural myth which exists at the heart of most successful brands. A myth is a sacred, heroic story of doubtful authenticity. In marketing terms, myths are associated with powerful brands such as Nike, Body Shop and Virgin.

As Alexander (1996) argues, the stronger the opposition, the stronger the myth and the stronger the brand positioning. The starting point for semioticians

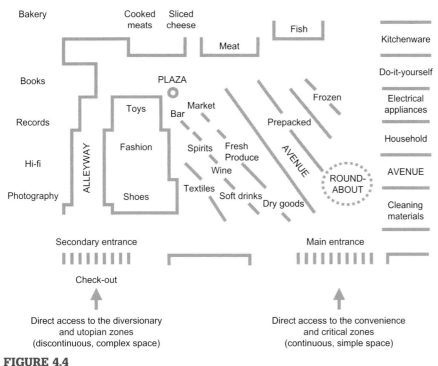

FIGURE 4.4

The hypermarket design proposed after semiotic analysis

helping marketers position brands, is to identify the attributes of the brand and at the same time to specify their opposites. They then take pairs of these oppositions and examine the resulting myth diagrams. For example, an attribute of a ready-prepared meal is real food and thus one opposition could be real food/junk food. Another attribute is homemade and the opposition would be homemade/commercially made. The myth diagram is then constructed from these two oppositions as shown in Fig. 4.5.

The opportunity for the myth explaining the brand positioning is in the two cultural contradiction boxes, i.e. commercially made real food and homemade junk food. Often only one of these quadrants is viable – not many would contemplate homemade junk food! Marks & Spencer focused on the opportunity presented by the cultural contradiction of commercially made real food and became the legendary myth of the store selling high quality ready-prepared meals.

FIGURE 4.5

The myth diagram for ready-prepared meals (*after Alexander 1996*)

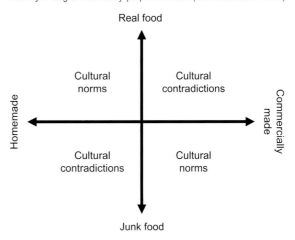

To assess the positioning of a brand through an advertisement, semiotics provides helpful insights about how the brand interacts with the cultural landscape. Lawes (2002) shows that there are several tools that can help decode the brand's positioning. These are:

- *Visual signs* — for example, a heart symbolising love or an umbrella symbolising security.
- *Linguistic signs* — the words and phrases used, act as metaphors. For example Golden Nuggets breakfast cereal evokes a sensation about exploration that can be rewarded.
- *Aural signs* — aspects such as tone of voice, regional accents and music tempo contribute to a cultural message.
- *The implied communication situation* — in any communication there is always an implied speaker and an implied recipient. By failing to appreciate consumers' views about these, brand marketers may be under the wrong impression about the way consumers perceive their brand.
- *Textual structure* — by looking at the way the text is constructed, further inferences can be drawn. For example is it telling a story, or implying a truthful report?
- *Information structure* — is the information presented as new versus taken-for-granted and are there features which are implied as having different degrees of importance?
- *Visual emphasis* — by examining how different elements are arranged relative to each other and their relative size, implications about relationships, order and relative importance emerge.
- *Binary oppositions and contrast pairs* — inferences can be drawn from appreciating how brands are described relative to other benchmarks. For example 'low-fat'.
- *Communication codes* — These seek to understand how different communication codes are being used. For example is white associated with purity and gold with luxury?

Bitoun (2006) provides an illustration of semiotics at work in perfume advertising. She presents findings from a semiotic audit which showed that, although advertisements were heterogeneous, they were characterised by three types of perfume representations, which also illustrated three different relationships between women and the perfume. The three classes of advertising are:

- *Class 1* — highlights the power of the product through the size and the compact, square shape of the bottle, the colour of the bottle relative to a black and white scene and the appearance of the model in the advertisement. She tends to have short or natural looking hair and wears little makeup and she conveys physical assertiveness, These cues suggest a fragrance for active, powerful women.

- *Class 2* − shows a metamorphosis of the perfume. It has greater use of colour, conveys more emotion and sensuality and depicts a smaller bottle with a rounded or curved shape. These cues suggest a fragrance for women who seek to charm others or be loved.
- *Class 3* − depicts a rich atmosphere, an imaginative world, with vivid colours. It features 'interiorised' glances which indicate an emphasis on the female character's own imagination. These cues suggest a fragrance for childlike women who want to enjoy themselves.

Semiotics were also critical to create associations during Barack Obama's campaign for the U.S. Presidency in 2008. The Obama logo uses the red, white and blue colours of the American flag. The logo depicts the sun rising over a ploughed field, symbolising a new dawn. The 'O' from the logo has become synonymous with Obama and has been used consistently on marketing devices. In addition, the logo is adaptable to numerous environments, which symbolise the adaptability of the person it represents.

CONCLUSIONS

This chapter has shown how brands perform a social and psychological role beyond that provided by their physical features. Consumers rely on brands to help them understand and communicate with different groups of people. The fact that consumers report greater pain relief after using a branded, rather than a generic version of the same analgesic, provides evidence of added values from brand images. Creative marketing has successfully positioned brands as effective problem solvers, with personalities that contribute to greater effectiveness.

Brands have the added values of symbolism − meanings and values over and above their physical constituents. Consumers look to brands not only for what they can do but also to help say something about themselves to their peer groups. Rolex watches are not worn just for their functional excellence but also to say something about who the owner is. To ensure that brands are effective, symbolic devices, it is crucial for marketers to communicate their capabilities to users and their peer groups, through advertising, public relations, packaging, merchandising, etc.

The symbolic aspect of brands makes them all the more attractive to consumers since they:

- enable consumers to convey messages about themselves and understand others better;
- help set social scenes and enable people to mix with each other more easily;
- act as ritual devices to celebrate specific occasions;
- provide a basis for a better understanding of the way people act;
- help consumers say something to themselves.

In effect, consumers are encoding messages to others by buying and using particular brands and are hoping that their target audience decodes the message the right way. Unfortunately, this is not always the case. For example, two friends meeting after several years may decide to meet for lunch. One opts for a low fat salad and no dessert, encoded to communicate their wish to take care of their health. The second person may decode this as 'time has dulled their sense of enjoyment'.

When consumers buy brands, they are making decisions about how well specific brands maintain or enhance an image they have of themselves. Just as consumers have distinct personalities, so do brands. Consumers take as much care choosing highly conspicuous brands as they do choosing their friends, since they like to be surrounded by like-minded personalities. Brands whose images match consumers' actual or ideal self-images are likely to be bought, while brands which reflect a lifestyle the consumer wishes to avoid are not likely to be consumed. When friends or colleagues admire someone's newly-bought brand, that person feels pleased that the brand reinforces their self-image and will continue to use the brand. The situation in which consumers find themselves will dictate, to some extent, the type of image that they wish to project. Through anticipating, and subsequently evaluating, the people they will meet at a particular event, consumers then seek brands to reflect the situational self-image that they wish to display.

Individuals' values are a powerful determinant of their brand choice behaviour. Much has been published showing that brands whose values reflect those of target consumers stand a greater likelihood of being bought. To make these value assessments, consumers interpret brands in humanistic terms and, through the metaphor of the brand as a person, are rapidly able to judge brands. Marketers are able to influence the personality of the brand through many routes, such as functional characteristics, packaging, advertising and sponsorship.

According to the type of personality clothing the brand, so there is a particular relationship between consumers and brands. Instead of just considering this from the perspective of the consumer, valuable insights about promotion strategies result from considering how brands perceive and react to consumers. A typology of relationships between brands and consumers has been reviewed and by considering the criteria necessary for an effective relationship, the strengths and weaknesses of the consumer—brand relationship can be assessed.

Semiotics, the scientific study of signs, can help brand development by assessing the cultural signs portrayed by different brands. For example, our culture brands the 07.00 train running ten minutes late as the 'late 07.00', carrying critical associations of inefficiency. However, in less developed economies, the train would be branded as the 07.10, portraying the triumph of mass transportation running against many odds in an under-resourced environment. Semiotics analyses a brand's communication capabilities at four different levels: utilitarian, commercial, socio-cultural and mythical. It can provide guidance about merchandising and displays. Brand advertising and design can benefit from checking communication briefs against the ways that consumers have interpreted the marketing activity as part of the social system. Furthermore, semiotics is a helpful tool in analysing the cultural myth at the heart of a brand to help develop a more powerful positioning.

MARKETING ACTION CHECKLIST

To help clarify the direction of future marketing activity, it is recommended that the following exercises are undertaken:

1. When did you last evaluate the added value of the image surrounding your key brands? If this has not been done within the past 12 months, it may well be advisable to assess this. One way of doing this is to identify the main competitor to one of your brands. Recruit a representative sample of consumers to try your brand and also that of your nearest competitor, seeking their comments about which brand they most preferred and why. This is a 'branded product test'. With another matched sample of consumers, repeat the product test but, this time, remove any branding and use identifying codes when presenting the brands. This is referred to as a 'blind product test'. Again, ask consumers which one of the two brands they most preferred, with their reasons. Calculate the proportions of the two groups who prefer each of the two brands on the branded and the blind product tests. Comparing the preference scores when the brands are assessed blind and then branded gives an indication of the value consumers ascribe to functional and emotional aspects of the brand.

2. Do you know what image surrounds your brand? If little is known about this, it would be wise to conduct some qualitative depth interviews with consumers. Ideally, this should be done by a skilled qualitative market researcher, preferably with a background in psychology, sociology or anthropology. Some of the ways of gauging the image associated with a brand are to ask consumers the following types of questions:
    'If the **brand** came to life, what sort of person would it be?'
    'If the **brand** were a person and they died, what would be written on their epitaph?'
    'If the **brand** were to be a car, what sort of car would it be?'
    'Tell me the first thing that comes to mind when I say '**brand**'.
    'What would a friend of yours most like about the **brand** and what would they most dislike about **brand?**'
    They could also be asked to role-play the way your brand solves a problem and then repeat this for a competitor's brand.

Once you have identified the image dimensions of your brand, it would then be useful to see how strongly your brand is associated with each of these image statements, comparing them against competitors' brands. This could be done using a questionnaire that asked respondents to use a five-point agreement–disagreement scale to state how well they felt that each of the given statements described each of the brands. By administering this to a representative sample of consumers, the image profiles of your brand and those of your competitors' brands can be assessed.

3. When did you last evaluate whether the characters portraying your brands are appropriate for today's consumers? If you feel your brands compete in a fashion-driven market, it would be advisable to undertake qualitative market research to assess the suitability of the people in your brand advertising.

4. To what extent do your key brands satisfy the following symbolic roles? Do they:
   - enable people to convey messages about themselves?
   - enable people to join new groups more easily?
   - help celebrate special events?
   - aid people to understand the actions of their peer group?
   - allow consumers to say something about themselves to themselves?

   Having undertaken this symbolic brand audit, evaluate how well your marketing activity helps support these symbolic roles.

5. If on exporting your brands you found a hostile consumer response, did you subsequently conduct qualitative market research to assess why your brands failed? Was any work undertaken to assess whether the symbols surrounding your brands meant something different overseas from its meaning in the UK? For example, putting your hand to your ear in the UK indicates that the person is talking too quietly but in Italy is taken as an insulting gesture.

6. How well-matched is your brand's image with the self-image of your target consumers? One way of assessing this is by comparing the image profile of your brand against the self-image profile of your target market. If you have no data on this, Question 2 in this section explains how to measure the image of your brand quantitatively. The same battery of attributes should also be administered to a representative sample of your consumers, asking them to use a five-point scale to assess how much they agree or disagree with each of the statements describing themselves. Compare the average brand image scores against the average self-image scores to assess how well your consumers' self-image matches that of your brand. Highlight the attributes showing the largest differences — these indicate areas where your brand does not meet consumers' expectations and should be investigated further.

7. How much is your brand the subject of situational influences? Do you know what supporting roles your brand plays when consumers use it in different situations? Does your marketing activity promote the appropriateness of your brand for particular situations?

8. If you are unsure about the values consumers perceive to be represented by your brand, consider undertaking the following exercise. Recruit a consumer and make it clear that you want to talk with them for about an hour. Show them your brand, ask them which other brands they would use besides your brand, then present them with all the brands they have mentioned. Explore the characteristics they believe your brand has. Having recorded these as they spoke, show the consumer the list of characteristics and ask them which one is the most important to them. For example, for a brand of crisps they may have said 'It has a very strong taste'. On a separate sheet of paper, record this characteristic; then ask them why this is important to them. They may have replied with the first consequence, 'I eat less when it tastes so strong'. Record this, then again ask, 'And why is that important to you?' They might now give the second consequence 'I don't get so fat'. Record this consequence and continue with this probing and recording until finally you arrive at the value associated with the base attribute. In our example it might be self-esteem. This is the first attribute-consequences-value ladder (similar to Fig. 4.2). Return to the list of characteristics and ask the consumer which is the second most important characteristic. On a separate sheet of paper record this, then repeat the questioning until you have arrived at the second value; i.e. you need to explore the second ladder. Repeat this process until you feel you have sufficient understanding of this consumer's perceptions of your brand's values.

   In this qualitative approach, we would encourage you to consider interviewing around ten consumers. This starts to give you a better appreciation of your target market's views. Due to the time and the expertise needed for these interviews, you may find it better to recruit a market research agency for this.

9. In Fig. 4.3, five types of brand values were identified. Use this framework to assess the extent to which your brand draws on each of these values. How does this profile of value importance vary between your light, moderate and heavy brand users? How is your brand strategy reflecting the importance of these values between these user groups?

10. Evaluate the information you have on consumers' perceptions of your brand. Does it also give an indication of how they think the brand thinks of them? How do you know that your consumers do not feel threatened, humiliated or otherwise negatively affected by the brand? What type of relationship have they developed with the brand? How can you enhance the relationship in the long-term?

11. Using Fournier's typology of relationships described in the section on 'Brand personality and relationship building', audit your consumer market research data for one of your brands and assess what type of relationship describes the consumer–brand bonding. Is this relationship congruent with the desired brand personality? If not, assess what changes are needed better to harmonise your brand strategy.

12. Take one of your recent brand advertisements and evaluate, with your colleagues, what the brand is communicating as a utilitarian sign, as a commercial sign, as a socio-cultural sign and as a mythical sign. Are these messages consistent at all four levels? Were the interpretations consistent across your team? Repeat the exercise with consumers and compare the findings between yourselves and the consumers. Any dissonant findings should be considered in more detail and corrective action taken.

13. If you feel that there may be a need to improve the position of one of your brands, consider the following exercise. With your marketing team, prepare a list of the attributes of your brand. Brainstorm these attributes to identify their opposites. Randomly choose two pairs of opposites and plot them on a diagram. Using the analogy of Fig. 4.5, identify the possible positioning that could characterise cultural contradictions. Evaluate each of these to assess new ideas for the brand's positioning.

STUDENT BASED ENQUIRY

1. Choose a consumer product category with which you are familiar. Select two brands within the category and compare and contrast their brand personalities. As you do so, identify the cues used by each brand to convey its personality.

2. Identify two brands you have purchased in the past month – one which reflects your *actual* self concept and one which reflects your *ideal* self concept. Assess how these brands appeal to your self-image.

3. Using Fournier's (1998) relationship typology, select brands which currently represent: i) a committed partnership, ii) a compartmentalised friendship, iii) a rebound relationship and iv) a childhood friendship for you, giving reasons for your brand choices.

4. Select a brand from the Interbrand Top 10 brands listing (see Chapter 2). Examine the brand's use of semiotics to convey brand meaning.

References

Aaker, J. L. (1997). Dimensions of brand personality. *Journal of Marketing Research, 34*(3), 347–356.

Alexander, M. (Oct. 1996). The myth at the heart of the brand. In *The Big Brand Challenge. Esomar Seminar, Vol. 203.* Berlin: Esomar Publication Series.

Bitoun, C. (2006). Semiotics, as a tool to understand and take action. *The Marketing Review, 6,* 111–121.

Blackston, M. (1992). A brand with an attitude: a suitable case for treatment. *Journal of Marketing Research Society, 31*(3), 231–241.

Bosnjak, M., & Brand, C. (2008). The impact of undesired self-image congruence on consumption-related attitudes and intentions. *International Journal of Management, 25*(4), 673–683.

Elliott, R. (1997). Existential consumption and irrational desire. *European Journal of Marketing, 34* (4), 285–296.

Elliott, R., & Wattanasuwan, K. (1998). Brands as symbolic resources for the construction of identity. *International Journal of Advertising, 17*(2), 131–144.

Fournier, S., Dobscha, S., & Mick, D. G. (Jan-Feb 1998). Preventing the premature death of relationship marketing. *Harvard Business Review*, 42–51.

Fournier, S. (1998). Consumers and their brands: developing relationship theory in consumer research. *Journal of Consumer Research, 24*(4), 343–373.

Gutman, J. (Spring 1982). A means-end chain model based on consumer categorisation processes. *Journal of Research, 46*, 60–72.

Knowles, J. (2001). The role of brands in business. In J. Goodchild, & C. Callow (Eds.), *Brands, Visions & Values*. Chichester: J. Wiley.

Kwak, D. H., & Kang, J. H. (2009). Symbolic purchase in sport: the roles of self-image congruence and perceived quality. *Management Decision, 47*(1), 85–99.

Lawes, R. (2002). De-mystifying semiotics: some key questions answered. *Market Research Society Annual Conference*. London: The Market Research Society.

Mahajan, V., & Wind, Y. (2002). Got emotional product positioning? *Marketing Management, 11* (3), 36–41.

Martin, W. E. (Sept/Oct 2003). A brand new you. *Psychology Today*, 72–95.

Mulyanegara, R. C., Tsarenko, Y., & Anderson, A. (2009). The big five and brand personality: Investigating the impact of consumer personality on preferences towards particular brand personality. *Brand Management, 16*(4), 234–247.

Ogilvie, D. M. (1987). The undesired self: a neglected variable in personality research. *Journal of Personality and Social Psychology, 52*(2), 379–385.

Schenk, C., & Holman, R. (1980). A sociological approach to brand choice: the concept of situational self image. In J. Olson (Ed.), *Advances in Consumer Research, Vol. 7* (pp. 610–615). Ann Arbor: Association for Consumer Research.

Sellers, D. (2008). *Apple's customers among 'sociologically elite'*. Available at: http://www.macsimumnews.com/index.php/archive/apples_customers_among_sociologically_elite. Accessed 22 July 2009.

Sheth, J., Newman, B., & Gross, B. (1991). Why we buy what we buy: a theory of consumption values. *Journal of Business Research, 22*(2), 159–170.

Further Reading

Apple's Branding Strategy. Available at: http://www.marketingminds.com.au/branding/apple_branding_strategy.html. Accessed 22 July 2009.

Apple Macintosh Superbowl Commercial. Available at: http://www.youtube.com/watch?v=OYecfV3ubP8. Accessed 22 July 2009.

Belk, R., Bahn, K., & Mayer, R. (June 1982). Developmental recognition of consumption symbolism. *Journal of Consumer Research, 9*, 4–17.

Biel, A. (1991). The brandscape: converting brand image into equity. *ADMAP, 26*(10), 41–46.

Birdwell, A. (Jan. 1968). A study of the influence of image congruence on consumer choice. *Journal of Business, 41*, 76—88.

Branthwaite, A., & Cooper, P. (May 1981). Analgesic effects of branding in treatment of headaches. *British Medical Journal, 16*, 282, 1576—1578.

Broadbent, K., & Cooper, P. (1987). Research is good for you. *Marketing Intelligence and Planning, 5* (1), 3—9.

Chisnall, P. (1985). *Marketing: A Behavioural Analysis.* London: McGraw Hill.

Combs, A., & Snygg, D. (1959). *Individual Behavior: A Perceptual Approach To Behavior.* New York: Harper & Bros.

Desbordes, M., & Chadwick, S. (2006). *Marketing and Football.* London: Butterworth-Heinemann.

Dolich, I. (Feb 1969). Congruence relationships between self images and product brands. *Journal of Marketing Research, 6*, 80—84.

Floch, J. (1988). The contribution of structural semiotics for the design of a hypermarket. *International Journal of Research in Marketing, 4*(3), 233—252.

Gallery, C. (April 10 2009). Yes, we can learn how to change from brand Obama. *Campaign*, 32—37.

Gordon, W., & Langmaid, R. (1988). *Qualitative Market Research. A Practitioners' and Buyers' Guide.* Aldershot.

Gower., Gordon W., & Valentine, V. (1996). Buying the brand at point of choice. *Journal of Brand Management, 4*(1), 35—44.

Grubb, E., & Hupp, G. (Feb. 1968). Perception of self, generalised stereotypes and brand selection. *Journal of Marketing Research, 5*, 58—63.

Heigh, H. L., & Gabel, T. G. (1992). Symbolic interactionism: its effects on consumer behavior and implications for marketing strategy. *Journal of Services Marketing, 6*(3), 5—16.

Hirschman, E., & Holbrook, M. (Summer 1982). Hedonic consumption: emerging concepts, methods and propositions. *Journal of Marketing, 46*, 92—101.

Landon, E. (Sept. 1974). Self concept, ideal self concept and consumer purchase intentions. *Journal of Consumer Research, 1*, 44—51.

Lannon, J., & Cooper, P. (1983). Humanistic advertising: a holistic cultural perspective. *International Journal of Advertising, 2*, 195—213.

Levitt, T. (July—Aug 1970). The morality of advertising. *Harvard Business Review*, 84—92.

Munson, J., & Spivey, W. (1981). Product and brand user stereotypes among social classes. In K. Munroe (Ed.), *Advances in Consumer Research, Vol. 8* (pp. 696—701). Ann Arbor: Association for Consumer Research.

North, W. (1988). The language of commodities: groundwork for a semiotics of consumer goods. *International Journal of Research in Marketing, 4*(3), 173—186.

Restall, C., & Gordon, W. (1993). Brands — the missing link: understanding the emotional relationship. *Marketing and Research Today, 21*(2), 59—67.

Ross, I. (1971). Self-concept and brand preference. *Journal of Business, 44*(1), 38—50.

Rokeach, M. (Winter 1968). The role of values in public opinion research. *Public Opinion Quarterly, 32*, 554.

Sirgy, M. (Dec. 1982). Self-concept in consumer behavior: a critical review. *Journal of Consumer Research, 9*, 287—300.

Solomon, M. (Dec. 1983). The role of products as social stimuli: a symbolic interactionism perspective. *Journal of Consumer Research, 10*, 319—329.

Wilkie, W. (1986). *Consumer Behavior.* New York: J. Wiley.

www.absolut.com. Accessed 22 July 2009

www.hearmeroar.com. Accessed 22 July 2009

Business-to-Business Branding

SUMMARY

The aim of this chapter is to consider the issues associated with the way brands can be developed in business-to-business markets. One way of appreciating this is through understanding how organisations buy brands. We open by making the point that brands play as important a role in business-to-business markets as they do in consumer markets. The nature of brand equity in organisational markets is addressed. The unique characteristics of business-to-business (B2B) marketing are considered, along with brand implications. We identify the people likely to be involved in B2B brand purchasing and discuss their roles. The stages involved in brand purchasing are presented, with consideration of the effort put in by the buyers. The importance of value in B2B brands is reviewed, focusing on the tangible and intangible components of four aspects of brand performance. The contribution of relationship marketing to B2B branding is considered, including the criteria used to assess the appropriateness

Creating Powerful Brands. DOI: 10.1016/B978-1-85617-849-5.10005-4

of potential partners, characteristics associated with successful relationships, how these vary over the duration of the relationship and factors that build brand loyalty in organisational markets. The rational and emotional factors affecting brand choice are reviewed. The traditional way that marketers present brand information is compared with buyers' views of the most useful sources and the two main routes by which buyers are influenced by sellers are considered. Finally, we address the important role played by corporate identity programmes and the corporate images perceived by buyers.

BRANDS AND ORGANISATIONAL MARKETING

The difference between consumer and business-to-business marketing

A distinction is frequently drawn between consumer and business-to-business (organisational, industrial) marketing. Consumer marketing is principally concerned with matching the resources of the selling organisation with the needs of consumers. It focuses heavily on those people at the end of the value chain who purchase brands to satisfy either their own personal needs or those of their families or friends. By contrast, organisational marketing is concerned with the provision of products and services to *organisations*. They are not the final consumers of the products and services. For example, Siemens Healthcare is actively involved in organisational marketing by supplying services such as communication and IT security, as well as IT project management to the hospital sector. This provides added value to hospitals, for example, greater energy efficiency, reduced costs and increased security; and also reduces potential violence or theft.

Emotional issues in business-to-business brands

While there are differences between consumer and organisational marketing, brands are just as important in both areas. Successful business-to-business branding is not a cold, stark form of branding which clinically targets companies. Rather it seeks to interest well-targeted business people who are very time-constrained, with vibrant propositions that are beneficial to them and their firms and that encourage particular types of relationships. As budgets for business-to-business campaigns are often smaller than those for consumer campaigns, marketers need to be more involved in the client's strategy, to understand how best their brand can help each client. It could even be argued that business-to-business branding is more interesting than consumer branding since one can only refresh consumers with Coca-Cola but IBM's IT Business Solutions can change people's lives.

Some argue that business-to-business buying is far more rational than consumer buying, yet emotional factors still play an important role.

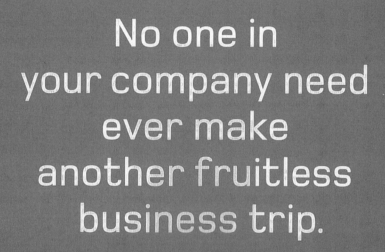

VIRGIN ATLANTIC

Virgin seek repeat business from the commercial traveller, through the rewards provided by its corporate loyalty programme.

Ward et al. (1999) noted that in high-tech markets, many managers mistakenly believe branding to be unimportant due to the supposedly rational purchase situation. It appears that the boundaries between high-tech products and consumer products are becoming blurred; the example is given of an ambitious MBA student purchasing an unnecessarily expensive notebook because of the success and status associated with its brand name. Similarly, managers are increasingly considering how the images associated with their purchases will reflect upon them personally, rather than just about how functionally appropriate the purchases will be to the business. In addition, new media such as social networks have increased the pace of user responsiveness, which in turn means that many business-to-business brands are more reactive to customer opinion leaders. For example, a recent Brandchannel debate explored market cynicism around 'greenwashing' by brands such as Exxon, who have chosen to adopt a green marketing strategy. Therefore, business-to-business brands need to be vigilant to capture emotion, while ensuring that emotion has ongoing relevance even at a consumer level.

Despite the importance of emotional factors, business-to-business brands are often characterised at the level of 'brand as reference' on Goodyear's consumerisation spectrum, discussed earlier. An examination of business-to-business advertisements shows that they often focus on the functional capabilities of the brand to guarantee its quality.

Brands permeate all areas of organisational marketing (e.g. Augmentin in pharmaceuticals, Pyrotenaz in fire-resistant cables, TelePresence in face-to-face communications, Novex in polyethylene, Brook Street in temporary personnel). They succeed because purchasers and users value the commitment of the suppliers behind their brands. Purchasers are proud to be associated with successful brands. For example, in the computer industry many firms are proud to promote the fact that their products are based on Intel microprocessors. In Chapter 2 we defined a successful brand as:

> *A cluster of functional and emotional values that enables an organisation to make a promise about a unique and welcomed experience.*

This definition is appropriate for both consumer and organisational marketing since in both sectors marketers are striving to enable their staff to behave in a way that is unique to their values and will aid the end users to achieve their desired experience. For example, in the business-to-business market, a strong brand creates value by enhancing shareholder confidence. Business-to-business brands also provide value to their business customers as they can enhance a firm's relationship with its employees. For instance, BT are offering Xero (http://www.xero.com/) as an online accounting package for SMEs. The brand offers the benefit of flexible accessibility, which creates value by enhancing employees' work-life balance.

Company or product/service's brand name?

Webster and Keller (2004) explain that, although we tend to think of brands in a consumer-marketing context, business-to-business brand names are among the strongest and most powerful brands. Chapter 2 presented the most recent Interbrand ranking, which shows that brands that serve organisational buyers, such as IBM, Microsoft and GE, are eclipsed in value only by Coca-Cola. Webster and Keller (2004) emphasise that these successful brands are psychological phenomena which take on meaning for customers. They argue therefore that a business-to-business brand must be considered in a broader context than its name, and instead, they recommend that the name would be considered a 'hook' upon which business-to-business firms can attach brand meaning.

It is more common to see organisational brand names bearing the name of the company. This enables a wide range of products from the same company to benefit from its corporate identity. As a consequence of this naming policy, many buyers see a brand's added values resulting from two factors. First, the added values from dealing with a particular firm and, second, the benefits from the specific product or service. As such, it is not uncommon in organisational marketing for buyers to talk about suppliers as brands (e.g. Xerox, GE, CISCO). In such situations the supplier has succeeded in augmenting their product with the added value of the firm's corporate identity. Consequently, the brand selection process for these buyers will be one of company selection first and then, at a later stage in the evaluation process, consideration of each company's brands. By contrast, where other firms have concentrated less on corporate brand endorsement and more on individual product branding, buyers will be less interested in the firm as a brand. For example in the SSRI class of the anti-depressants market, Prozac, from Eli Lilly and Seroxat from GlaxoSmithKline compete against each other without excessively relying on the added values of their corporate origins.

Are brands important in business-to-business marketing?

Organisational marketers who think that brands have no role to play are ignoring a powerful tool. More often than not, in a rather blinkered sense, they perceive brands to be little more than 'commodity items with a name stuck on' or they attach to them a very strong emphasis on functional capabilities with minimal presentation of any type of brand value. Ward et al. (1999) note that the majority of high-tech products are fast becoming commodities due to the similarities between products and the fact that

SINGAPORE AIRLINES:
The airline appeals to the business traveller through its promise of a good night's sleep

innovative features do not remain innovative for long. They call for a change in managerial attitudes from a 'product-centric' to a 'promise-centric' business model. Auh and Shih (2009) emphasised the importance of branding in business-to-business marketing. In two studies, they found that brand name type was critical for multi-generational product introduction, as it positively affected perceived technological improvement, product differentiation and the degree to which customers were willing to pay more for the more recent version of the technology.

As several case histories show, the general lack of attention to branding is naive and is an ineffective use of resources. An American study published late in the 1980s reported that established producers of wood and plywood panels were facing increasing competition from new entrants. Producers felt that the best way to counter this challenge was to brand their products. In this case all they did was to develop names for their lines, with the prime objective of differentiating themselves from competitors. Several months after the adoption of this so-called branding strategy, interviews were conducted with timber merchants. They were asked what criteria they used when deciding between wood suppliers. In the majority of cases, the first consideration was price. The buyers clearly regarded the competing products as commodities and not brands. If they had perceived any changes, they would have recognised the competing items as being differentiated because of their added values, for which price premiums could be charged. A further irony of this study was that the timber merchants were highly critical of the consumer confusion caused by these 'branding strategies'. The producers had used names that were only appropriate for distributors. They had ignored the fact that consumers could not relate the brand names to the performance capabilities of the different types of wood panels.

This is reinforced by the work of Mitchell et al. (2001) in the UK, who found a belief amongst managers in industrial firms that providing product identity and a consistent image were two of the most important functions of an organisational brand name. Examples of other functions considered to be important included aiding communication and market segmentation (see ranking in Table 5.1). They also found that on average, firms believed brand names to be important to realising corporate success.

Do organisational brands have notable equity?

The role of brand equity in organisational markets has been the subject of some debate over recent years. The first challenge is defining what brand equity means in a business-to-business market. van Riel et al. (2005) explored this challenge of conceptualising of B2B brand equity. Just as consumer literature would say that the power of the brand resides in the

Table 5.1 Perceptions of Industrial Brand Name Benefits Ranked by Importance (*from Mitchell et al. 2001*)

Importance Ranking	Benefit from Industrial Brand Name
1	Provides product identity
2	Provides image consistency
3	Valuable to marketing success
4	Major asset to firm
5	Confers uniqueness
6	Provides competitive edge
7	Helps product positioning
8	Aids communication
9	Helps market segmentation
10	Makes buying easier
11	Of value to customers
12	Provides legal protection

minds of the customer (Keller, 1998), van Riel et al. (2005) suggest that business-to-business brand equity can also be measured from the industrial buyer's perspective. They explain that although industrial buyers are often thought to be more rational and price driven, they sometimes make decisions based around the brand, for example where product failure would carry high performance or personal risk or where the product needs substantial support.

The benefits of brand equity have also been presented in the literature. Schultz and Schultz (2000) explain that a strong brand allows the business-to-business company to command a premium price, gain a greater market share and be perceived as higher quality, even in what they term 'rational and negotiated purchasing categories' including oil, computers and medical imaging.

Overall, research has suggested that organisational brands do have notable equity. In a study of 450 firms across 47 industries, Gregory and Sexton (2007) found that corporate brand equity was responsible for 7% of average stock performance. This figure can be as high as 20% for managed brands. They advocate that firms should seek the maximum brand equity value possible, as a portion of a percent in brand equity can equate to a value of hundreds of millions of dollars.

While there are similarities between the way brand equity has been conceptualised in organisational and consumer markets, there are also some differences (Kim et al. 1998). Mitchell et al. (2001), in their work on business-to-business brands, examined 10 major components of brand equity from the literature. In organisational marketing, the most important components of brand equity were considered to be perceived quality and a recognisable image. Related to these, market leadership, a differentiated position, a significant market share and the ability to attract a price premium were also considered important. On average or of some importance, were longevity, global presence, extendibility of the brand and difficulty of imitation.

Hutton (1997) carried out research in the USA that provides further evidence for the existence of notable organisational brand equity. He found that organisational buyers exhibit significant 'brand equity behaviour', in that for their favourite brand, they were prepared to pay a price premium, were willing to recommend the brand to others and were willing to give special consideration to another product with the same brand name (the 'halo' effect). It was also discovered that the extent to which respondents perceived their favourite brand to be well-known by others correlated significantly with these three indicators of brand equity.

Brands and the value chain

To succeed, brands in organisational markets must take into account the needs of everyone in the value chain. In the man-made fibres market, suppliers thought that it was more important to stress brand names without carefully relating the name to the uses of the fibres. As a consequence, in a market rich with competing brand names (e.g. Dacron, Terylene, Acrilan, Celon) but poor in brand explanation, users were confused about the capabilities of different fabrics. A more effective strategy would have been to identify the different customers and influencers in the value chain (e.g. the weavers, knitters, designers, manufacturers, distributors). Promotional strategies suitable for each group should have then been developed, unified by a corporate theme to clarify the unique capabilities of the different fibre brands.

Our concern in this chapter is to show how an appreciation of the differences between organisational and consumer marketing can help in developing successful brands. Some of the key differences are addressed in the next section.

THE UNIQUE CHARACTERISTICS OF ORGANISATIONAL MARKETING

More people involved

In consumer marketing, brands tend to be bought by individuals, while many people are involved in organisational purchasing. The business-to-business brand marketer is faced with the challenge of not only identifying which managers are involved in the purchasing decision but also what brand attributes are of particular concern to each of them. The various benefits of the brand, therefore, need to be communicated to all involved, stressing the relevant attributes to particular individuals. For example, the brand's reliable delivery may need to be stressed to the production manager, its low level of impurities to the quality control manager, its low life cycle costs to the accountant and so on.

Lower prices and costs involved

Consumers are generally faced with relatively inexpensive brands which, therefore, they do not spend much time evaluating. By contrast, organisational purchasing involves large financial commitments. To reduce the risk of an inappropriate purchase decision, organisations involve several managers from different departments to help in the evaluation process. More recently, firms are allowing their business-to-business customers and potential customers to become part of the product design and product testing process. Online communities of software testers, for example opinion-leader blogsites, can help companies improve their software before launch. Such activities reduce potential problems and costs and encourage trial and adoption through online endorsement.

More time involved

It is not unusual to observe consumers making a brand choice with only a short deliberation time. By contrast, organisational buying generally involves much longer deliberation periods. Salespeople in organisational marketing are more frequently regarded as technical advisers compared with salespeople in consumer marketing. They often expect to have several meetings with potential purchasers before the firm feels sufficiently confident to make a purchase decision. The implication of this is that the effectiveness of brand support needs to be assessed over a much longer period of time than in consumer marketing.

More loyalty involved

Whilst consumers are generally loyal to particular brands, from time to time they like to experiment with new brands. In organisational marketing, however, it is much more common for purchasers to seek long-term buying relationships with particular suppliers. They have invested a considerable amount of work in the selection process and have learnt the idiosyncrasies of working with their suppliers. To experiment by trying out new supplier's brands has implications throughout the whole organisation (e.g. delivery, quality control, production, invoice processing, etc.) and is not lightly entertained.

Segmentation is important

Just as certain groups are prone to buy certain types of brands in consumer markets, some clearly identifiable segments are also apparent in organisational purchasing. The characteristics of these segments, however, differ from those seen in consumer markets. For example, marketers of computer software brands (e.g. SPSS, Agresso) find it beneficial to segment potential purchasers according to their computing experience. Compared with more experienced users, novices appreciate a service that includes a help desk and training workshops to learn how to use the software.

Buyers are more rational

Organisational buying is generally more rational than consumer buying, even though emotional considerations still influence the final decision. They may not be a particularly dominant force, but they can still be part of the brand selection criteria. For example, some buyers like to be treated as very important people by sales representatives and others question whether they can 'get on with the people in that firm'. If these 'less rational' issues are not satisfied, the competitor may succeed.

Greater risks are involved

Chapter 3 showed that perceived risk helps us to understand better why consumers select certain brands. Likewise, an understanding of perceived risk is useful in organisational buying. Buyers perceive risk when buying a new brand for the first time and look for ways of reducing it. One way is to involve more people in the evaluation process. One pharmaceutical company took heed of perceived risk in its marketing strategy. Its sales representatives had an estimate of the extent to which each GP is risk averse. When launching a new drug they directed their calls initially towards those GPs with a high threshold for risk. Once these GPs have prescribed it and are satisfied with the results from the new drug, they are then encouraged to talk about the new drug with their colleagues. This is facilitated by the company hosting meetings at which, for example, there may be a lecture by a hospital consultant on a novel surgical technique.

It is important that marketers do not overlook the risk dimension in career advancement. Some buyers may be particularly ambitious to gain rapid promotion and keen to show the improvements they have achieved for the firm. If approached by a new supplier with a particularly attractive brand proposition, they are likely to be receptive to change. By contrast, the purchasing manager who has gradually achieved promotion may be more cautious and therefore less receptive to a new brand proposition.

In conclusion, there are differences between consumer and organisational markets. These differences, however, are subtle. Consequently, marketers do not need to get to grips with a whole new theory of branding. Rather they have to appreciate how to fine-tune the techniques widely practised in consumer marketing.

NATURE OF BRANDS IN ORGANISATIONAL MARKETS

Ward et al. (1999) define a brand as 'a distinctive identity that differentiates a relevant, enduring and credible promise of value associated with a product, service or organisation and indicates the source of that promise' (88). This definition focuses on the issue of value unlike our broader perspective, which addresses resulting experiences (e.g. production, throughput, etc.). The brand's promise can be based upon a variety of things; for example, innovative technology or superior service and customer support. The promise must be relevant to the intended customers of the brand and may differ slightly between customer groups in order to more exactly meet their needs. The credibility of the brand promise hinges upon persistence and consistency in its delivery; if this can be achieved, this represents a significant competitive advantage.

Based on material developed by Larry Light, Ward et al (1999) show how the brand pyramid can be used to build a powerful high-tech brand. The brand pyramid is shown in Fig. 5.1.

The first two levels of the pyramid represent the elements of product competition rather than brand competition. However, if a company can raise its

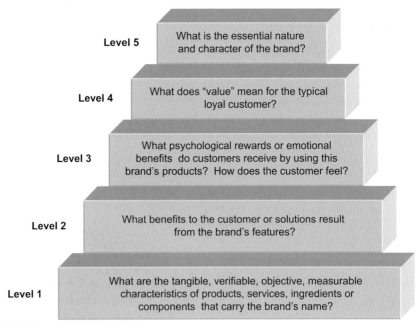

FIGURE 5.1

The brand pyramid (*Ward et al. 1999*)

offering to encompass level three, emotional rewards can offer competitive advantage. The top two levels of the pyramid represent the stages reached by powerful brands, where the brand is recognised in terms of its personality and values.

WHO BUYS BRANDS?

Unlike consumer brand buying, brand buying in organisations typically involves several people. A buying centre, sometimes also called a decision making unit, is a group of people from different departments who are involved in the evaluation and selection of a particular brand. When buying a particular brand of capital equipment, for example, it is likely that representatives from engineering, purchasing, finance, manufacturing, marketing and site services will be involved. It is not unusual, particularly when larger firms buy expensive and complex brands, to see as many as 20 people involved in the buying centre. This typically occurs when the organisation has little experience of a new brand or when they perceive a high degree of risk in purchasing a complex and expensive item. One of the reasons for such a wide range of skills and functional backgrounds is the greater sense of confidence amongst the decision makers.

Where the user has a continual need for specific items, such as rubber seals for car doors, the buying will typically be left to the purchasing manager. Even here, however, the user is likely to produce and update the specification for the purchasing manager.

In situations where the user has expert knowledge of a particular area and is not involved in major financial expenditures, they alone will make the brand decision. An example of this is the market research manager deciding which consultant to employ for a specific project.

The matrix in Fig. 5.2 is a useful guide when anticipating who is likely to be involved in the organisational brand purchase decision. Composition can be predicted from the commercial risk facing the organisation and its perception of the degree of product/ service complexity. When, for example, the firm feels there is a considerable financial commitment (both in the initial purchase cost and in running and maintenance costs)

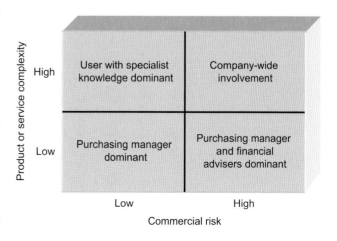

FIGURE 5.2

Predicting who will be involved in brand purchasing

in buying a brand that is difficult to assess functionally because it involves unfamiliar features, the organisation will feel at its most vulnerable. In such a situation there is likely to be at least one person from each of the interested departments. After they have evaluated alternative brands, it is probable that recommendations will then be channelled to more senior levels for final consideration.

As the matrix indicates, the purchasing manager rarely makes brand purchase decisions alone. In general, buying centres rather than individuals become involved in brand selection when:

- the size of the purchasing firm becomes larger;
- the firm has little experience of buying or using the brand;
- there is weak loyalty towards a supplier;
- the brand is regarded as being an important part of the production process;
- the financial size of the order increases;
- individuals perceive risk of any kind in buying the brand.

Membership of the buying centre changes as more information becomes available. For example, if in the early stages of the evaluation process it was learnt that one of the competing brands could be either bought or leased, then the team would be supplemented by a financial adviser able to evaluate these options.

Brand marketers need to put themselves in the organisation's position and consider which departments will be most affected by their brand. If the brand offers a significant opportunity for the organisation to cut production costs but after-sales service will deteriorate for a few months, it is likely that production, marketing, customer service and finance will be involved in the buying decision. The relative importance of any member in the buying centre varies according to the type of product being bought. For example, plant managers and engineers are more influential than purchasing managers when buying technically sophisticated capital equipment. Plant managers are the most active information seekers in buying centres and members of the buying centre often refer to them rather than going to external sources for information.

Having identified the members of the buying centre, the brand marketer needs to monitor it to identify any shifts in terms of who the key deciders are. X-ray film used to be sold to hospitals on the basis that the deciders in the buying centre were the radiologists and the technicians. However, with the government's review and changes to the Health Service, this led to a greater involvement of administrators in the buying process. It has been argued that brands in the medical X-ray film market succeeded because marketers recognised the changing composition of the buying centres and altered their presentations to reflect the increasing importance of administrators.

ANTICIPATING THE ROLE OF BUYING CENTRE MEMBERS

Often, recommendations about brand purchases are referred to senior management or directors because the evaluating team has a limit on how much they can spend. Yet even after this highly rational process, the decision may be overruled for an emotional reason. We learnt about one organisation which employed an IT consultant to work with its managers in evaluating and recommending which computer to buy. A recommendation was made to the board about two possible brands. The decision was overruled in favour of the more expensive and less technically sophisticated option. The chairman thanked the team for its work but felt that it was safer to stay with a well known brand, even though he had very little knowledge of IT. If the brand marketers had more insight into the roles played by the purchasing team, this outcome may have favoured their brand.

One of the most widely employed ways of understanding the different roles of members of the buying centre is based on the five categories: users, influencers, deciders, buyers and gatekeepers.

Users

Users are those people in the firm who will be using the brand. These people usually start the buying process and write the requirement specifications. Problems are sometimes created for the marketer when there are two or more groups of users, with conflicting objectives. For example, a chemist in a laboratory may want a particular brand of spectrometer because of its high resolution capabilities but the R&D manager, who would make less frequent use of the spectrometer, is concerned about the lack of space to house the equipment. The shrewd marketer needs to identify who the primary and secondary users are and find the right balance in appealing to each group.

Influencers

Influencers are sometimes difficult to identify as they can either exert influence directly, by defining brand criteria requirements, or indirectly, through informally providing information. For example, a manager evaluating particular brands of oscilloscopes would seek information from potential suppliers but in a chance corridor meeting may learn the views of a colleague in a different department. Influencers are not just those people inside the organisation, but can include external consultants. Often consultants are employed to write brand specifications or to help evaluate the competing brands. They may also include individuals working in competing firms who, because of their perceived expertise, are approached through the informal networking system.

Deciders

Deciders have the power to make the final decision about which brand should be bought. Ironically, it is sometimes difficult to identify these people. For example, the user may have written the specification requirement in such a way that only one brand can be bought. Or, in the final debate conducted in a closed session, the managing director may make the decision but leave the purchasing manager to place the order. Thus, the purchasing manager may appear to be the decider, when in fact it is the managing director.

Buyers

Buyers are those with the formal authority for arranging the purchase. While the purchasing manager may appear very forceful in negotiations, often the objectives are specified by others in the organisation. For relatively routine, low-cost purchases, the purchasing manager will proceed without recourse to company-wide discussion.

It should be appreciated that purchasing managers are keen on maintaining and improving their status within the firm. To do this, they employ several tactics. For example, some are deliberately rule oriented. Regardless of who approaches them, they insist on working by the book and take no actions until in receipt of formal notification, even though verbal decisions were reached earlier. Such an approach causes frustration amongst all those who need to work with purchasing. Another tactic seen is that of favouring a few colleagues. Just for these few individuals, purchasing managers project an aura of friendship and willingness to help, expecting favours to be done for them in return.

Gatekeepers

Gatekeepers are individuals who control the flow of information into the buying centre. They may be the managing director's secretary, opening the daily emails and post and deciding which attachments and circulars should be seen and by whom; or the purchasing manager, insisting to the receptionist that any sales people seeking new business with the firm should always be directed to the purchasing department. Gatekeepers tend to exert their influence at the early stage of the buying process when the full range of competing brands needs to be identified.

It must be appreciated that the same person can perform several of these roles. The challenge facing the brand marketer is to identify who is playing which role and when any one member of the buying centre becomes more influential. Evidence suggests that the purchasing department becomes more influential when:

- commercial considerations, e.g. delivery or terms of payment, are seen to be more important than technical considerations;
- the item is routinely bought;

- the purchasing department is highly regarded within the firm because of their specialised knowledge of suppliers;
- the technology underpinning the brand has not changed for some time, neither have the evaluation criteria.

When members of the buying centre meet, individuals from different departments with different motivations are brought together. Group dynamics may cause tension and covert attempts are made by some members to gain a more influential position. In a large buying centre, people sense that those who have expert knowledge are more powerful influencers. Surveys have shown that these individuals are also powerful when the firm does not feel it is under pressure to arrive at a rapid brand decision.

Armed with a better understanding of who is likely to be involved in the buying centre and of the roles that these individuals are likely to play, brand marketers should be better able to decide how to position their brand to appeal to the different participants. They should also be able to anticipate where the influencing power lies, and thus, where more effort should be directed. By also appreciating the stages of the buying process, they should be able to assess when more effort needs to be put behind their brand, an issue which is considered next.

THE ORGANISATIONAL BUYING PROCESS

An eight-step process

In common with the way we model consumer buying behaviour as a process, starting with problem recognition and progressing through to post-purchase evaluation, so the same logic satisfactorily describes organisational buying. The seminal work of Robinson and his team of researchers (1967) resulted in a model that charts the organisational buying process as an eight-stage process, shown in Fig. 5.3.

The process starts when the firm becomes aware of a problem. For example, their product is outdated, an opportunity for a new line has been identified, a piece of capital equipment has broken and so on. Someone within the organisation recognises this problem and starts to involve others. They would consider how their particular problem should be solved. A detailed specification would be drawn up at the third stage and, following internal discussions, this would be redrafted until it reflected a consensus view. If marketers know who will draw up the specification, they can target their brand's commercial and technical promotion. At the fourth stage, the organisation searches for potential suppliers and qualifies these. The criteria for qualification are considered later in this chapter. Some of the possible suppliers will be eliminated at this stage.

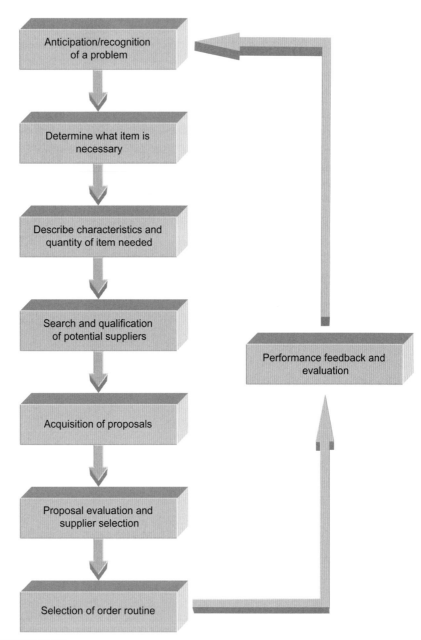

FIGURE 5.3
The organisational buying process (*after Robinson et al. 1967*)

The screened suppliers would then be invited to submit their brand proposals. This normally entails a series of meetings to ensure that the supplier fully understands the company's needs. Each of the proposals are analysed against the agreed evaluation criteria and a brand purchase decision made. The purchasing manager is then given the authority to place the order and undertake any negotiations about terms and deliveries. Finally, once the brand has been used, an internal review takes place and assesses how well the brand and the supplier are performing against the evaluation criteria. This would usually only be done formally if the brand proved unsatisfactory. However, it is likely that there would be an informal review, where individuals would talk amongst themselves about the brand's capabilities.

While this eight-phase model describes the buying stages passed through, the amount of time and effort devoted to any one phase depends on the type of purchase. Robinson and his team identified three types of purchases: **new task**; **modified rebuy**; and **straight rebuy**. These three types of purchases give some indication about the amount of effort undertaken by the buying centre.

New task

In the new task purchase, the organisation has no previous experience of buying the product or service. In this situation, the purchasing firm will put in a lot of work and will seek a considerable amount of information about different brands. Marketers have to work hard, explaining how their brand can solve the purchaser's problem. The buying centre feel that there is a lot of risk in this purchase and the marketers should present their brands as a low-risk option, describing how other firms have successfully benefited from using the brand.

Modified rebuy

In the modified rebuy situation, the firm has experience of buying brands in the product field but feels it is time to consider whether significantly better brands are available. An example of this would be the office services manager becoming aware of the benefits of a new class of laser printers, which print faster than their current printers. The organisation knows that there will be a lot of work involved in finding a new brand, as well as in adapting its internal processes to absorb the change. It feels, however, that the benefits from change warrant the review. For the firm currently supplying the organisation, news of a re-evaluation of alternative suppliers should alert them to the dangers of complacency. A thorough and rapid review should be instigated to assess:

■ how the supplier is currently working with the purchaser (e.g. Has there been a change in personnel causing friction? Have deliveries been on time? Are there any problems with quality?);

- what market changes are occurring?
- what competitive brands are available and how these compare against their brand.

A meeting should be convened with the purchasing firm to identify their revised needs fully and a programme of change instigated. One lesson that can be learnt from this is that organisations often seek relationships with suppliers based on an expectation that they will always be trying to improve their brands in order to give their customers an even better deal. We are aware of an industrial advertising agency that holds regular seminars for its clients about new issues in marketing. These seminars are valued by its clients, who feel that not only do they have a good advertising agency but they are also being kept up to date and are able to become more effective.

Straight rebuy

The straight rebuy situation involves repeat purchasing of previously bought brands. In such cases, the purchasing process is relatively fast and simple. It is handled on a routine basis, increasingly using electronic reordering. The purchasing organisation has a lot of relevant experience about the brand and, apart from the reordering mechanics, little other effort is expended. In some industries (e.g. packaging), it is not unusual for the purchaser to split the order between two suppliers. If problems occur with one supplier, the purchaser immediately increases the order with the second supplier. It is clearly in the interest of the 'in' supplier not only to ensure customer satisfaction with their brand but also to make the reordering procedure as easy as possible.

Once the customer starts to order the brand routinely, it may well prove profitable for the marketer to invest resources in an automated reordering system. Otherwise, the 'in' supplier may face the problem of technological developments in purchasing, making it easier for buyers to change brands. With the internet facilitating the role of purchasing departments through comparison evaluations and electronic auctioning, the chance of the purchaser switching may not be insignificant.

BRAND VALUES IN BUSINESS-TO-BUSINESS BRANDING

Although many competing organisational brands have similar physical specifications and performance capabilities, in each market only one of these brands achieves and maintains the dominant market position. One of the reasons for this is that the brand leader has been well differentiated and customers perceive greater value, anticipating the leading brand to be superior in some way. Mudambi and her colleagues (1997) developed a framework which helps us better understand the nature of brand value in industrial marketing. They argue that brand value is a function of the expected price and the expected performance of four components. These components are the product itself, the company itself, the distribution of the brand, the services supporting **branding**. Each of these four performance elements has a tangible component (e.g. the physical quality of the product) and an intangible component (e.g. the reputation of the company). When seeking to differentiate an industrial brand through its value to customers, this framework, displayed in Fig. 5.4, therefore helps managers to assess how each of the four performance elements is adding value to both the tangible and intangible components.

Product performance is the core of the brand's value. In the case of a computer, the tangible product performance characteristics include its processing speed and the size of its memory. A model may be preferred due to the intangible component of perceived reliability. **Distribution performance** includes ease of ordering, availability and speed of delivery. Office stationery retailers assess their suppliers through tangible measures, such as lead times and range of

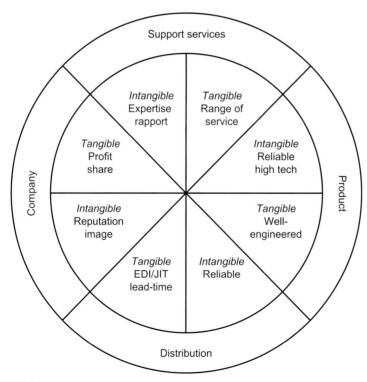

FIGURE 5.4
The pinwheel of industrial brand value (*after Mudambi et al. 1997*)

products available. However, the final choice will also be influenced by intangible factors, such as the willingness and ability to respond rapidly to an unusually large and urgent order. **Support services performance** involves issues such as technical support, training and financial support. Many car manufacturers assess suppliers on the tangible basis of whether they are able to provide technical support during the research and design stages. Some of these suppliers may be chosen because, on the intangible dimension, they also tailor their communication systems to meet the manufacturer's request. **Company performance** covers all aspects of the company. There is an underlying assumption that industrial customers prefer to deal with companies that are reliable and successful. The tangible evaluation of a management consultancy could include its size, geographical coverage and financial stability. In the final decision, intangible aspects will also be considered, such as its reputation and perceived experience in different industries.

This framework was tested in the UK precision bearings market, where interviews were undertaken with manufacturers, distributors and purchasers. The results indicated that although price is rated overall as the most important criteria, it is considered less important at the early screening stage. Price plays a less dominant role during the purchase of original equipment, as opposed to that of replacement parts. The research highlighted the importance of intangible product and company attributes in adding value, as shown in Table 5.2. This research demonstrated that intangible factors are important in rational and systematic decision making. These factors can provide a key source of differentiation, especially when it is difficult to continue to compete on product quality or price. There is a limit to how much brands can be improved through enhancing their physical performance. Once this limit is exceeded, the brand becomes criticised as being over-engineered.

Table 5.2 Summary of Perceived Sources of Value in Precision Bearings (*after Mudambi et al. 1997*)

Product	Distribution	Support	Company
Tangible	*Tangible*	*Tangible*	*Tangible*
Precision	Stated availability	Design advice	Financial stability
Load bearing	Stated lead times	Product testing	Years of experience
Dimensions	EDI and JIT	State support	Global coverage
Intangible	*Intangible*	*Intangible*	*Intangible*
Innovation	Ease of ordering	Understands our	World class
Fit for purpose	Reliable delivery	needs/business	Technical leadership
Well-engineered	Emergency response	Troubleshooting	Global perspective

BRANDS AS RELATIONSHIP BUILDERS

A weakness of the Robinson model is that it does not address the relationships between buyers and sellers sufficiently well. Building on the Robinson model, Johnston and Lewin (1996) analysed 25 years of literature exploring organisational buying behaviour. From this, they emphasise that organisational buying behaviour is affected by buyer-seller relationships. They highlight power/dependence, behaviour/performance monitoring, cooperation/trust, adaptability and commitment as factors that are used to examine buyer-seller relationships in the literature. Our understanding of organisational brand buying has increased as a result of international projects undertaken by the IMP Group. They found that both buyers and sellers were seeking close, long-term relationships. Many of their interviews with different industrialists showed a desire for stability. The broad implications for brand marketers are that they should look at brand marketing not just in terms of marketing resource management, but also in terms of employing the right interpersonal skills and managing relationships through the most appropriate negotiating style.

Buyers were reticent about switching between competing brands because:

- they did not want to keep on spending more time in finding and evaluating alternative brands;
- they were worried about technical problems in adapting their production processes for a new brand;
- they might have had internal production problems which could be easier to resolve by involving a loyal supplier.

Many purchasers were of the opinion that they had a very good working relationship with their suppliers. As such, any brand alternative had to be extremely good to warrant any thought about change.

The IMP Group showed that the relationship between the brand marketer and the purchasing organisation was influenced by four factors: the interaction process; the organisations involved; the atmosphere affecting and affected by the interaction; and the environment within which the interaction took place.

The interaction process

In the interaction process, a series of 'episodes' take place between an order being placed and delivered. For example, the brand exchange, information exchange, financial exchange and social exchange. Over a period of time, these exchange episodes lead to institutionalised expectations about the respective roles of buyer and seller. For example, there are unwritten rules about which party will hold stock. Many case histories showed that stable relationships were characterised by frequent social and information exchanges.

Organisations

In looking at the characteristics of two organisations, there is some indication of the likely relationship. Technical issues are often important indicators of the likely buyer-seller relationship. Ultimately the interaction process is concerned with matching the production technology of the seller to the application technology of the buyer. Where two organisations are at different stages of technological development, their working relationship will be different from that where two firms have a similar level of technical expertise. Likewise, where two firms are of a different size, have little experience of working together or have individuals with differing backgrounds, then the relationship between buyer and seller will take a lot of work to ensure harmony.

Atmosphere

The relationship between two firms is affected by the overall atmosphere, which itself can be characterised by several factors. For example, the firms' mutual expectations, the overall closeness or distance of the relationship, whether there is a sense of conflict or cooperation and whether the dominant firm is trying to use its power over the weaker partner. Where there is an atmosphere of overall closeness, cost advantages can be gained through a variety of sources, such as more efficient negotiation and administration, joint work on redesigning existing brands and more effective distribution. By contrast, some atmospheres may be characterised by a power-dependence relationship. For example, to ensure that brand deliveries are scheduled primarily for the convenience of the powerful purchaser, the purchaser will take advantage of the supplier's dependence on his firm.

Environmental issues

Wider environmental issues such as social systems, channel structure and market dynamism have an effect on the interacting relationship. The buyer and seller have to appreciate the type of social system they are working in. For example, the brand supplier has to be aware of any nationalistic buying preferences they are facing. The relationship will also be affected by the type of channel used — an electronic components producer may sell to an actuator producer who in turn sells a range of actuators to another firm working on aircraft systems. The relationship between any two firms in this extended channel will be influenced by the relationships between other members of the channel. Highly dynamic markets, characterised by frequent new brand launches, make suppliers and purchasers aware of the need for a large number of relationships that are not as intense as may be the case with much more stable markets.

By taking account of these four influencing factors, brand marketers can better appreciate the basis for their relationships with purchasing organisations. They should make all of the brand team aware of the institutionalised activities that each purchaser takes for granted and protect these from being cut in recessionary periods. Making changes to institutionalised activities without thorough negotiation is likely to sour the working relationship.

The IMP research shows that the same brand support teams cannot work as effectively with every customer. The relationship of trust and mutual respect will be nurtured when people are selected because their backgrounds and personalities are ideal for sustaining a long-term working relationship with certain customers. This research makes the point that both rational and emotional factors influence brand selection, an issue that we consider in more detail.

Research by Mitchell et al. (2001) provides more insight into the factors that build brand loyalty in organisational markets. They found that that quality (in 65% of cases), reliability (in 61% of cases), performance (in 60% of cases) and service (in 49% of cases) were the primary precursors of brand loyalty. Availability, familiarity, relationship with the sales team, price and advertising were found to be secondary factors, in descending order of importance.

Lynch and de Chernatony (2007) also highlight the role of the sales team in building brand relationships. As brands consist of both functional and emotional values, they explain that the salesperson personifies the brand values. In addition, the salesperson has an opportunity to be adaptive and can revise their approach to suit the buyers' communication style and information processing mode. This adaptability ensures that the brand is perceived to have relevance for the business-to-business buyer.

Managers have little time to continually explore alternative long-term relationships with new partners. The pressure to achieve higher quality standards, implement JIT and reduce the time taken to launch new brands has led them to work more closely with a lower number of partners. Wilson (1995) has developed a useful matrix to reflect which potential candidates could become effective relationship partners, as shown in Fig. 5.5.

Those firms who represent a low operating risk as partners and who show high value added to the buyer's business are ideal relationship partners. Relationships between buyers and sellers have always

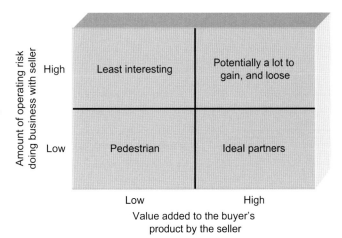

FIGURE 5.5

Classifying potential partners (*Adapted from Wilson 1995*)

existed but today's increasing competition, pressure to reduce costs and the need for profitable alliances have transformed these relationships into key strategic tools which improve a company's performance.

Wilson has expanded on the four factors of the IMP research and identified variables that can predict the success of a relationship between buyer and seller. These are:

- *commitment*: the desire and the effort of both parties to continue the relationship;
- *trust*: the belief that each partner will act in the best interest of the other;
- *cooperation*: complementary actions taken to achieve mutually beneficial outcomes — if the outcomes are beneficial to only one partner, this bodes badly for the future of the relationship;
- *mutual goals*: the degree to which partners share goals;
- *interdependence and power*: the degree of power balance between partners, which determines the degree of dependence of one partner on the other;
- *performance satisfaction*: the degree to which firms can meet or exceed the expectations of the partner;
- *structural bonds*: systems, such as shared technology, interlinking the two companies so closely that it becomes difficult to end the relationship;
- *comparison level of the alternatives*: the number of choices a company has among available high-quality outcomes. The fewer high-quality partners a company has, the more dependent the company will be on the current ones;
- *adaptation*: the degree of modifications one company undergoes to accommodate the needs of the partner;
- *non-retrievable investments*: the amount of resources a company commits specifically to that relationship;
- *shared technology*: the degree to which one partner uses the technology contributed by the other partner to the relationship;
- *social bonds*: the degree of personal relationship that develops between the firms.

These factors measure the degree of dependence between the two parties in a relationship and help anticipate the likelihood of the relationship continuing. Their importance varies according to the stage of the relationship. For example, trust has more importance in the early stages than later on. There are five stages in the relationship:

- *search and selection* of an appropriate partner;
- *definition of the purpose* of each partner, so that a satisfactory balance of power can be developed;
- *setting boundaries*, which consider where each partner's organisation ends and the hybrid exists; and when they can make legitimate claims on each other;

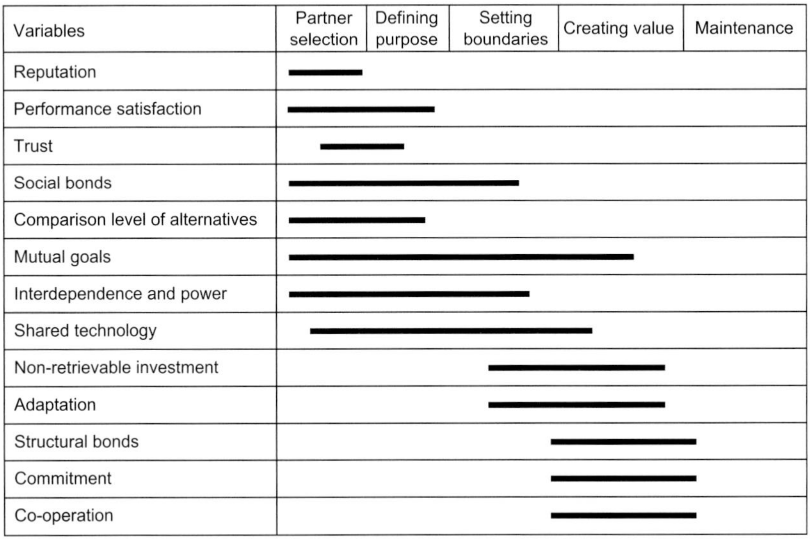

Variables	Partner selection	Defining purpose	Setting boundaries	Creating value	Maintenance
Reputation	▬▬				
Performance satisfaction	▬▬▬▬▬				
Trust	▬▬				
Social bonds	▬▬▬▬▬▬				
Comparison level of alternatives	▬▬▬▬				
Mutual goals	▬▬▬▬▬▬▬▬				
Interdependence and power	▬▬▬▬▬▬				
Shared technology	▬▬▬▬▬▬▬				
Non-retrievable investment			▬▬▬▬▬		
Adaptation			▬▬▬▬▬		
Structural bonds				▬▬▬	
Commitment				▬▬▬▬	
Co-operation				▬▬▬▬	

FIGURE 5.6
The changing nature of business relationships (*after Wilson 1995*)

- *creating value*, which involves consideration of the processes which will add greatest value to each partner and the best way to share the value;
- *maintaining the relationship*, when both parties strive for the long-term survival of their bond. At this stage most of the factors become latent.

Fig. 5.6 shows when the different relationship factors (including the new factor of *reputation*) become more important over the five stages of the evolving relationship.

Wilson's framework indicates that the character and content of a buyer-seller relationship evolves over time, though these changes are not necessarily sequential or as clearly defined in practice. Each stage is an opportunity for both partners to assess the relationship and decide whether to maintain, broaden or curtail it.

One of the key drivers for developing relationships is the chance to create more value. However, assessing whether extra value has resulted from the relationship can be difficult and a dogmatic calculation of individual benefits may damage the relationship. Nevertheless, the relationship exists in order to add value or reduce costs to the parties engaged. Therefore, some attempt needs to be made to agree upon a reasonable system to determine the degree of value created by each party and their share of the profits. This will necessitate an 'open-book' policy, where both parties provide full details about how they have helped to enhance value.

There clearly are costs involved in developing a relationship and these should be identified, then balanced against the extra value that the relationship might bring. Some of the extra costs include the effort to coordinate the activities of both parties, the time necessary for each party to learn about the other organisation and the work to blend both production, selling and administration systems.

FACTORS INFLUENCING BRAND SELECTION

When choosing between competing brands a thorough evaluation, particularly for brands new to the firm, takes place, often with an agreed list of attributes. This reflects the views of all members of the buying centre. But a more covert assessment also takes place. This is based on social ('Can I get on with this rep?') and psychological ('Will I be respected if I'm seen to be buying from that firm?') considerations. Let us now examine some of these rational and emotional issues.

Rational brand evaluation criteria

In business-to-business marketing, considerably more emphasis is placed on the use of resources which appeal to buyers' rational, rather than emotional considerations. For instance, a survey amongst firms marketing high technology brands showed that considerable importance was placed on having state-of-the-art technology, employing effective salespeople, backing the brand with a strong service capability, being price competitive and offering a complete product range. Much less importance was placed on engendering a favourable attitude between the buyer and the sales representative. It is not possible to generalise about the kinds of functional components of a brand that might appeal to buyers, since this depends on several factors, some of which include:

- the different requirements of members of the buying centre;
- the type of industry buying the brand;
- the type of product being bought.

It is unlikely that **all members of the buying centre** will be equally interested in the same attributes. People work in different departments, with different backgrounds and different expectations of the brand. In fact, the brand marketer may well face a situation where different members have opposing views about relevant brand criteria. A chief chemist may be particularly concerned about the purity of a brand of solvent, while the purchasing manager's sole concern is keeping costs down. In a study considering the marketing of solar air conditioning systems, it was shown just how diverse the evaluation criteria were amongst members of the buying centre. Plant managers were more attentive to operating costs, whilst general managers were more concerned about the modernity of the brand and its potential for energy saving.

If the same brand is being sold to **different industries**, it is unlikely that they will be using the same evaluation criteria. In a study comparing the evaluation criteria used by manufacturing companies and hospitals, it was found that while there were some similarities, there were also major differences; both manufacturers and hospitals regarded reliability and efficiency as being very important issues, but hospitals saw after-sales service as a key factor, whilst the manufacturers rated the technical capabilities of the brand as very important.

The greater the similarity between purchasing industries, the greater the likelihood of similar evaluation criteria being used. When comparing the use of 20 brand evaluation criteria between electric power generating industries and electronic manufacturers, there were only four criteria not considered to be of the same importance (repair service, production facilities, bidding compliance and training aids).

The **type of product** also influences brand evaluation criteria. One particularly informative study was able to classify industrial products into four categories and showed that similar attributes were considered according to the category. Specifically, the four most important considerations for each category were:

1. for frequently ordered products that pose no problems in use: reliable delivery, price, flexibility and reputation;
2. for products requiring training for use: technical service, ease of use, training offered and reliability of delivery;
3. for products where there is uncertainty about whether the product will perform satisfactorily in a new application: reliability of delivery, flexibility, technical service and information about product reliability
4. for products where there is considerable debate amongst the buying centre, price, reputation, information on product reliability and reliability of delivery.

There is an erroneous belief amongst some business-to-business marketers that brands succeed if they offer an attractively low price to purchasers. This is not so. A team of researchers examined the buying records of large manufacturing companies. They focused on 112 purchases of capital equipment. From this database they found that, on average, three competing brands were evaluated before a purchase decision was made and, in 41% of the purchases, and the successful brand was *not* the lowest priced bidder. The buyers paid a price premium for:

■ interchangeability of parts;
■ short delivery time;
■ working with prestigious suppliers;

- full range of spare parts rapidly available;
- lower operating costs;
- lower installation costs;
- higher quality materials.

Only when competing brands are perceived as being very similar, does price become important. A study investigated buyers' perceptions of different brands of electrical devices (oscilloscopes, switches, resistors, etc.). When buyers perceived *little* brand differentiation, the three main choice criteria were price, specifications and delivery. By contrast, when buyers perceived *significant* differences between competing brands, price was not one of the 10 criteria considered. A similar finding resulted in another study amongst buyers purchasing undifferentiated brands of industrial cleaners, lubricants and abrasives. Price was viewed as being one of the key choice criteria when choosing between mainly undifferentiated brands.

Even when price is a dominant choice criterion, different aspects of price are considered by the buyers. The purchase price of the brand may appear to be high but when taking into account longer-term economies resulting from the brand, such as lower defect rates in production, the buyer may well look at the purchase in terms of the long-term savings and quality improvements. Buyers also consider the brand's life cycle costs – the total costs likely over the lifetime of the brand. Mercedes-Benz trucks were once advertised using the slogan 'Are you buying a truck or an iceberg?', underneath which was an iceberg depicting the fact that over the vehicle's life the buying price represented 15% of the cost and the running costs 85%. Another example is the online advertising for Mac computers. Despite their premium cost, Mac emphasise their advantages with the slogan 'Why your next PC should be a Mac'. The site highlights the functional ('brains') and aesthetic ('beauty') advantages of the Mac relative to the PC, as well as benefits for business users, such as security, wireless capabilities and screen sharing.

Going with too low a price makes the buyer wonder what has been cut. We are aware of one consultant who lost a project because his price was too low compared with the other bidder. The client could not understand how the consultant could do a sufficiently thorough piece of work at such a low price. It transpired that as he worked alone, he had much smaller overheads and was content to operate at lower margins than the more expensive firms; yet his low price lost him the contract. Thus, in common with consumer brand marketing, there exists a feeling of 'you get what you pay for'.

There are other reasons why pitching the brand at a low price may not be wise. Some buyers are evaluated on their ability to negotiate discounts and pricing low allows little leeway for negotiations. There are also some buyers who seek the satisfaction of always being able to negotiate a better price.

In conclusion, functional issues are important components of organisational brands that appeal to the rational side of the buyer. But buyers are not solely concerned with quantitative measures to assess technical and commercial performance. The next section explores some of these non-rational criteria.

Emotional brand evaluation criteria

The individuals involved in the brand selection process enjoy the challenge of finding the best solution to the firm's problems. They are also motivated by more personal issues such as job security, a desire to be well regarded by colleagues inside and outside the firm, the need for friendship, ego enhancement, aspirations of career advancement, loyalties based on their beliefs and attitudes and a whole host of other social and psychological considerations.

In a study we undertook for a tin packaging manufacturer, we became very aware of these emotional brand selection issues. Our brief was to find out why purchasers only awarded small contracts to this firm. In a depth interview, one purchaser told us that he used to place large orders but felt that the quality standards were too variable and only used this supplier for small runs. When asked why he still bothered to use them, he spoke about the good social relations he had built with the firm's managing director. He viewed our client

> 'as an old friend, whose company you value because of the years you've been together. But as you get older, you develop particular idiosyncrasies, which as a good friend you just accept and try to still enjoy their company.'

He clearly valued his friendship with the managing director and, at a lower level of commitment, was still prepared to buy tin packaging, albeit for jobs where tin quality was not a critical issue.

The way buyers make decisions influenced by rational (objective) issues and emotional (subjective) factors can best be summarised in Fig. 5.7.

Clearly a power station with significant capital costs, running costs and environmental impact, will be purchased primarily on objective/functional grounds. However, there is often little to choose between competing products and services such as lubricants, ball bearings, insurance and the like. In such cases, it is personal

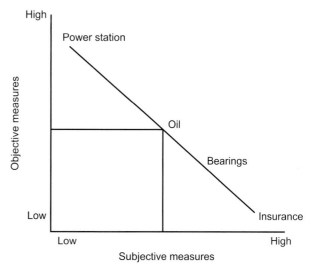

FIGURE 5.7

The impact of objective and subjective issues on decisions

relationships and the perceived quality of the supplying company that make the difference. In these situations subjective measures become more important.

Buyers, as emotional individuals, take account of how the brand will affect them socially and psychologically. Having a brand delivered late not only affects production schedules but also causes personal anguish. Late delivery is interpreted by some buyers as 'you're not that important to us'. It is a broken agreement, which is read as an attitude of complacency and hurts their pride. They feel that the supplier is not serious about their business and start to wonder whether they can trust someone who shows such disinterest.

Buyers like to deal with prestigious suppliers as they feel this increases their status within the firm. They are proud to tell their colleagues that they are using particular suppliers as they believe they gain more credibility and authority, particularly when the supplier's corporate identity has clearly communicated associations of excellence.

The size of a supplier is not considered just in terms of production capabilities but also as indicative of the type of relationship. Larger suppliers are sometimes viewed as impersonal, unapproachable, self centred, bureaucratic firms, who are unlikely to be flexible. By contrast, smaller firms are perceived as warmer, friendlier, more attentive and more flexible in responding to the supplier's problems.

One purchasing manager was part of a team deciding whether to buy glass or tin packaging for a new grocery brand. In spite of the fact that technical data was obtained and evaluated, he recommended tin because he personally found glass aesthetically unappealing. To get around this personal dislike, the glass packaging manufacturer appealed to his emotional instincts by leaving a small glass statue on his desk. He hoped that the sight of this expensive statue on his desk would change the purchasing manager's attitude. In this case, functional issues were less important to the buyer than emotional considerations.

This is but one of many examples showing the importance of positioning organisational brands to satisfy the emotional needs of buyers. For example, a surgical instrument manufacturer succeeded when they spoke about their brands not in cold, medical terms but using instead the terms surgeons would use — 'smooth and elegant'.

In this case the manufacturer won through blending a technical approach to the communication with a more personal tone.

Understanding the psychological concerns of organisational buyers can give the brand an edge if other competitors are focusing solely on functional issues. In another study to understand why firms continued to purchase computers from the same organisation, we became aware of the importance of psychology. A buyer was approached by another supplier whose computer virtually matched the performance and cost characteristics of the incumbent. A few days later the buyer broke his leg and rang the incumbent computer firm to reject their invitation to a Grand Prix because of his accident. He was impressed by the way they sent a luxurious car to chauffeur him and provided special facilities in their hospitality room at the race. This feeling of 'being special' helped maintain his loyalty.

As was pointed out earlier in this chapter, managers perceive risk when buying brands. By understanding managers' concerns about brand buying, marketers should then be able to devise ways of presenting their brand as a risk reducer. One study (Hawes and Barnhouse, 1987) investigated purchasers' perceptions of personal risk when choosing between competing brands in a modified rebuy situation. Their concerns, after feeling personal remorse due to purchasing incompetence, were that relations with the internal user would be strained, the status of the purchasing department would decrease and there would be a less favourable annual career performance review with less chance of promotion. In order of importance, these buyers felt that the most effective way of reducing personal risk was to visit the supplier's plant and then to ask some of the supplier's customers about their opinion of the supplier. Clearly, personal contact and openness are two important ways of reducing perceived risk.

As organisational buyers' perceptions of risk increase, they are more likely to consult informal, personal sources of information. Henthorne and his colleagues (1993) examined how organisational buyers' perception of risk is influenced by talking to people inside and outside their firm. A conclusion from their study is that suppliers should identify those sources who reduce buyers' perceptions of perceived risk, then consider whether they can be used to minimise other buyers' concerns in different firms. Specifically, marketers could provide names of satisfied customers and independent experts who can attest to the quality of the purchase. They found that the credibility of external advisers is often higher than that of internal staff.

Brand marketers can develop strategies that position their brand as an effective risk reducer. The Caterpillar Corporation is but one of many firms who benefitted from understanding buyers' perceived risk. They were faced with the problem of firms producing inferior quality spare parts for after-sales service. These looked similar to the genuine Caterpillar part but sold at a considerably lower price. Their lifetime was less than the genuine part and premature failing while in use could cause considerable engine damage. To counter this threat, Caterpillar developed a series of leaflets for its customers showing dice next to Caterpillar equipment. The headlines were 'Don't gamble with it', 'Don't play games with it', 'Don't risk it'. Each brochure showed photographs of the damage caused by using pirated parts. These successfully communicated the functional excellence of the brand, as well as resolving the buyer's personal risk about using only genuine parts.

PROVIDING ORGANISATIONAL BUYERS WITH BRAND INFORMATION

In business-to-business brand marketing, buyers undertake a thorough evaluation of competing brands. Due to the more complex and expensive nature of brands, they rely much more on personalised messages. As such, more emphasis is placed on using sales representatives to present brand information. Their task, particularly when approaching new firms, is that much easier when the brand is recognised as emanating from a well-respected organisation. Corporate advertising, reinforcing a clear corporate identity, can pave the way for a more effective sales presentation. The sales representatives don't have to spend long reassuring the buyer about the company and can devote more time to explaining the brand's benefits.

This promotional push, using advertising to give reassurance to customers and to enable salespeople to focus on brand capabilities, is seen in many industrial sectors. A survey (Traynor and Traynor 1989) amongst high technology firms showed that suppliers felt it was most important to promote their brands using the sales force, then advertising in trade magazines, followed by display at Trade shows.

The work of Schmitz (1995) sheds some light upon the different ways in which organisational buyers are persuaded or influenced by the signals given out by sellers, including promotions. Schmitz identified two distinct routes by which attitudes are formed in response to any signals communicated by the seller. The first route, known as the central route, is an active process of thinking and drawing conclusions. The second, the peripheral route, is more of an instinctive and subconscious processing of cues that bypasses active thought. Central route persuasion occurs when the recipient is motivated and able to process the signal, which may for example be the information and product details contained in an advertisement or promotional literature. The peripheral route is used when either motivation or the ability to process the signal are absent and peripheral cues may include the style of dress of the sales person, the number of arguments used, the way in which information was presented, etc. Attitudes formed via the central route are more enduring, persistent and related to behaviour than they would have been if formed via the peripheral route. By understanding the buying group's motivational levels and their abilities to appreciate information about the brand, the sales person can better tailor their interactions with different members of the DMU.

SIEMENS

Siemens develop their corporate brand through the message of Innovation. This is in keeping with the brand slogan 'Global network of innovation'

Smarter business for a Smarter Planet:

What's the window of opportunity on an opportunity?

We live on a planet that is now generating more than 43,000 gigabytes of data per day. Think about the opportunity that comes with that ocean of data. The computing models and advanced analytics we have today actually allow us to use that information, not merely to sense and respond but to predict. So data isn't just telling us what's going on in the world—it's telling us where the world is going. Recognizing patterns. Crystallizing trends. Using information to make smarter decisions and to apply the right insights to the business. Today, IBM is helping companies do just that. From detecting fraudulent behavior before insurance claims are paid to improving customer retention by spotting the patterns of people who are likely to defect to competitors.

A smarter business needs smarter thinking.
Let's build a smarter planet. ibm.com/analytics

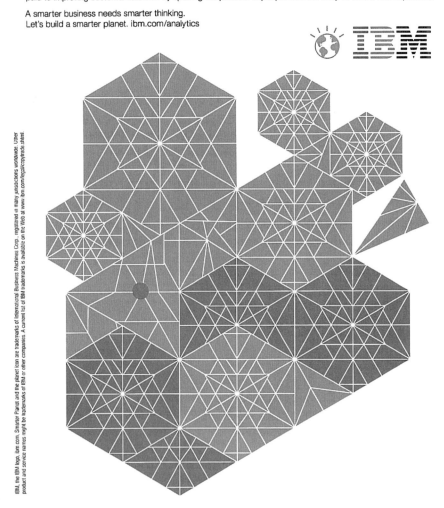

IBM (INSIGHTS BY NUMBERS AND WINDOW OF OPPORTUNITY)

IBM advertise their corporate brand by using a series of print messages which emphasise their abilities to process data, and create opportunities for business. Their slogan 'Smarter Business for a Smarter Planet' is reinforced through tangible proof in their claim that (in 2008) '80% of the top 20 Fortune 500 Companies used IBM for its financial management expertise'. (Reproduced by kind permission of IBM)

A study (Dempsey, 1978) amongst purchasers of capital equipment and component materials found that buyers rated their purchasing records as the most useful source of information. This was followed by visits from salespeople and then discussions with their colleagues in other departments. Marketers who have a **range** of brands may find it easier to widen their business with their existing customers, since they can reassure themselves about their track record from their purchasing records. These results also indicate that brand marketers should build a favourable relationship with other departments in the firm to give the purchaser a greater sense of confidence when they talk with other colleagues. The fourth most useful way of finding out about the brand was by visiting the supplier's factory and glancing through the relevant brochures.

The key points that the organisational brand marketer should not lose sight of are that buyers and their colleagues in the buying circle may well:

(a) perceive the contents of the brand message differently from the marketer;
(b) perceive the content of the brand message differently between themselves.

As Chapter 3 discussed, people have finite mental capabilities and to protect themselves from excessive amounts of information, their perceptual process filters much out. An example of this is an advertisement once used for Zantac in Germany. The brand team wanted to communicate the anti-ulcer capabilities of this drug to GPs in Germany and took a one-page advertisement in a respected medical journal. At the bottom of the advertisement there was a detailed explanation of the brand's pharmacology. Subsequent research showed that none of the doctors read the detailed explanation about the drug. Most read the caption, skimmed over the brand name but took little further notice of the advertisement.

Loctite is a good example of an organisation that took steps to avoid the problem of different perceptions amongst members of the buying centre. It was concerned about the way different types of engineers in the buying firm interpreted their trade advertisements for its range of industrial adhesives. They conducted some market research and found that design engineers felt strong pressure on themselves to be 'right' and not make a bad decision. They were risk avoiders who liked to see diagrams, charts and graphs in brand advertisements, with explanations of how the brand worked. By contrast, plant engineers saw themselves as fixers, who kept things running by being creative in solving problems. They felt more comfortable with advertisements showing photographs of products rather than graphs. With this understanding of the different needs of members of the buying centre, Loctite then developed trade advertisements to appeal to specific types of engineers. A consequence of this was a greater similarity of perception amongst engineers about Loctite's industrial adhesives and this in turn led to greater sales success.

THE COMPANY AS A BRAND

In any purchase, organisational buyers will remember their image of the company longer than any detailed product information. Industrial buyers always question a potential supplier's credibility, particularly whether they really are specialists in their field. However, as we have discussed, as business decisions are taken by people, about people, there are also emotional factors involved and the supplier needs to ensure they have a well-conceived identity underpinned by a respected personality.

The purpose of corporate branding is to:

- make the company name known, distinct and credible in the mind of potential buyers;
- facilitate the building of relationships with buyers and suppliers;
- portray the benefits it offers to the buyer;
- embody the value system of the corporation.

Business-to-business companies have become more focused on building their corporate brands. In research among organisational buyers of technology products, Kuhn et al. (2008) found that the corporate brand had more influence than individual brands in overall choice. Therefore, they explain that the role of the corporate brand is particularly important in the business-to-business context. Corporate branding is also important for this sector as it faces unique challenges. For example, Wise and Zednikova (2009) explain that conventional marketing and communications efforts may not be effective for business-to-business firms in sectors such as chemicals or mining, as they do not have large marketing budgets relative to other industries. In addition, those firms may find it more difficult to attract new employees than other industries would. Therefore, such companies are now thinking about the way they are perceived by their stakeholders and managing external perception by ensuring that they have a strong corporate brand.

In order to better understand the business-to-business corporate brand, Aspara and Tikkanen (2008) differentiated three aspects of the business-to-business organisation that influence the customers' perception of its brand. These are:

- *Offering related perceptions* – these are customer perceptions about product range and delivery; perceptions about services, such as installation and repair, or solution/systems offerings, such as the understanding of the customer's business and experience of dealing with similar customers.
- *Personal contacts related perceptions* – these involve customer perceptions about the experience, competence and likeability of customer-facing staff.
- *Network role related perceptions* – these include customer perceptions about the organisation's willingness to co-operate and partner with a business customer.

They emphasise that business customers' trust, loyalty, repeat business and recommendations will be influenced by these perceptions. Managers have an opportunity to examine how their corporate brand is positioned along these dimensions. Once they do this, they can identify the dimensions that may be strengths and could provide differential advantage relative to their competitors.

To present themselves in the most favourable way, firms should develop a **corporate identity** programme, ensuring that all forms of external communication are coordinated and presented in the same way. The problem is that corporate identity is akin to 'branding as an input process', as discussed in Chapter 2. Owing to the problems created by the buyer's perceptual process, the resulting perception of the firm and its **corporate image** may well be different from what was intended.

Corporate identity is a valuable asset which, if efficiently managed, can contribute to brand success. As such, any firm needs to manage its corporate identity programme in such a way that all members of a particular buying centre perceive a similar corporate image, encouraging a feeling of trust and confidence in the supplier. Each member of the buying centre could have a different line of contact with a particular supplier. Without a cohesive approach to managing the corporate identity, each member of the buying centre may perceive a different corporate image — a result which bodes badly for the brand. Buyers are impressed by the consistency with which the firm presents itself. Their increased confidence places the brand in a more favourable light.

Any corporate identity programme is supported by a myriad of resources; for example, the firm's name, its structure, employees, offices, letterhead paper, promotional activity, core values, culture, logo, promotional work and even the way the telephone is answered. It is wrong to think that corporate identity equates to the logo — this is but the tip of the iceberg.

Suvatjis and de Chernatony (2005) developed the 'six station model' of Corporate Identity, which they successfully tested among industry practitioners. Strong corporate identity is linked to competitive advantage. Therefore the model featured in Fig. 5.8, depicts key stations or components that interrelate to form corporate identity.

As one moves through the stations, there is greater contact between stakeholders and the company. These stations are:

■ Head Station — which is the firm's top management and its vision and values;
■ Strategy Station — which incorporates the firm's brands, products or services and its strategy and marketing;

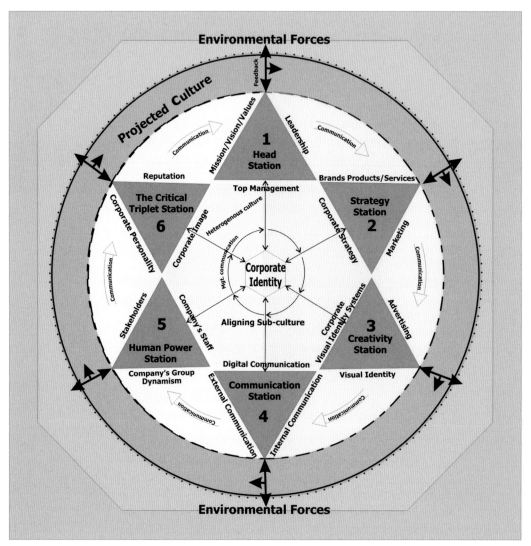

FIGURE 5.8
The six-station corporate identity model (*Suvatjis and de Chernatony. 2005*)

- Creativity Station — which develops the firm's visual identity and advertising;
- Communication Station — which includes internal and external communication;
- The Human Power Station — which includes stakeholders and staff;
- The Critical Triplet Station — which encompasses corporate reputation, image and personality.

In a business-to-business context, the model works in the following way: the Head Station sets out the company's mission and values. These are translated into strategies in the Strategy Station and consensus is achieved with each strategic unit in the firm. The Creativity Station then uses visual representation to enable stakeholders to understand the firm's objectives, mission and values. The Communication Station operates as a two-way communication channel, exchanging information for both internal and external environments. The role of the Human Power Station is then to harness human resources to communicate the corporate identity through their behaviour. Finally, the Critical Triplet Station builds on the previous five stations to create a corporate personality, which is judged favourably by external stakeholders.

To maximise the assets of the company as a brand, the more enlightened firms expect their employees to act as 'ambassadors' for the firm. Most employees come in contact with other firms and they must be able to present a personality of knowledgeable helpfulness. The personnel director should no longer be concerned just with an *internal* focus on strategic human resource management but should also take an *external* orientation, looking, for example, at recruitment in terms of the individual's abilities to 'sell' for the firm. Training programmes are increasingly being devised to give employees the skills to talk knowledgeably with all external contacts. Employees need to take 'ownership' of problems and structure their departments around customers' needs.

There are many advantages to adopting a well-thought-through corporate identity programme. The first advantage is coping with shorter brand life cycles. The dynamic nature of markets and the continual pressure on performance improvement is resulting in shorter brand life cycles. To succeed, the brands must be adaptable. Uncompromising, staid brands blur the firm's image.

A second advantage of corporate identity is that it sustains a real point of differentiation. It is more often the case that functional advantages are soon surpassed by technology leapfroging. This is particularly so in financial services, where it is only a matter of days before new 'look-alikes' follow the innovator. Where, however, the point of difference is based on an emotional, rather than a functional discriminator, buyers normally perceive extra value and competitors take longer to copy this. For example, the buyer may constantly receive fast responses from a technical representative who takes an active interest in their business and, on the odd occasion when the rep is unobtainable, the buyer may continue to receive a similar level of service. The emotional benefits from customer service are valued by buyers, who are unaware of the fact that this is one of the core values that the supplier has worked hard to instill internally (e.g. communication systems and training). This aspect of corporate identity is not just apparent from the promotional campaign but from the way that *all* points of contact with the customer are geared to deliver customer service.

A third advantage is that when media inflation exceeds retail price increases, it can be a cost-effective means of communicating the broad values to which the company subscribes. The presence of the letters IBM with the individual brands instantly enables the buyer to form an image about what the brand might be like.

We do not believe, however, that the promotions budget should just be directed towards corporate communication. Instead, a process has to be developed whereby corporate advertising helps communicate the corporate identity, allowing each brand to benefit from the corporate goodwill, yet not stifling the individual brand's personality.

Corporate identity can be a powerful tool helping business-to-business marketers promote their brands. However, its power is limited by the extent to which all of the component parts are coordinated and whether they reinforce each other. The supplier needs to identify the different ways in which it comes into contact with all members of the buying centre, e.g. brochures, staff, delivery vehicles, stationery, etc. For each of these elements, it then needs to assess whether a unifying device (e.g. logo) should be displayed and whether each of the elements supports the corporate identity objectives. The corporate identity objective of communicating concern for quality may be well supported by impressive brochures and smartly presented, knowledgeable sales staff – but it will fail when dirty lorries deliver the products, with impatient lorry drivers thrusting badly prepared invoices at the goods inward clerk.

It should now be apparent that the logo is not the sole basis for such programmes, although it is a useful device to communicate corporate objectives. Particularly when the firm is responding to a changed environment, it can be a very visible way of communicating change to all interested publics.

An example of a change in brand elements to achieve a strategic goal is Andersen Consulting, who were renamed Accenture on 1 January 2001. The name change severed its link with its former parent, Andersen Consulting. The firm explained that after an iterative process, the name was screened for trademark and URL availability as well as cultural sensitivities and local pronunciation. The name change achieved a strategic objective. In an interview with BBC, Andersen Consulting's Chief Executive, Joe Forehand, explained that the name Accenture 'puts an access or emphasis on the future, just as the firm focuses on helping its clients to create their future'.

VARYING RECEPTIVITY TO BRANDING

While the importance of brands within the business-to-business market has been discussed throughout this chapter, one should be cautious in assuming that brands are equally important to all customers in all purchase situations. Mudambi (2001) addressed this issue in her research within the UK in the precision bearings market, carrying out depth interviews and a postal questionnaire with various people involved in purchasing bearings. She identified three distinct 'clusters' of firms who buy bearings.

Highly tangible firms – To these firms, the tangible aspects of the offer such as price and physical product properties were perceived as more important than the intangible aspects. These firms constituted 49% of the firms investigated.

Branding receptive firms – Firms in this cluster placed the most emphasis upon branding. The branding elements included how well-known the manufacturer is (brand name awareness), the general reputation of the manufacturer (brand image/reputation) and the number of prior purchases from the manufacturer (brand purchase loyalty). These firms also saw the service aspects of the offering as significantly important. The branding receptive cluster represented 37% of the purchasing firms.

Low-interest firms – This cluster of firms saw none of the attributes to be more important than others. To these firms, bearings represent an unimportant purchase.

Branding can be seen to play an important role for a notable percentage of business-to-business buyers. The implications of segmentation based upon the perceived importance of branding, such as that discussed above, has implications for organisational marketers. For example, when faced with low-interest customers, marketers could use brand communications to increase customers' perceptions of the importance of the purchase decision. When faced with highly tangible firms, marketers might choose to combine excellent information about the tangible benefits of the product with some information about how customers might evaluate the more intangible aspects. Business with branding receptive organisations could be increased or improved by focusing upon the unique nature of each product, its physical attributes, the reputation of the manufacturer, the services that augment the product and also the emotional and self-expressive functions of the brand.

CONCLUSIONS

This chapter has shown that branding plays as important a role in business-to-business marketing as it does in consumer marketing. The organisational buyer has the encouragement to assess rationally competing brands, yet emotional considerations also influence brand buying. Not only do buyers seek competitive features and benefits, but they are also influenced by emotional aspects such as the prestige associated with specific brands or the reassurance of an established relationship with the organisational brand. Organisational brands have been found to have notable brand equity, with organisational buyers exhibiting significant 'brand equity behaviours', such as being willing to pay a price premium or recommend the brand to others. However, caution should be exercised when making assumptions about the overall impact of brands in an organisational setting, as certain firms, under certain conditions, will be less receptive to branding than others.

Aspects of organisational marketing are different from consumer marketing and these issues need considering when developing organisational brands. For example, several people from different departments are likely to be involved in the brand selection decision, each having different brand expectations. On the supplier side, the salesperson may have a critical role in adapting their approach to meet the communication styles and needs of different business customers.

Non-routine purchasing in organisations typically involves about five people, but increases for more complex purchases. One way of anticipating which departments are likely to be involved in the brand purchase is through a consideration of the organisation's view about the commercial risk and technical complexity of the brand. High technology brands from a well-respected supplier will attract considerable interest from technical specialists but only a small amount of interest from company accountants.

Further insight to the challenges facing the brand can be gained by identifying which members of the buying centre will be the users, influencers, deciders, buyers and gatekeepers. Brand marketers must work to ensure that brand information is not blocked by gatekeepers. Brand presentations then should not just be directed at the needs of the decider but also at the needs of key influencers, such as architects influencing the choice of office heating systems. Attention also needs to be given to the nature of the persuasion process between members of the buying centre and the seller, as the two main forms of persuasion (the central route and the peripheral route) have different implications for marketers.

An eight-stage model of the buying process enables marketers to anticipate the way organisations go about deciding which brand to select. The amount of work undertaken by members of the buying centre will vary according to how much previous experience they have of buying the particular product. With little experience, they seek a considerable amount of information and undertake a detailed review of alternate brands. As they gain more experience, they demand less data from suppliers and eventually they place considerable value on the ease of rapid reordering. Just as consumers are keen on buying the best value brands, so are industrialists. Their concern is the trade-off between performance and price. One way to enhance a brand's value is to increase its performance, while holding price constant. Four aspects of performance present opportunities for increasing value — the product itself, supporting systems, distribution and the company's reputation. By looking at the tangible and intangible aspects of these four performance characteristics there is greater scope for better matching the value of the brand with buyers' needs.

The IMP Group have shown that many buyers seek a long-term relationship with their suppliers, due in no small part to the joint benefits to be gained from the supplier continually improving their portfolio of brands. Partnerships are more likely to occur when a new supplier can show that they can add more value and there are few risks in working with them. Factors that influence the success of a relationship were reviewed and the importance of each of these varied according to the stage of development of the relationship.

To facilitate the buying centre's evaluation of both rational and emotional aspects of the brand, personal visits by sales representatives are of considerable value. However, buyers also place a lot of importance on the supplier's track record with the firm, in addition to discussing matters with colleagues and visiting the supplier's factories. A strong corporate brand is essential in creating customer trust, loyalty and invaluable customer recommendation. Managers can generate strong brands by considering the dimensions of their firm that offer advantages to customers and are differentiating relative to competitors. The brand purchase decision is more confidently made when the buyer favourably associates the supplier's brand with a well respected corporate image. To achieve a strong image, firms should ensure that all aspects of the business combine to produce a consistent brand message. The 'Six Station Model' of corporate identity illustrates that business-to-business firms comprise key stations which interrelate together. When these stations are integrated successfully, all aspects of the business combine to produce a corporate personality that is judged favourably by external stakeholders.

MARKETING ACTION CHECKLIST

To help clarify the direction of future brand marketing activity, it is recommended that the following exercises are undertaken:

1. Write down the criteria that you believe your customers use to evaluate your brands. If you only have rational reasons listed, discuss this with your colleagues and identify the emotional issues that your customers take into account. Does your marketing programme take account of your customers' rational and emotional needs?
2. For one of the contracts that your firm is trying to win, work out with the rest of the team the composition of the buying centre. Do you know who is playing the roles of user, influencer, decider, buyer and gatekeeper? How are you tailoring your brand presentation to appeal to

the different members of the buying centre? Have you made sure that *all* members of the buying centre have relevant brand information?

3. For a contract that you recently lost, work with your colleagues to identify who was in the buying centre and the roles they played. With hindsight, were you correctly targeting and tailoring your brand presentation?

4. For a new contract that you are bidding for, assess whether the buying centre's experience makes the purchase a new task, modified rebuy or straight rebuy. With this assumption made explicit, use Fig. 5.3 to map out the stages that the buying centre is likely to pass through and estimate where it will devote most effort. For each of these stages identify the action you need to take to help the buyers.

5. What actions are you taking to ensure that your customers can place repeat orders? Have you investigated developing a computerised reordering process for each of your customers?

6. When did you last audit the way that all members of your organisation interact with your major customers? For each major account, do you know whether:
   - there have been any organisational changes?
   - there is any personal friction between your staff and the buying centre?
   - deliveries are always on time, with the correct product mix?
   - the buyers are satisfied with the consistency of quality?
   - there are market changes occurring that will result in the buyer changing his brand purchasing?

7. What are you doing with your major customers to make them feel that you are always trying to improve your brands and help their business grow?

8. For a contract that you are currently bidding for, prepare a table comparing your brand against the other competing brands showing:
   - purchase price;
   - installation costs;
   - operating costs;
   - regular maintenance costs;
   - depreciation costs.

   If any of this information puts your brand in a favourable perspective, how could you incorporate this into your brand presentation?

9. Choose a brand that is under pressure from competition. By looking at the four performance factors of the product, supporting services, distribution and your company, consider what tangible and intangible value you are providing. By repeating this exercise for your competitor's brand, identify on a relative basis, where you are offering more or less value. Through considering how your customer uses your brand, follow through this analysis to identify how more value could be delivered.

10. For each of your partners with whom you believe you have established a relationship, use the two-dimensional matrix in Fig. 5.5, to plot where

they lie in terms of the amount of value they add to your brand and the operating risk of being associated with them. Why are you dealing with partners who do not fall in the high value added/low-risk quadrant?

11. Using the factors shown in Fig. 5.6, which are associated with successful relationships, audit each of your partners to assess the well-being of each relationship. For those relationships where over half these factors receive low scores, identify the value these partners are adding to your brand and whether the cost of maintaining the relationship warrants this.

12. For a new contract that you are bidding for, consider how confident the buying centre is about your brand. Have you identified previous customers and independent experts that could attest to the quality of your brand and thereby enable you to present your brand as a low-risk purchase?

13. When was a market research study last undertaken to assess buyers' views of your brands and those of your competitors? If this was longer than two years ago, it may be worthwhile commissioning a new study. This should identify buyers' evaluation criteria and their assessment of competing brands. It should also investigate those aspects of risk that buyers perceive when buying your brand; and their preferred ways of reducing risk.

14. Do you know which sources buyers most value when seeking information about competing brands? If not, interviews should be undertaken with key buyers and your promotional strategy should be adjusted accordingly.

15. Consider how the six-station model of corporate identity might apply to your company. Which stations add greatest value? How might you increase consistency between the corporate identity you project and your stakeholders' views about your company?

STUDENT BASED ENQUIRY

1. Select a business-to-business brand you are familiar with. Distinguish between the functional and emotional messages communicated by that brand. Which messages do you think are most effective?

2. Consider the eight-step organisational buying process presented by Robinson et al. (1967). For each stage of the process identify one action that a business-to-business brand might take to improve the likelihood that it would be selected as a supplier.

3. Select three business-to-business brands of your choice. Using only the information available on their corporate websites, explore the extent to which they communicate their brand values. Evaluate whether you believe that they translate those values into action and how they do so.

4. Using the Interbrand list of leading brands, select a corporate brand that serves business-to-business customers. Discuss and evaluate the brand elements that combine together to create that corporate brand.

References

Aspara, J., & Tikkanen, H. (2008). Significance of corporate brand for business-to-business companies. *The Marketing Review, 8*(1), 43−60.

Auh, S., & Shih, E. (2009). Brand name and consumer inference making in multigenerational product introduction context. *Journal of Brand Management, 16*(7), 439−454.

Dempsey, W. (1978). Vendor selection and the buying process. *Industrial Marketing Management, 7*, 257−267.

Gregory, J. R., & Sexton, D. E. (March 2007). Hidden wealth in B2B brands. *Harvard Business Review*, 23.

Hawes, J., & Barnhouse, S. (1987). How purchasing agents handle personal risk. *Industrial Marketing Management, 16*, 287−293.

Henthorne, T., LaTour, M., & Williams, A. (1993). How organizational buyers reduce risk. *Industrial Marketing Management, 22*, 41−48.

Hutton, J. (1997). A study of brand equity in an organizational-buying context. *Journal of Product and Brand Management, 6*(6), 428−439.

Johnston, W. J., & Lewing, J. E. (1996). Organizational buying behavior: towards an integrative framework. *Journal of Business Research, 35*, 1−15.

Keller, K. L. (1998). *Strategic Brand Management: Building, Measuring and Managing Brand Equity*. New Jersey: Pearson Prentice Hall.

Kim, J., Reid, D., Plank, R., & Dahlstrom, R. (1998). Examining the role of brand equity in business markets: a model, research propositions and managerial implications. *Journal of Business-to-business Marketing, 5*(3), 65−89.

Kuhn, K. A., Alpert, F., & Pope, N. K. L. (2008). An application of Keller's brand equity model in a B2B context. *Qualitative Market Research: An International Journal, 11*(1), 40−58.

Lynch, J., & de Chernatony, L. (2007). Winning hearts and minds: business-to-business branding and the role of the salesperson. *Journal of Marketing Management, 23*(1/2), 123−135.

Mitchell, P., King, J., & Reast, J. (2001). Brand values related to industrial products. *Industrial Marketing Management, 30*, 415−424.

Mudambi, S., Doyle, P., & Wong, V. (1997). An exploration of branding in industrial markets. *Industrial Marketing Management, 26*(5), 433−446.

Robinson, P., Faris, C., & Wind, Y. (1967). *Industrial Buying and Creative Marketing*. Boston: Allyn and Bacon.

Schmitz, J. (1995). Understanding the persuasion process between industrial buyers and sellers. *Industrial Marketing Management, 24*, 83−90.

Schultz, D. E., & Schultz, H. F. (2000). How to build a billion dollar business brand. *Marketing Management, 9*(2), 22−28.

Suvatjis, J. Y., & de Chernatony, L. (2005). Corporate identity modeling: a review and presentation of a new multi-dimensional model. *Journal of Marketing Management, 21*(7), 809−834.

Traynor, K., & Traynor, S. (1989). Marketing approaches used by high tech firms. *Industrial Marketing Management, 18*, 281−287.

van Riel, A. C. R., de Mortanges, C. P., & Streukens, S. (2005). Marketing antecedents of industrial brand equity: an empirical investigation in specialty chemicals. *Industrial Marketing Management, 34*, 841−847.

Ward, S., Light, L., & Goldstine, J. (July-Aug 1999). What high-tech managers need to know about brands. *Harvard Business Review*, 85−95.

Webster, F., & Keller, K. L. (2004). A roadmap for branding in industrial markets. *Brand Management, 11*(5), 388–402.

Wilson, D. (1995). An integrated model of buyer-seller relationships. *Journal of the Academy of Marketing Science, 23*(4), 335–345.

Wise, R., & Zednikova, J. (2009). The rise and rise of the B2B brand. *Journal of Business Strategy, 30* (1), 4–13.

Further Reading

Anderson, J. (1995). Relationships in business markets: exchange episodes, value creation, and their empirical assessment. *Journal of the Academy of Marketing Science, 23*(4), 346–350.

BBC News. (2007). *Andersen Consulting Renamed.* Available at: http://news.bbc.co.uk/2/hi/business/992946.stm. Accessed 15 September 2009.

Berkowitz, M. (1986). New product adoption by the buying organization: who are the real influencers? *Industrial Marketing Management, 15*, 33–43.

Blois, K. (1996). Relationship marketing in organizational markets: when is it appropriate? *Journal of Marketing Management, 12*, 161–173.

Brandchannel.com. (2009). *How should brands react to manic public opinion?* Available at: http://www.brandchannel.com/forum.asp?bd_id=112. Accessed 9 September 2009.

Chisnall, P. (1985). *Strategic Industrial Marketing.* Englewood Cliffs: Prentice Hall.

Choffray, J., & Lilien, G. (April 1978). Assessing response to industrial marketing strategy. *Journal of Marketing, 42*, 20–31.

Dichter, E. (Feb 1973). Industrial buying is based on same 'only human' emotional factors that motivate consumer market's housewife. *Industrial Marketing*, 14–18.

Diefenbach, J. (1987). The corporate identity as the brand. In J. Murphy (Ed.), *Branding: A Key Marketing Tool.* Basingstoke: Macmillan.

Hakannson, H. (Ed.). (1982). *International Marketing And Purchasing Of Industrial Goods.* Chichester: J. Wiley.

Hill, R., & Hillier, T. (1986). *Organisational Buying Behaviour.* Basingstoke: Macmillan.

Hutt, M., & Speh, T. (1985). *Industrial Marketing Management.* Chicago: The Dryden Press.

Ind, N. (1990). *The Corporate Image.* London: Kogan Page.

Kelly, J., & Coaker, J. (1976). The importance of price as a choice criterion for industrial purchasing decision. *Industrial Marketing Management, 5*, 281–293.

King, S. (1991). Brand-building in the 1990s. *Journal of Marketing Management, 7*(1), 3–13.

Kiser, G., & Rao, C. (1977). Important vendor factors in industrial and hospital organizations: a comparison. *Industrial Marketing Management, 6*, 289–296.

Kohli, A. (July 1989). Determinants of influence in organizational buying: a contingency approach. *Journal of Marketing, 53*, 50–65.

Lehmann, D., & O'Shaughnessy, J. (April 1974). Difference in attribute importance for different industrial products. *Journal of Marketing, 38*, 36–42.

Mattson, M. (1988). How to determine the composition and influence of a buying centre. *Industrial Marketing Management, 17*, 205–214.

McQuinston, D. (April 1989). Novelty, complexity and importance as casual determinants of industrial buyer behavior. *Journal of Marketing, 53*, 66–79.

Naumann, E., Lincoln, D., & McWilliam, R. (1984). The purchase of components: functional areas of influence. *Industrial Marketing Management, 13*, 113–122.

Parket, I. (1972). The effects of product perception on industrial buying behavior. *Industrial Marketing Management, 3,* 339–345.

Patton, W., Puto, C., & King, R. (1986). Which buying decisions are made by individuals and not by groups? *Industrial Marketing Management, 15,* 129–138.

Pearson, S. (1996). *Building Brands Directly.* London: Macmillan.

Reed, K. (2009). *BT launches into SME online accounting market.* Accountancy Age. Available at: http://www.accountancyage.com/accountancyage/news/2244896/bt-launches-sme-online. Accessed 8 September, 2009.

Saunders, J., & Watt, F. (1979). Do brand names differentiate identical industrial products? *Industrial Marketing Management, 8,* 114–123.

Sheth, J. (Oct 1973). A model of industrial buyer behavior. *Journal of Marketing, 37,* 50–56.

Shoaf, F. (May 1959). Here's proof – the industrial buyer is human. *Industrial Marketing, 43,* 126–128.

Sinclair, S., & Seward, K. (1988). Effectiveness of branding a commodity product. *Industrial Marketing Management, 17,* 23–33.

Stewart, K. (1990). Corporate identity: strategic or cosmetic? In A. Pendlebury, & T. Watkins (Eds.), *Marketing Educators Conference Proceedings.* Oxford: MEG.

Webster, F., & Wind, Y. (1972). *Organizational Buying Behavior.* Englewood Cliffs: Prentice Hall.

Woolfson, K. (Oct 1990). British Telecom plans a name with a new ring. *The European,* 5–7.

www.accenture.com/Global/About_Accenture/Company_Overview/Advertising/default.htm. Accessed 8 September 2009.

http://www.apple.com/getamac/whymac/. Accessed 10 September 2009.

www.medical.siemens.com/. Accessed 8 September 2009.

Service Brands

OBJECTIVES

After reading this chapter you will be able to:

- Identify the reasons why service brands are different to product brands.
- Understand the implications of services' characteristics for brand building.
- Discover how traditional brand building models might be applied in the services context.
- Identify the stages involved in creating a successful services brand.
- Understand the roles played by front line employees and customers in creating strong services brands.
- Describe the main ways in which service managers build service brand equity.

SUMMARY

The purpose of this chapter is to examine the issues associated with the creation and development of powerful service brands. It opens with a discussion about the nature of services branding. We outline the challenges in creating service brands and explore in particular those characteristics which make services uniquely challenging for managers. We outline the differences between goods and services branding and examine how the 'fast moving consumer goods' (FMCG) approach to branding services needs to be adjusted. We present a model for developing successful service brands. Particular emphasis is placed on the intangible nature of services and how problems created by intangible offerings can be overcome. The chapter next explores the roles of employees and customers in the delivery of service brands and discusses how these roles can be strategically designed to strengthen service brands. Finally, service brand equity is examined and the four ways firms can build strong service brands are presented.

205

Creating Powerful Brands. DOI: 10.1016/B978-1-85617-849-5.10006-6

NEW WAYS OF THINKING ABOUT SERVICES

This chapter is dedicated specifically to the branding of service firms. In doing so, we must first ask: 'Are services different and therefore worthy of unique consideration?' Traditionally, the answer would have been 'yes' but we will discuss the characteristics that differentiate physical products from services. These characteristics are well-known to many as intangibility, inseparability, heterogeneity, perishability and ownership. Theorists suggest that, as these characteristics are unique to services, services require special consideration in theory and practice and these considerations extend to services branding. Marketing academics suggest that there is a broader marketing mix for services. In addition to the '4Ps' of marketing for traditional products, services involve People, Processes and Physical Evidence. Understanding the characteristics of services has real value in creating powerful brands and we explore their relevance further at a later stage of this chapter.

First however, we consider work by Vargo and Lusch (2004) which suggests that ALL businesses are in service. Their central premise is that a firm's success depends on intangible resources, co-creation of value with customers and relationships. Vargo and Lusch (2004) present a new 'dominant logic' for marketing that places the provision of services, not goods, at the heart of economic exchange. They present eight 'foundational premises' or 'FPs' that encapsulate their new logic. This new perspective has been hailed as a new worldview of marketing. We therefore briefly address each FP to explore how this new logic might influence service brands.

> *FP 1: The application of specialised skills and knowledge is the fundamental unit of exchange.* People derive benefits or utility from their consumption of goods and services. What this means is that marketing is no longer about goods or services but about adding value through benefits. Therefore, a consumer buying a car brand, such as Toyota, is really buying the service of Toyota's skills and knowledge.
>
> *FP 2: Indirect exchange masks the fundamental unit of exchange.* This means that all employees should focus on all customers (internal and external). In traditional service firms, the role of internal marketing is vital for the adoption of the brand and the front line employee has been identified as the service brand ambassador. We shall address the role of the front line employee in brand building later in the chapter.
>
> *FP 3: Goods are distribution mechanisms for service provision.* People buy goods for emotional as well as functional reasons. In Chapter 2, we defined the brand as a 'cluster of functional and emotional values that enables organisations to make a promise about a unique and welcomed experience'. FP 3 reinforces this, as we see that consumers buy 'end states' such as happiness and security, as well as functional benefits such as

performance and reliability. In services, the intangible nature of the service brand provides the firm with an opportunity to create positive 'end states' for the consumer.

FP 4: Knowledge is a fundamental source of competitive advantage. Through knowledge, firms can differentiate themselves from competitors. In service brands, firms can differentiate themselves through unique processes or unique information. We see an example of this in the case of Apple stores, where consumers interact with extremely knowledgeable staff.

FP 5: All economies are service economies. Services are becoming more apparent in each economy as, in addition to the service sector's contribution to economic measures such as GDP and employment, adding a service component helps to differentiate manufactured goods. For example, IBM extends this advantage with their Service Management product offering, which allows firms translate assets into value.

FP 6: The customer is always a co-producer; in other words, the customer helps to produce value from products. Customers must learn to use and repair goods in order to create value from them. We see other examples of co-production in the rise of social media such as Facebook and websites such as Gizmodo, where customer bloggers advise other users about the benefits and best techniques for using products and services.

FP 7: The enterprise can only make value propositions. The customer decides on the value of a service and participates in creating value through co-production. Services need customers in order to have value and those customers create value through interaction with the service provider. Service brands must therefore differentiate themselves on the extent to which the relationship with their consumers creates value for those customers. For example, O2, the mobile phone network, forms relationships with their customer base by offering priority concert tickets to customers for concerts held in O2 arenas.

FP 8: A service-centered view is customer oriented and relational. The ongoing relationship between the firm and its customer is as important as a single transaction. Therefore, service brands must seek to keep their customers and evolve with them to sustain valuable customer relationships.

Vargo and Lusch (2004) advise that firms leverage their competencies for the benefit of the consumer and hold this as the central mission of the firm. Therefore, firms should identify their competencies and, in order to create competitive advantage, position them as adding value.

Our original question remains: if all firms are in service, are services different and should services brands require special consideration? It is important to distinguish between firms that are in service and those that are services firms. Brodie et al (2007) explain that the service brand concept is critical whenever 'service is super-ordinate to the branding of goods and/or services'.

Vargo and Lusch (2004) suggest that the primary goal of the organisation is value creation. In service firms, brands play a broader role because they involve all those who create value: this incorporates the customer as well as employees and stakeholders. In services, Brodie et al. (2007) explain that customers interactions with employees delivering the service brand determines brand meaning. Therefore, in services, the company is the primary brand. In this chapter, therefore, when we discuss the service brand, we are specifically exploring the brands of those companies who are predominantly in services businesses. We next explore the challenges faced by services in building strong brands.

THE CHALLENGE OF SERVICES BRANDING

The increased competition in services markets has made many companies realise that a strong service brand is an essential part of their competitive advantage. Unfortunately, the understanding of service branding has not kept pace with the growth of the services sector. For example, an overview of the top 10 brands in the world (Interbrand, 2009) shows that just three — McDonalds (number 6), Google (number 7) and Disney (number 10) — are 'pure' service brands. Further examination of these brands also shows that the product component of their offering has helped to build brand strength: for example, Interbrand credits the McCafé and healthy food offerings provided by McDonalds in opening new target markets and creating sales growth for the brand; Google's android phone software has helped to sustain the brand's innovative image; and Disney spin-offs such as video games and theme parks have helped to retain the brand's position. We must therefore ask: 'Why are service brands struggling and what are the unique challenges faced by services brands?'

Brands are even more important for services than for goods since consumers have no tangible attributes to assess the brand. It is harder to communicate the values of service brands. An effective route to convey the values of a service brand is through 'the way the company does things'; therefore the company's culture acts not only as a source of differentiation but as a key communicator of values. This means that a brand personality cannot just be communicated through an internet or television advertisement but it also depends very much on everyone in the company, from the CEO to anyone who has contact with consumers, since staff are an integral part of service brands. It is therefore important to train staff to ensure a greater likelihood of consistent delivery of the service brand. Building and sustaining brands needs to be undertaken by everyone in the firm and involves a profound understanding of every aspect of the interaction between consumers and the company.

Just as the consumer doesn't want Persil washing powder but clean clothes, so the brands of financial institutions should reflect the fact that consumers purchase them not as an end in themselves but rather as a means to achieve other goals. After all, customers booking a weekend at Ashford Castle in Ireland (http://www.ashford.ie) do not simply want overnight accommodation. They want the romance and exclusivity of a five star hotel set in fairytale surroundings.

The lack of powerful brands in the financial services sector is indicative of new challenges and the need for a new mindset when developing service brands. A service brand has to be based on a clear competitive position, which in turn has to be derived from the corporate strategy. This requires a holistic approach involving everyone in the company. Only when the brand's positioning and benefits have been communicated to staff, who understand it and have been trained and are therefore capable and confident in delivering the brand's promise, should these be communicated to consumers.

Furthermore, to successfully develop and maintain service brands, marketers should take heed of Conrad Free's (1996) contention that an effective brand strategy must reflect a true competitive advantage, encompassing factors such as:

- *High quality top management* – The commitment of high calibre management is fundamental to guarantee excellent service brand delivery.
- *Vision* – All employees need to understand and be committed to the brand vision. Long-term rather than short-term plans are required to ensure the development of meaningful relationships with consumers.
- *Results driven* – The vision should be translated into clearly defined goals for all staff.
- *Competitiveness* – The company should benchmark its performance against best practice, both inside and outside the sector.
- *Use of technology* – Effective exploitation of new technologies is a fundamental source of sustainable competitive advantage.
- *Consumer focus* – The consumer needs to be regarded as central to everything the organisation does.

The development of a services brand must also be approached in the context of the challenges posted by the nature of services themselves. We address those characteristics: intangibility, heterogeneity (non-standardisation), inseparability of production and consumption and perishability (Zeithaml et al., 1985), and explore how they each influence the service brand.

INTANGIBILITY

Bateson (1979) viewed intangibility as *the* single biggest difference between goods and services. It is long acknowledged that tangibility is not limited to goods and intangibility is not limited to services. Shostack (1977) illustrated

MANGO

Mango use celebrity endorsement to target their stores at young stylish consumers. Here, the actress Scarlett Johansson is associated with the brand. (Reproduced by kind permission of Mango)

goods and services on a continuum, with physical goods such as salt on the 'tangible dominant' side of the spectrum and services such as education and management consulting on the 'intangible dominant' side. 'Pure' services therefore have greater intangibility and this creates problems for brand building. Services cannot be felt, tasted, touched or seen in the same way as goods; however, there are always some tangible components which help consumers evaluate services. On a flight, for example, the total service experience is an amalgam of many disparate components, such as the style of welcoming on the aircraft, the nature of the service on board and the in-flight entertainment, to mention but a few components. There are clearly many tangible elements during the flight but this is hardly comparable with buying a television or a suit, where the total product can be seen, examined and evaluated.

Since the nature of a service is inherently abstract, the development of a brand image should enhance the reality of the offering through the representation of tangible clues. Brady et al. (2005) explain that intangible services have fewer cues which a customer could use to select a service brand. The two types of cues customers use are intrinsic and extrinsic cues. Intrinsic cues are taken from the product and are therefore more difficult to change. Extrinsic cues 'surround' the product and can be changed more easily. Brady et al. (2005) examined student purchases of computers (very tangible), hotel accommodations (less tangible) and mutual funds (very intangible). They found that intrinsic cues were more important for intangible dominant services, for example investment services, whereas extrinsic cues were more important for more tangible services and products. Managers should take care to develop their brand messages according to the level of intangibility of their service offering.

As shown in Fig. 6.1, there are those offerings which are high in 'search qualities', i.e. attributes consumers can identify and assess before making a decision. These include price, size, shape, colour, smell and feel. Physical goods are typically high in search qualities. A second category of offerings are those which are characterised by 'experience qualities' since the differentiating attributes can best be evaluated while being used — for example, a meal in a restaurant or a holiday using a particular tour operator. The third category, typifying numerous services, covers those offerings high in 'credence qualities' where consumers have insufficient knowledge to fully evaluate these services, even after consumption; for example, many people would find it difficult to fully assess the quality of major surgery.

Organisations need to assess the position of their services on the evaluation spectrum when designing effective communication strategies to help consumers evaluate service brands. The difficulties consumers have in evaluating services forces them to rely on any available clues and processes. Thus,

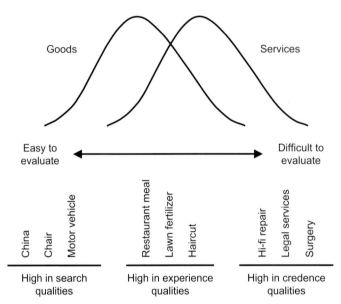

FIGURE 6.1
Distinguishing between search, experience and credence qualities

while the lawyer may be extremely qualified, the abrupt secretary may strongly influence the potential client's perceptions. The more difficult it is to evaluate a service, the greater the perceived risk associated with it and therefore the more important the need to use the brand as a risk reducing device.

For service brands that are high in search qualities, a popular strategy involves encouraging consumers to recommend the brand to other customers through promotions such as 'introduce a friend and you will receive a month's free subscription'. Word of mouth acts as a credible source of information. For service brands high in experience qualities, demonstrations of the service help people make judgements. For example, some travel companies offering ocean cruises invite potential consumers to social evenings where there are video presentations of their cruises. Service brands high in credence qualities need to communicate their brand's qualities more creatively. For example, many customers rarely check their mortgage rates and tend to forget that it is the mortgage which enables them to enjoy living in their own houses. In this situation it may be more appropriate for marketers to communicate once a year that their houses have appreciated in value (when economic conditions bring this about) and that they are achieving this increased prosperity through a mortgage with a well-managed firm. Likewise the publication of an easy to understand annual statement about the performance of investment products can be an easy way to bond customers, particularly when they are not bombarded with vast amounts of facts that are difficult to understand. Brands that are high in credence values should convey a strong message of trust, almost along the lines of 'Don't even think about whether it will be good — of course it will. Just trust us. Now relax and enjoy the service.' The work of Srinivasan and Till (2002) effectively demonstrates how brand names can influence consumers' evaluation of the search, experience and credence attributes of an offering and how this is moderated by trial. They found that consumers view well-established brands as being of a higher quality than generics, but this effect was reduced when consumers had trial opportunities. Before trial, established brands were viewed more positively than generics on experience and credence attributes but not on search attributes. Trial moderated this advantage in terms of experience attributes, but consumers' evaluations of credence claims remained unchanged in favour of established brands.

HETEROGENEITY

Service based brands are characterised by multiple interactions as consumers frequently have to talk with several people in a service firm. One has only to consider the traveller at an airport and the numerous interactions they have with different people as they pass from checking in to security to passport control to ticket checking prior to boarding the airplane and then with the cabin crew. Each interaction gives a message about the brand of the airline. Due to the different people having contact with the traveller, the chance for conflicting messages about the service brand increases, unless managers have worked with staff to ensure a consistent style of behaviour.

Since services are predominantly performed by people, it is more difficult to ensure that the same standard of service will be delivered by two different people in the same firm. Members of the staff who represent the service in the eyes of consumers may not only deliver the service differently from each other, but the service may differ from day to day. For example, a member of the cabin staff on a long haul flight may provide different service levels between the outward and return flights because of the extent to which he or she adjusted to the various time zones. The perceptions of the service quality may also vary because consumers differ in their individual demands, expectations and their own assessment criteria.

Many organisations try to standardise staff performance through careful planning, control and what can almost be considered automation. While they achieve a high degree of homogeneity in service delivery, they increase the risk of being inflexible and their staff may react poorly to unforeseen problems. Some organisations regard heterogeneity as an opportunity for strengthening their brand by customising their service to better serve the needs of individual consumers through greater empowerment. If organisations follow the latter approach of empowering their staff, they need to shape their employee relations and their internal communication channels to motivate and retain these customer-focused employees.

Whether goods meet quality standards can be checked before they leave the supplier, whereas the quality control of services needs to be assessed when consumers come in contact with staff. This is more problematic to check. The implication for service brands is that staff training needs particular attention, as every member of staff plays a major role in ensuring the quality of the service. Regular training 'refreshers' are useful in ensuring that staff are continually providing high-quality service. Furthermore, organisations need to place more emphasis on closer collaboration between members of the staff so that they can learn from each other's experiences and build on best practice.

INSEPARABILITY OF PRODUCTION AND CONSUMPTION

While goods are generally produced, then sold and finally consumed, most services are sold first, then simultaneously produced and consumed. The consumer is present while the service is being produced and needs to participate in the production process. This presents an opportunity to influence customers' perceptions of the service during its delivery since, through close contact, staff can better sense consumers' views and either amend their approach or suggest to consumers how their own actions can improve the outcome. For example, a dentist performing a wisdom tooth extraction can sense the patient's nervousness and adapt their behaviour accordingly, by explaining the procedure and offering reassurance.

Consumers are often actively involved in co-creating the service, either by undertaking a lot of the service themselves, as in a launderette, or by cooperating with the staff, as in a hair salon. Consequently, it is important that consumers are made aware of the roles they are expected to undertake as they can affect the service delivery. For example, consumers at a KFC drive-through are expected to know what they would like to order, drive to a booth, place an order, pay and collect their order, without entering a KFC premises. On the other hand, in a restaurant, customers need to take a seat and wait to be served. Consumers need to be informed about changes in the nature of a service, which affect the role they play in the service delivery process. Aer Lingus introduced touch-screen, self-service check-in to facilitate passengers with early check in, seat selection and self-printing boarding pass facilities. Aer Lingus was aware of the need for customers to 'learn the system'. Following its launch, ground crew were deployed to assist users in their use of the new technology.

There is now an increasing acceptance of the importance of emotion and mood within the services context. As the service encounter is essentially an interpersonal interaction, it follows that emotion and mood, as vital factors in human relationships, play a significant role. Mattila and Enz (2002) considered the effect of emotion and mood in consumers' responses to brief, non-personal service encounters, specifically interactions between service staff and consumers at the front desks of a number of first class hotels. They found that the consumer's mood immediately prior to the service encounter and the emotions displayed by the consumer during the service encounter (measured through facial expressions, bodily gestures, tone of voice and body language) were strongly linked to the consumer's eventual assessment of the service encounter and the organisation in general. However, they also found that the consumer's evaluation of a given service encounter was unlikely to reflect that of the employee providing the service. The suggested implication of this for service organisations is that consumers' evaluations of a service may be improved by training employees to be more responsive to consumers' moods and displayed emotions, particularly because there may be a 'gap' between consumers' and employees' perceptions.

Hennig-Thurau et al. (2006) examined emotional contagion and emotional labour during the service encounter. Emotional contagion is the extent to which the customer 'catches' the mood of the service employee. Emotional labour is the employee's display of expected emotions. For example, employees may be encouraged to smile and wish the customer a good day. Emotional labour can involve 'surface' (fake) or 'deep' (genuine) acting of emotions. In their study of a movie rental service, they found that an employee's emotional display affects the customer's affective or emotional state. In addition, when employees were perceived to be authentic in their emotional display, this had a greater effect on the customer's emotions. Overall, the study found that the change in customers' emotional states following the service encounter affected their satisfaction with the service. Frei (2008) suggested that service managers might hire service employees for their attitude and then train them to deliver the service. It would seem that managers should seek out those employees who can deliver a genuinely positive demeanour.

Services can also be characterised by the degree to which consumers come in contact with other consumers — for example, minimal contact in legal services yet high consumer-consumer contacts at a rock concert. In service brands, consumers' appreciation and enjoyment may be adversely affected not only by the poor performance of the service provider but also by their contact with other consumers. For instance, a couple going out to an expensive restaurant may appreciate the polite and formal style of the waiter but be appalled by the loud remarks of a group nearby who have had too much wine. Brands which are designed to appeal to different consumer segments need systems to ensure they are kept separated. For example, curtains separate business travellers from tourists in the economy class of most airlines. Furthermore, consumers within some segments may wish to reduce their contact with other consumers. Airlines respond to this by having wider seats and bigger gaps between passengers, for business class travellers.

PERISHABILITY

Unlike goods, a service brand cannot be stored. An hour of an accountant's time that is not used cannot be reclaimed and used later. This perishability highlights the importance of synchronising supply and demand. An unused seat on an aircraft's return journey shows no profit to the airline when the aircraft takes off. Likewise, sudden increases in the numbers of consumers, such as students arriving at a cafeteria at the end of a lecture, are difficult to satisfy because there is no 'inventory' available for back-up. Service organisations need strategies to cope with fluctuating demand or to smooth demand to match capacity more closely. Failure to address this problem not only leads to increased costs and lost revenues but also to weakened brands. Long queues at checkouts can ruin the brand image of a supermarket chain, unless ways are found either to reduce waiting times or to entice consumers to shop at off-peak periods.

Associated with the fact that services cannot be stored, is the issue that many services are bought long before they are experienced, as with pensions. Thus, many service brands face a double challenge: first, they need to develop an image and a reputation to attract consumers; secondly, they must then retain these consumers as competitors try to attract them away, even though they have yet to experience the actual service.

For most services, consumers are prepared to wait only for a limited time before receiving the service. Since organisations cannot always ensure no waiting time, they need to find the best compromise to satisfy consumers. When queues occur, companies should first analyse the operational processes and possibly redesign them to remove any inefficiencies. If the queues are inevitable, long waiting times can be alleviated through a reservation system; for example, at the dentist's office, a pleasant and efficient receptionist can help shift demand to off-peak times. Another strategy is to differentiate between consumers according to their importance or the urgency of their requests. As a last resort, consumers' tolerance of queues can be increased by making the waiting time enjoyable or at least more tolerable, through such actions as:

■ *Entertaining consumers* – company switchboards often play music when they put callers on hold. Cinema Omniplexes have TVs displaying upcoming movie trailers to occupy guests as they queue for their popcorn.

- *Starting the process* — reading a menu in a restaurant or filling out forms at the beautician are perceived as activities which are part of the service and not waiting time.
- *Reassuring consumers* — that they are waiting for the right service; for instance, sign in an airport, which point out the flights that are being checked in by the specific lane.
- *Informing consumers* — when they will be served or at least reassuring them that they have not been forgotten; for example, when someone is 'on hold' in a telephone queue, stopping the music every 30 seconds to play a message apologising for the delay and explaining that the staff are busy but are aware that the caller is waiting.
- *Explaining* — the reasons for the delay, for example, through automatically updated arrival-screens in airports and railway stations.
- *Establishing a first-come first-served rule* — The allocation of waiting numbers to consumers helps avoid jostling. Nevertheless, consumers are prepared to accept apparent inequality in exceptional cases, such as hospital emergencies.

MOVING BEYOND THE FAST-MOVING CONSUMER GOODS MODEL

The previous section has highlighted various difficulties associated with service branding and in particular, the danger of applying the traditional fast moving consumer goods model without adapting it to the characteristics of the service sector. One service sector that has been given consideration is the financial services sector, due to its size and importance. Depth interviews among brands' teams in UK financial services organisations (de Chernatony and Cottam, 2006) revealed specific problems for brand building:

- Some brands are too firmly rooted in the past. Managers also accept that some employees confuse 'brand' with 'logo' and are concerned with the logo's colour, rather than the strength of the brand's promise.
- It is not unusual for a company to emphasise financial performance rather than customer orientation.
- Banks with weaker brands have inadequate brand leadership.
- Managers found it hard to find ways to differentiate their brands.
- Weaker brands could be strengthened by more focus on and understanding of branding.
- HR practices and policies do not always support the brand.
- Culture and brand values are vital to the success of a financial service brand.

The following two cases illustrate how the goods branding concept can be adjusted for the development of successful service brands. Brookes (1996) is but one amongst many who argue that goods brands and service brands can be developed through broadly similar process, i.e.:

- Setting clear brand objectives.
- Defining a clear positioning.
- Selecting appropriate values.

Levy (1996) provides a further example of how successful service brands can be developed by adapting the FMCG model. Their analysis showed the need to adopt the FMCG model, taking into account services differences such as:

- *Product definition* — Financial services are not as well-defined as FMCG brands. A current account is not a single physical entity but consists of a collection of several elements, from the cheque book to the conversation with the cashier.
- *Brand differentiation* — FMCG marketers strive to differentiate their brands by communicating a competitive edge and by avoiding launching 'me-too' products. Marketers in financial services, however, focus on building long-term relationships with their consumers through having a broad product range to satisfy all their needs. Since the maintenance of existing relationships lies at the core of their strategy, emulating competitors' products to fill a gap in the firm's portfolio is more often considered than it would be in the goods sector.
- *Consumer motivation* — While in FMCG markets consumer loyalty is hard to win and retain, consumers in the financial sector are reluctant to switch between companies because they perceive the differences as negligible and not worth the possible disruptions.
- *Measurement of brand strength* — Measuring awareness levels and the propensity to repurchase provides a reasonable assessment of brand strength in the FMCG domain. With financial brands the concept of repeat purchase is less meaningful as it needs to be measured over years. Likewise, brand loyalty is hardly relevant as inertia prevents consumers from switching. Thus marketers of financial services draw more on qualitative data about the interactions with existing customers.
- *Product benefits* — Although product features can help attract consumers, they are not the obvious brand discriminator. Furthermore, in financial brands, extra features can confuse already reluctant consumers about the benefits of the brand and provide them with an excuse to reject it.

A challenge within the financial service sector is in creating and sustaining consumer trust. Brands have a significant role in developing or regaining customer confidence. Gill (2007) explains that branding activities are critical for banks to retain the role of 'trusted advisor'. Consumers are moving away from traditional media; for example, Gill (2007) cites the number of persons who visit financial service sites online as 36% of the total online population or over 16 million customers. In Chapter 8 we will also see how consumers take matters into their own hands through online communities and social networking. Complicating this further, Gill (2007) explains that financial service brands are now monitored on broader aspects of their business, so consumers 'no longer take a company's promise at face value' (Gill, 2007, p.149). Banks must therefore strive to create collaborative advertising that engages their audience and extend their use of the internet beyond functionality to create imaginative ways of spreading brand messages online. For example, ING, the Dutch bank, promotes itself through its website (www.ing. com). ING's mission is 'To set the standard in helping our customers manage their financial future'. The website is a tool to enhance consumer trust. As Mannung-Schaffel (2006) identifies, ING has fewer 'bells and whistles' than other banks, thus promoting its image of transparency; and to engage with its publics, it highlights its corporate responsibility activities, provides a link to careers and provides an 'eZnomics' section, where customers can learn to haggle a bargain, consider their food miles or learn interview tips. By utilising the website in this way, ING promotes its brand identity to its diverse publics.

THE PROCESS OF BUILDING AND SUSTAINING POWERFUL SERVICES BRANDS

Recognising that little was known about the actual *process* of building and sustaining powerful services brands, de Chernatony et al (2003) conducted a series of interviews with experts in the field of services branding in order to investigate this issue further. While a number of models exist that deal with the building of goods brands, this research resulted in one of the only dedicated services brand building models. The experts interviewed were brand consultants (the majority of whom were at the level of Chairmen and Directors) and a management journalist, as it was believed that these were the people with the most knowledge and experience in the field. As a result of the interviews, a 'cog wheel' model evolved as shown in Fig. 6.2. The cog wheel reflects the fact that, within the process, it was found that feedback resulted in closed loop planning. While the majority of respondents viewed brand building as a planned process, there was a small minority who spoke of coincidence, luck and post-rationalisation. These elements should certainly be taken into account as in any walk of life, as should the fact that not all brands will go through the stages in the same order or go through every single stage.

> *Identify external opportunities* — Initially, a market opportunity must be identified. Extensive research (both qualitative and quantitative) is useful, encompassing competitors, resource availability, supply and demand, cost-benefit analysis, the political and economic environment and current segmentation within the sector.

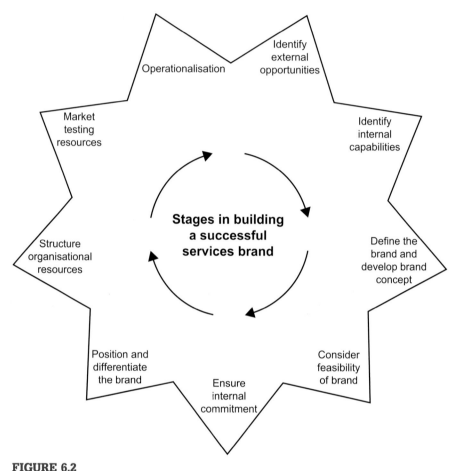

FIGURE 6.2
The cog wheel model for building and sustaining services brands (*de Chernatony et al. 2003*)

Identify internal capabilities – The identification of the organisation's key competencies is essential and benchmarking may be used to facilitate this. *Define the brand* – The brand concept is developed, which may involve the use of creative judgement, visual representations, symbolic representations and language. It may be helpful to identify the 'brand essence', the meaning of the brand expressed in just a few words. Another possibility is that the brand may be defined in terms of values and consumer beliefs. de Chernatony et al. (2004) advocate identifying brand values internally and recommend that those values should be organic so that they can evolve over time. *Consider feasibility of brand* – Practical considerations such as scale, timing and the firm's existing track record will need to be taken into account. A key issue at this stage is the availability of financial resources.

Ensure internal commitment — There needs to be genuine belief in the service brand throughout the organisation. If employees 'live the brand', the brand's values will be more effectively communicated to the consumer. de Chernatony et al. (2004) found that service brands must be internalised by employees and employees should be encouraged to be proud of their brand.

Positioning and differentiation — This allows the service brand to gain a meaningful competitive advantage over its competitors and to appear unique to the consumer.

Structure organisational resources — In a services context, the most important organisational resource will almost always be staff. It is important that the right people are recruited and are then trained in the most effective way. As well as considering the skills and abilities of staff, however, it is also important to factor in personalities. As an interaction between individuals, the service encounter will be strongly influenced by the personalities of both the consumer and the service employee. Also important is staff motivation — examples of techniques used include giving employees a share in the organisation or attempting to imbue staff with genuine excitement about the brand.

Market testing — As service brands inherently involve public exposure, there is a need to pilot test and fine-tune the brand prior to full service roll-out. Unlike with goods brands, it is very difficult to perfect the brand behind closed doors.

Operationalisation — Communications and delivery often merge at this stage in a services context. It is vital to have a consistent, strong brand message across all media. Staff views on brand communications need to be taken into account, as it is the staff who will ultimately be delivering the promises being made and the service delivery process must match or even exceed the brand promise.

In their research on financial services brands, de Chernatony and Cottam (2006) advocated the following factors for successful brands:

- The brand is a holistic experience, therefore everything experienced by the customer is the brand.
- The brand is based on excellent, personalised customer service, where customer service is the central ethos of the brand.
- The brand challenges the norm, as mediocrity will fail to differentiate service firms.
- The brand is responsive to change, through openness and organisational learning.
- Organisational members need to become more brand literate as they need to better understand the brand and what it means for their jobs.
- There is a synergy between brand and culture, where the organisation's espoused values, employees' values and brand values are congruent.

STARBUCKS

Starbucks emphasise the quality of their coffee by offering it for sale in retail outlets. The slogan 'It's not just coffee. It's Starbucks' seeks to emphasise the superiority of Starbucks coffee

Frei (2008) had an interesting suggestion for growing service brands. He proposed that firms should choose what *not* to do well. Instead of struggling to provide a full service package which is not differentiated in any way, the service should compromise and seek to excel in specific areas of strength. Frei offers the example of Wal-Mart, whose customers do not value ambience and sales help as much as low prices and wide selection. Therefore, Wal-Mart targets segments of the customer population that need low prices and wide selection and those customers self-select Wal-Mart because it suits their needs. This approach means that the service brand may cede superiority on some aspects of service to the competition but it also allows the service firm to offer excellence on specific service attributes. An example of this is Ryanair, 'The Low Fares Airline', whose competitive advantage is embedded in their brand promise of low fares. Companies should therefore explore ways they can create specific tangible benefits through branding.

BRANDING TO MAKE TANGIBLE THE INTANGIBLE

One of the most problematic aspects associated with service brands is that consumers have to deal with intangible offerings. In an attempt to overcome this problem, marketers put a lot of emphasis on the company as a brand, especially in sectors such as financial services, since this is one way of making the service more tangible. Research in the financial services sector has shown that consumers know little about specific products, often they do not want to know more and they are content to assume that the best-known companies have the best financial products. This further compounds the problem of service branding! However, brands are critical in this sector. Colton and Oliveira (2009) explain that corporate brands are one of the main ways banks can build trust with customers.

Because of their intangible nature, service brands run the risk of being perceived as commodities. To overcome this problem, strong brands with a clear set of values, which result in positive perceptions amongst consumers, are essential. However, these common and consistent perceptions amongst consumers are difficult to establish for intangible offerings. Services brands need to be made tangible to provide consumers with well-defined reference points.

An effective way to make brands tangible is to use as many physical elements as possible that can be associated with the brand, such as staff uniforms, office décor and the type of music played to customers waiting on the telephone. A service brand can project its values through physical symbols and representations, as Virgin Airlines has so successfully done with its vibrant red colour reflecting the dynamic, challenging position being adopted. The first points of contact with a service organisation, such as car parking, design of building and appearance of the reception area, all interact to give consumers clues about what the service brand will be like. Other ways that brands communicate with consumers are through tangible elements such as stationery, the way employees dress and brochures. One major retailer had a full length mirror that all the members of its staff passed as they left the canteen to go into the store. Above the mirror was a sign saying, 'This is what the customer sees', to draw their attention to the importance of thinking about the tangible cues consumers are presented with, as representative of the brand. However, as well as using physical cues to communicate the brand to the consumer, service organisations are increasingly recognising that they also play an important role in communicating to employees, affecting their motivation, enthusiasm and mood. For example, one leading services organisation provides toiletries and perfumes expressly for the use of employees, in cloakrooms and toilets. Employees are aware that this is to reward them for their excellent performance and they feel more valued as a result.

Package design plays an important role for branded goods and in service brands, it represents an opportunity for more effective differentiation. McDonald's boxes for childrens' meals, for instance, have been shaped as toy houses to reflect the playful element associated with the experience of a lunch in the fast-food chain. The tangible elements surrounding the brand can also serve to facilitate the service performance. For example, the use of PayPal on eBay can offer reassurance to a customer using the website for the first time. The setting within which the service is delivered may either enhance or inhibit the efficient flow of activities. The yellow and blue stripes in IKEA stores, for example, not only allude to the Scandinavian tradition of the company but also guide consumers through the different sections. The design of the surroundings also plays a socialisation function, informing consumers about their expected behaviour, the roles expected of staff and the extent to which interactions are encouraged between them. Club Med dining facilities are structured so that customers can easily meet and get to know each other. Finally, the design of the physical facilities may be used to *differentiate* the service brand from its competition. HMV used store design to reposition itself from a music store to a multi-channel social gateway. For example, the retailer now incorporates technology in stores to create a social space, in response to consumer trends in gaming and online music file sharing (Design Council, 2008).

The tangible elements of a service brand encourage and discourage particular types of consumer behaviour. For example, a 7-Eleven store played classical symphonies as background music to retain their 'wealthier' customers, while driving away teenagers who tended to browse the shelves rather than spend any money. Different aromas can elicit emotional responses and thereby influence consumer behaviour — some food retailers pump the fragrance of freshly baked bread into their stores, evoking a more relaxed, homely feeling and stimulating appetites!

The previous examples have shown different ways in which service organisations can make their brands more tangible. The approach adopted when 'tangibilising' the service brand must be consistent with the service and should not promise more than the service will actually deliver. The elegant uniforms worn by sales assistants at Marks & Spencer clearly communicate the brand message of striving for excellence. If the physical evidence of the brand is unplanned, inconsistent or incompatible with the message of the added values that the brand aims to convey, consumers will perceive a gap and reject the brand. The following questions can help marketers to assess the extent to which they are capitalising on tangible cues to support their brand strategy:

- Do all the elements of the physical evidence convey a consistent message?
- Do the physical evidence and the conveyed message appeal to the target market?
- Does the physical evidence appeal to employees and motivate them to develop the brand?
- Are there additional opportunities to provide physical evidence for the service?
- What are the roles of the surrounding elements of the service? How does each tangible element contribute to the development of the brand?

CONSISTENT SERVICE BRANDS THROUGH STAFF

Even though the service organisation may have developed a well-conceived positioning for their brand and devised a good communication programme, the brand can still flounder because of insufficient attention to the role the staff plays in producing and delivering the service. The service encounter usually involves an interaction between the customer and a service employee during the delivery of the service. Earlier in this chapter we learned that customers co-produce the service with the service employees. It is this interaction during the service delivery that will form the customer's impression of the service. Frei (2008, p.75) explains that 'companies often live or die on the quality of their workforces' and because services are people intensive, good employees can provide a real competitive advantage.

In particular, the following factors can compromise the success of the brand:

- Ineffective recruitment.
- Conflict in the duties staff are required to perform.
- Poor fit between staff and technology.

For example, Frei (2008) argues that if the company's systems and processes require 'heroism' by employees to keep customers happy, the service is flawed by design. While employees are trained to deliver the brand promise, they are not expecting to continually be 'heroes' to compensate for poor systems and processes. Managers should ensure that people and processes are aligned so that the best service can be provided to customers.

Earlier, we explained that heterogeneity is a characteristic of services. As the quality and process of every service performance can vary and often requires the involvement of several employees, empowering employees and building a culture based on teamwork is likely to enhance consumer satisfaction.

Staff embodies the service brand in the consumer's eyes. In many cases the service staff is the only point of contact for the consumer and by thoroughly training staff and ensuring their commitment to the brand, its chance of succeeding is greater. The success of the Disney brand results from the firm's insistence that employees recognise they are always 'on stage' whenever in public, encouraging them to think of themselves as actors who have learnt their roles and are contributing to the performance and the enjoyment of visitors. Frei (2008) suggests that employees should be hired for their attitude, which would mean that even lower aptitude employees will perform well and deliver a positive service.

However, it must be recognised that despite best intentions, not *every* consumer will experience the service encounter positively *every* time. At this point, the organisation's commitment to 'service recovery' becomes vital, will determine whether the consumer abandons or stays with the brand and may prevent any potentially damaging negative publicity. The 'recovery paradox' is the name given to the phenomenon of dissatisfied consumers being so impressed by the service recovery they experience, that they become even more loyal and satisfied than they were originally. For example, the customer who complains about the temperature of their meal and receives a bottle of wine with the chef's compliments, may be more satisfied with the service than they would have been if nothing had gone wrong. However, research has provided conflicting evidence regarding the actual validity of this and so organisations would be wise not to use it as a deliberate strategy. Wallace and de Chernatony (2007) have shown that some services, such as banks, avoid explicit apologies and service recovery as they fear it could encourage customer litigation! Once a consumer has experienced some form of service failure, there are a number of key factors important in effective service recovery:

Fairness— A basic expectation of consumers is that they should be treated in a fair manner.

Promptness — If the organisation 'drags its feet', consumers will only become more annoyed.

Accountability — Consumers should be provided with, and informed about, appropriate channels for any complaints they may have. These should be easy to access and use and consumers should feel that the organisation welcomes their feedback.

Learning — The organisation should be prepared to use a service failure situation positively, in that they learn from the experience and take steps to prevent it recurring.

Also, the staff of a service organisation can positively enhance the perception consumers have of the service quality. We use the dimensions of SERVQUAL (Parasuraman et al, 1988) to illustrate this:

- *Reliability* — For example, Facebook published the site's 'uptime' as 99.92% in 2008, highlighting that it was down for only 7.2 hours in that year. This provides reassurance to users about the reliability of the site.
- *Responsiveness* — A Tesco employee may accompany a customer to the correct aisle when they are looking for a specific product.
- *Assurance* — O2 billing provides information for customers about each category of spend and provides text alerts about bill amounts.
- *Empathy* — A manager at a Dublin Superquinn grocery store encouraged his employees to know their local elderly customers by name. He noticed these customers entering the store daily for a small number of items. Although they did not create large value through weekly trolley shops, he understood that these women were often widowed, lived alone and visited the store daily for company. Therefore cashiers were encouraged to address the customers and make them feel welcome. This sense of community enhanced staff morale.
- *Appearance* — The uniform worn by the Delta Airlines crew is designed by Richard Tyler, which adds to their perception as sophisticated and perfectly tailored. The uniform design helped to refresh the airline's image, as well as boosting staff morale.

Failing to take heed of these factors can have a negative impact on the perceived quality of the service.

While front line employees have an excellent opportunity to support the brand through their actions, their behaviour can also be potentially detrimental. Harris and Ogbonna (2006) show that employees can sometimes act as 'saboteurs'. This is acting in a manner that is 'deliberately deviant or intentionally dysfunctional'. They caution that up to 95% of employees can be dysfunctional and this is costly to firms because it effects service quality perceptions and ultimately, the survival of the firm. Their research focused on customer contact employees in the food and beverage sector of the hospitality industry. They found that greater sabotage occurred among risk takers, those with a greater need for social approval, employees who perceived they were not monitored, those who did not intend to stay with the organisation, employees who perceived a fluid labour market and those who perceived low levels of cultural control. Harris and Ogbonna (2006) also looked at the outcomes of sabotage. They found that sabotage increased self-esteem and team spirit for those employees who engaged in acts of deviance but those actions reduced rapport between employees and customers, reduced functional quality and reduced the financial performance of the firm. Managers must be careful therefore, to make sure that the culture of the firm does not encourage sabotage. In the banking sector, Wallace and de Chernatony (2008) explored the nature and causes of sabotage in a service environment where staff enjoy greater job security, where there is reduced turnover and where managers are not always able to 'fire' a saboteur. In banking, sabotage is broader than deviance, as managers also interpret underperformance or non-compliance to service standards as forms of sabotage. Managers are sympathetic to saboteurs, as they recognise that fears about change, institutionalisation, lack of opportunity and life demands are partly responsible for the saboteur's behaviour. In addition, managers caution that bank financial metrics might not always support customer orientation, therefore the staff is unsure about the best behaviour to adopt. For example, should a mortgage advisor in a branch sell an additional mortgage and meet their financial metric if it is not in the best interests of the customer? If services are to create strong brands, service and customer orientation should be at the heart of the business agenda.

To ensure that the staff is willing and able to deliver high quality services, organisations should motivate their staff and encourage a customer-orientation culture by considering the following:

- *Recruit the right people* — Successful companies such as Google or Apple are regarded as preferred employers. The brands of the major consultancy firms, such as McKinsey, are built on their policy of hiring only the best people. Some firms explore the values of potential staff to assess whether these are similar to the values of their firm. Where values are strong, those values will attract like-minded employees, which will ultimately strengthen the culture of the firm.
- *Train staff to deliver service quality* — The consistency of the McDonald brand is ensured by the formal education employees undergo at the famous 'Hamburger University'. To enhance their brands, companies such as Federal Express invest heavily in the training of their employees. Nordstrom, the American retail store, is a good example of a brand built on staff empowerment: the only rule staff are taught at Nordstrom is 'use your judgement'. Furthermore, members of the staff need to regard each other as internal customers within the service process, enhancing the quality of the brand through teamwork and cooperation. Perhaps the most unusual code of conduct is Google's 'Don't be evil'. This code is further explained by

guidelines such as 'serve our users', 'respect each other', 'preserve confidentiality' and 'obey the law'.

■ *Provide support systems* — Appropriate technology and equipment are essential to support staff in delivering quality service. For example, bank clerks need to have easy and fast access to up-to-date customer records if they are to deliver accurate and prompt customer service. Supermarket cashiers need a reliable Electronic Point of Sale (EPOS) system to allow them to smoothly check through groceries and make sure prices are accurate.

■ *Retain the best people* — Although many organisations are aware of the importance of recruiting the best people, there are instances where firms do not put as much effort into retaining them as they could. A high staff turnover usually translates into low consumer satisfaction and poor service quality. In order to retain their employees, organisations need to involve them in the company's decision-making process and devote as much attention to them as to their customers. Moreover, they should reward employees for good service delivery through financial and non-financial measures, such as McDonald's scheme 'The employee of the month'.

By addressing some of these issues, a company can establish a more customer-focused service culture, which is a prerequisite for delivering consistently high-quality services and for building successful service brands. A customer-centred focus should pervade the whole organisation so that the commitment to customers becomes second nature for all employees. The development of a genuine service culture is neither easy nor quick but companies that have overcome this challenge have been duly rewarded. Ultimately, service firms should seek to cultivate 'Brand Ambassadors'. These employee types, proposed by Ind (2004) are storytellers who spread the brand idea. These individuals may be senior managers or front line employees who bring the brand message to life through their performance with customers and through their actions which support employees working at the front line. A good example is Sir Richard Branson, who founded Virgin in 1970 and has created over 200 companies worldwide with 50,000 employees. Virgin stands for value for money, quality, innovation, fun and a sense of competitive challenge and Branson reflects this through his own adventures, including his work on the Virgin Galactic space tourism project, his membership of The Elders environmental group, and the friendly tone of his 'Ask Richard' Q&A section of the Virgin website.

SERVICE BRANDS WITH THE OPTIMUM CONSUMER PARTICIPATION

While the previous section has shown how service organisations can enhance their brands by building on the role employees play during the service delivery, this section focuses on how consumers contribute to the development of the service brand. The way consumers evaluate a service brand depends largely on the extent to which they participate in the delivery of the service. If a yachting enthusiast did not get on too well with an instructor at Club Med, this interaction would affect their view of the brand. When subsequently hiring a dinghy at the end of the course and having difficulties rigging the sail, they may complain to their friends about ageing equipment. Yet the real reason for their problems, lies in their not paying proper attention to the instructor.

If the service performance requires a high degree of consumer involvement, it is vitally important that consumers understand their roles and are willing and able to perform their roles; otherwise their inevitable frustration will weaken the brand. Large, easy to read signs and displays at the entrance of IKEA stores inform consumers how they are supposed to take measurements, select pieces of furniture and collect them.

The level of consumer participation varies across services. In service sectors such as airlines and fast-food restaurants, the level of consumer participation is low, as all that is required is the consumer's physical presence and the employees of the organisation perform the whole service. In sectors such as banking and insurance, consumers participate moderately and provide input to the service creation through providing information about their physical possessions. When consumers are highly involved in the service, for example participating in WeightWatchers, they need to be fully committed and actively participate.

Consumers can be regarded as productive resources and even as partial employees of the service organisation or 'co-producers', because they provide effort, time and other inputs for the performance of the service. They are also contributors to the quality and value of the service, thereby influencing their assessments about the service brand. Consumers who believe they have played their part well in contributing to the service tend to be more satisfied. The IKEA brand is built on the principle that consumers are willing to be involved in 'creating' the service, not just in consuming it. Since actively involved consumers feel that the responsibility is theirs when the service turns out to be unsatisfactory, they are particularly pleased when the service provider attempts to redress the problem.

To involve consumers in the service-delivery process, organisations can implement different strategies that are based on the following three factors:

- Defining the role of consumers;
- Recruiting, educating and rewarding consumers;
- Managing the consumer mix.

The organisation needs to determine the level of customer participation by defining the consumer's 'job'. Some strong brands, such as Federal Express and

MASTERCARD

Mastercard's 'World Masterclass' offers a loyalty scheme comprising points for purchases. Here, the tangible benefits of points is evident in the suggested additional luxury such as Preferred Seating or Room Upgrade which are gained by members

DHL, are built on low consumer involvement, as consumers rarely see the service provider's facilities and have only very brief phone contact with its employees. In these cases, as consumers are minimally involved in the service delivery process and their role is extremely limited, strong service brands can be developed through standardised offerings and precisely defined procedures. On the other hand, for service organisations like business schools and health clubs, there are higher levels of consumer participation and more tailored offerings can be developed.

Effective consumer participation may require that consumers go through a process similar to a new company employee – a process of recruitment, education and reward. In telephone banking, consumers are first recruited and then they receive formal training and information about the service. Only then will they be rewarded with easier access to financial services. Brands such as First Direct have been successful because they have effectively communicated the benefits consumers can gain from their participation. Service brands can be strengthened through an effective management of the mix of consumers who simultaneously experience the service. All major airlines, for example, are aware of the need to separate different segments.

However, a potential downside of increasing customer participation is beginning to receive more attention. In the same way that an employee's misbehaviour will damage the services brand, so will the misbehaviour of a consumer. Lovelock (2001) describes such consumers as 'jaycustomers', and recognises that the customer is not necessarily always right. Depending upon the type of behaviour being displayed, there are a number of potential remedies available. Reynolds and Harris (2006) found that front line employees create coping mechanisms at three stages of the service encounter: 'pre-incident' such as mental preparation and observing patrons; 'during the incident' such as emotional labour or eliciting support from other patrons; and 'post-incident' such as talking to colleagues or gaining revenge. Managers must always ensure that there are sufficient support systems for employees under stress, to avoid the potential for employee retaliation.

In observing Google, Jarvis (2009) offers positive advice to service managers: 'Your worst customer can become your best friend.' Jarvis (2009) explains that Web 2.0 has given a greater voice to the customer and companies can be hit by 'blogstorms'. Jarvis (2009) gives the example of 'Angry Jim', a customer who is disappointed with the service, sets up a website with his own angry blog and a timer which monitors the response time of the company. In addition, Angry Jim shares every correspondence with the company with his audience. He can create a YouTube video and if it is funny or controversial, it will spread. Jarvis (2009) advises that companies should see such customers as helpers and involve them in the resolution of their original problem. He advocates that the

CEO should share the problem and the dialogue with his customers — and thank Angry Jim for his help!

BUILDING SERVICE BRAND EQUITY

Brand equity represents the differential attributes underpinning a brand that give increased value to the firm's balance sheet. Berry (2000) proposes one of the few services-specific models for cultivating brand equity. The model (Fig. 6.3) is comprised of four main ways in which services companies can build strong brands, and he notes that the strongest brands tend to employ all of them.

Dare to be different — This should be a conscious effort to differentiate the brand from others. Berry uses the example of the Starbucks coffee chain, in that while it could fit more people into its shops, this would detract from the real offering: 'a respite and a social experience'. Today, Starbucks keeps in contact with customers through its own Facebook site, and an Apple iPhone app., which allows customers to find an outlet and lets Seattle customers pay for their coffee in advance.

Determine your own fame — Strong service brands should mean something important and represent a valuable offering to their target market. In Berry's words, they 'enhance the customers' experience by doing something that needs doing'. They perform this service better than competitors do and communicate the fact effectively to consumers.

Make an emotional connection — The strong services brand will evoke feelings of closeness, affection and trust in the consumer. In order to do this, its values must reflect the core values of the consumer, so that the consumer identifies with the brand. For example, in a hectic world, Twitter allows users to enjoy an 'always on' virtual presence. Through 140 character 'tweets', users can update friends and 'followers' can keep up to date with celebrities, politicians and businesses who respond to the simple question 'what are you doing?'.

Internalise the brand — Because of the centrality of service employees to the consumers' experience of the service, it is essential that employees internalise or 'live' the brand's values and ideas. If the brand has been internalised, its delivery will be more

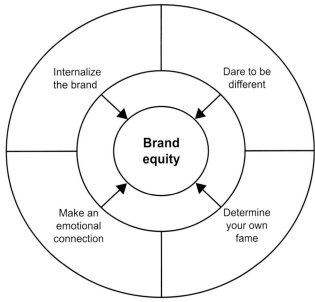

FIGURE 6.3

Building service brand equity (*from Berry. 2002*)

consistently in line with its values and therefore the values will more effectively be communicated to consumers.

CONCLUSIONS

The growing importance of the services sector has made firms aware that the creation and development of service brands represents a source of sustainable, competitive advantage. While all firms have a service component, services are predominantly intangible and their very nature creates challenges in creating strong service brands.

Despite similarities between the principles of branding for goods and services, the specific nature of services therefore requires tailored approaches and models. The case studies of the insurance and financial services sectors have illustrated some of the challenges marketers face when establishing service brands.

We have argued for a fine-tuning of the existing branding theories, as opposed to the creation of a whole new theory, and examined the distinctive differences between goods and services to help managers fine-tune the brand development process. Since every service is based on a series of performances, service brands run the risk of being perceived as commodities. To overcome this problem, service brands need to be made tangible to provide consumers with a favourable set of perceptions. An effective way to do this is to use the physical components associated with the service. These representations, which need to be compatible with the added-value message the brand aims to convey, can be used to inform staff and consumers about their expected roles and to differentiate the service brand.

The consumer's appreciation of service brands depends on a variety of factors such as the role played by the staff, the role consumers play, the interaction between consumers and the consumer's mood at the time of the service encounter. All employees, as they embody the organisation in consumers' eyes, can influence perceptions of the service brand. Front line employees are particularly critical as ambassadors of the service brand, as their performance during the service encounter brings the service brand message to life. Marketers therefore need to carefully consider their recruitment processes, the role staff are expected to play and their technical support to ensure they are able and motivated to deliver high-quality services. Managers may wish to consider whether employees who have a genuinely positive attitude and pleasant demeanour are preferable to those who may have stronger aptitude but poor interaction skills. Managers may also identify those employees who are champions or saboteurs and strive to harness the brand champions to ensure brand supporting behaviour. Given the ubiquity of sabotage, managers must be alert to cultural norms which encourage deviance and underperformance.

Consumers are co-creators of the service and can therefore contribute to the development of a service brand. However, marketers also need to be aware of the potential threat caused by consumer misbehaviour and the possible ways of deflecting this. In particular, the increasing use of online media for customer dialogue provides a forum for customers to become serious threats to the service brand and access a wide audience. Such customer complaints should be considered opportunities for managers to address and use to improve the quality of the service. Finally, service brands can create equity by differentiation, building the brand internally, making emotional connections and determining their own fame by doing something important.

MARKETING ACTION CHECKLIST

To better help determine brand marketing activities, it is recommended that the following exercises are undertaken:

1. For each of your service brands, list the tangible elements that are used to convey a consistent message of the brand's added value.
   - How clearly do these elements communicate the brand benefits to customers?
   - What other functions, for example, socialising or facilitating, do they have?
     With your marketing team, identify opportunities to use these elements to differentiate your brand from competitors.
2. From the list of tangible elements in the previous exercise, assess whether these elements encourage only the target consumer segment to purchase the service and whether they discourage other segments. Is this discouragement intended because the other segments are incompatible with the primary target segment? Does the target segment need this 'protection' from other segments?
3. If you are planning to launch a new service:
   - How well does it fit with your existing products and consumer base?
   - Will the existing consumers relate to the new brand?
   - Will they be confused by contrasting messages?
   - Does the new brand need a product-specific name or could it build on the current corporate identity?
4. For any of your service brands, how easy it is for consumers to evaluate the quality of the service delivery? Where do they gather information about brand benefits and will these satisfy their needs? Do consumers perceive a high risk associated with purchasing the service? What strategies could be used to minimise this perceived risk?
5. Are the benefits of your brands delivered consistently by all employees? What are you doing to ensure consistency between your staff members and

over time? Are the principles of service recovery understood by staff and are they implemented when necessary? What would be the implications for your brands if your company decided to take a more empowerment-orientated approach?

6. Are your consumers informed about the role they are expected to play during the service delivery? Does your brand require a low, medium or high degree of consumer involvement? What guidelines do consumers receive prior to or during the delivery of the service to be aware of their expected behaviour? What, if any, guidelines are in place to deal with instances of consumer misbehaviour?

7. As the service offered by your brand cannot be stored, what strategies have been implemented to smooth fluctuating demand and supply? When waiting time cannot be avoided, how can consumers still be satisfied?

8. Examine the role of your staff during the production and delivery of the service brand, according to factors such as:
   - reliability;
   - responsiveness;
   - assurance;
   - empathy and appearance.

   Do these factors offer further opportunities to enhance consumer perceptions of your brand?

9. The long-term success of the service brand depends on the ability and motivation of your staff.
   - Is the recruitment process geared to provide the organisation with employees who share the corporate vision and the message that the brand aims to convey?
   - Does your organisation provide regular training for employees to ensure a consistent delivery of the service over time?
   - Are staff trained to tailor the service offering in response to the differing moods and displayed emotions of consumers?
   - What technical support do employees receive to deliver the brand benefits?
   - How does your company strive to retain good employees?

10. What role do consumers play in the delivery of your service brand? Do they receive sufficient information to understand this role and perform it effectively? How can you enhance their participation to make the brand stronger?

STUDENT BASED ENQUIRY

- Choose a service brand that you use regularly. Apply Vargo and Lusch's (2004) eight foundational premises to identify the brand's strengths and weaknesses. Based on the eight premises, how might the service improve its brand?

- Select a well-known service brand you are familiar with. Identify how each of the characteristics of the service presents a challenge for brand building. For each challenge you identify, outline three ways service managers might overcome this to build a strong brand.
- Recall a service experience you had in the past month. Using the five components of SERVQUAL discussed in the section 'Consistent service brands through staff', describe how the service employees you encountered influenced your perception of the service.
- Select an online service brand you are familiar with. Identify and discuss three ways the brand could increase the tangibility of its service.

References

Bateson, J. E. G. (1979). Why we need service marketing. In O. C. Ferrell, S. W. Brown, & C. W. Lamb, Jr. (Eds.), *Conceptual and Theoretical Developments in Marketing* (pp. 131–146). Chicago: American Marketing.

Berry, L. L. (2000). Cultivating service brand equity. *Journal of the Academy of Marketing Science, 28* (1), 128–137.

Brady, M. K., Bourdeau, B. L., & Heskel, J. (2005). The importance of brand cues in intangible service industries: an application to investment services. *Journal of Services Marketing, 19*(6), 401–410.

Brookes, J. (1996). Awaking the 'sleeping dinosaur': a case study of the Liverpool Victoria Friendly Society Ltd. *The Journal of Brand Management, 3*(5), 306–312.

Colton, S., & Oliveira, P. (2009). *Banking on it: the role of the corporate brand in rebuilding trust.* Available at: http://www.interbrand.com. Accessed 29 June 2009.

de Chernatony, L., & Cottam, S. (2006). Internal brand factors driving successful financial services brands. *European Journal of Marketing, 40*(5/6), 611–633.

de Chernatony, L., & Cottam., S. (2006). Why are all financial services brands not great? *Journal of Product and Brand Management, 15*(2), 88–97.

de Chernatony, L., Drury, S., & Segal-Horn, S. (2004). Identifying and sustaining services brands' values. *Journal of Marketing Communications, 10,* 73–93.

de Chernatony, L., et al. (2003). Building a services brand: stages, people and orientations. *The Services Industries Journal, 23*(3), 1–21.

Design Council. (2008). *HMV Group: design to overcome a downturn case study.* Available at: http://www. designcouncil.org.uk/Case-Studies/All-Case-Studies/HMV-Group. Accessed 2 December 2009.

Frei, F. X. (April 2008). The four things a service business must get right. *Harvard Business Review,* 70–80.

Gill, C. (2007). Restoring consumer confidence in financial services. *International Journal of Bank Marketing, 26*(2), 148–152.

Harris, L. C., & Ogbonna, E. (2006). Service sabotage: a study of antecedents and consequences. *Journal of the Academy of Marketing Science, 34*(4), 543–558.

Hennig-Thurau, T., Groth, M., Paul, M., & Gremler, D. D. (2006). Are all smiles created equal? How emotional contagion and emotional labour affect service relationships. *Journal of Marketing, 70,* 58–73.

Ind, N. (2004). *Living the Brand: How to Transform Every Member of Your Organization into a Brand Champion*. London: Kogan Page.

Jarvis, J. (2009). *What Would Google Do?* New York: Harper Collins.

Levy, M. (1996). Current accounts and baked beans: translating FMCG marketing principles to the financial sector. *The Journal of Brand Management, 4*(2), 95–99.

Lovelock, C. H. (2001). *Services Marketing*. Upper Saddle River: Prentice Hall International.

Mattila, A. S., & Enz, C. A. (2002). The role of emotions in service encounters. *Journal of Service Research, 4*(4), 268–277.

Parasuraman, A., Zeithaml, V. A., & Berry, L. L. (1988). SERVQUAL: A multiple-item scale for measuring customer perceptions of service quality. *Journal of Retailing, 64*(1), 12–40.

Reynolds, K. L., & Harris, L. C. (2006). Deviant customer behaviour, an exploration of frontline employee tactics. *Journal of Marketing Theory and Practice, 14*(2), 95–111.

Shostack, G. L. (1977). Breaking free from product marketing. *Journal of Marketing, 41*(4), 73–80.

Srinivasan, S. S., & Till, B. D. (2002). Evaluation of search, experience and credence attributes: role of brand name and product trial. *Journal of Product and Brand Management, 11*(7), 417–431.

Vargo, S. L., & Lusch, R. F. (January 2004). Evolving to a new dominant logic for marketing. *Journal of Marketing, 68*, 1–17.

Wallace, E., & de Chernatony, L. (2007). Exploring managers' views about brand saboteurs. *Journal of Marketing Management, 13*(1–2), 91–106.

Wallace, E., & de Chernatony, L. (2008). Classifying, identifying and managing the service brand saboteur. *The Service Industries Journal, 28*(1–2), 151–166.

Zeithaml, V. A., Parasuraman, A., & Berry, L. L. (1985). Problems and strategies in services marketing. *Journal of Marketing, 49*(2), 33–46.

Further Reading

Allen, D. (2009). *Starbucks app. uses your iPhone to pay for lattes*. Available at: http://gizmodo.com/ 5366650/. Accessed 3rd December 2009.

Boyd, W. L., et al. (1994). Customer preferences for financial services: an analysis. *International Journal of Bank Marketing, 12*(1), 9–15.

Brodie, R. J., Whittome, J. R. M., & Brush, G. J. (2008). Investigating the service brand: a customer value perspective. *Journal of Business Research, 62*(3), 345–355.

Camp, L. (1996). Latest thinking on the optimisation of brand use in financial services marketing. *The Journal of Brand Management, 3*(4), 241–247.

CNN.com. (2006). *Plane clothes investigation*. Available at: http://edition.cnn.com/2006/TRAVEL/ 01/19/fashion.uniform/index.html. Accessed 23 November 2009.

de Chernatony, L., & Dall'Olmo Riley, F. (1996). Experts' views about defining service brands and the principles of service branding. *Paper presented at 9th UK Services Marketing Workshop, University of Sterling*.

Denby-Jones, S. (1995). Retail banking: Mind the gap. *The Banker, Vol. 145/828*, 66–67.

Free, C. (1996). Building a financial brand you can bank on. *The Journal of Brand Management, 4* (1), 29–34.

Google (2009). *Google code of conduct*. Available at: http://investor.google.com/conduct.html. Accessed 3 December 2009.

IMB (2009). *Integrated service management*. Available at. www.ibm.com/ibm/servicemanagement/ index.html. Accessed 1 December 2009.

Inside Facebook. (2009). *Social network downtime study shows Facebook's reliability improved as 2008 progressed*. Available at: www.insidefacebook.com/2009/02/23/social-network-downtime-study-shows-facebooks-reliability-improved-as-2008-progressed. Accessed 3 December 2009.

Lovelock, C. H. (1996). *Service Markets*. Upper Saddle River: Prentice Hall International.

Lovelock, C. H., Vandermerwe, S., & Lewis, B. (1996). *Services Marketing: A European Perspective*. London: Prentice Hall Europe.

Manning-Schaffel, V. (2006). *ING safe*. Available at: http://www.brandchannel.com/features_webwatch.asp?ww_id=270. Accessed 10 December 2009.

New York Times. (2009). *Twitter*. Available at. http://topics.nytimes.com/top/news/business/companies/twitter/index.html?scp=1-spot&sq=twitter&st=cse. Accessed 23 November 2009.

Office for National Statistics. (2002). *The UK Services Sector*. London: Office for National Statistics.

Onkvisit, S., & Shaw, J. J. (Jan–Feb 1989). Service marketing: image, branding and competition. *Business Horizons*. 13–18.

Watters, R., & Wright, D. (1994). Why has branding failed in the UK insurance industry?. In *ESOMAR Seminar: Banking and Insurance* (pp. 137–151) Amsterdam: ESOMAR.

Zeithaml, V. A., & Bitner, M. J. (2003). *Services marketing*. Boston: McGraw-Hill.

www.ashford.ie/index.php. Accessed 3 December 2009.

www.virgin.com/about-us. Accessed 3 December 2009.

www.virgingalactic.com. Accessed 3 December 2009.

Retailer Issues in Branding

OBJECTIVES

After reading this chapter you will be able to:

- Understand the importance of the retail store as a brand and outline the critical success factors of leading retail brands.
- Examine the challenges and opportunities store brands present for retailers and manufacturers.
- Explore the strategic options available to retailers who wish to optimise store brand market share.
- Describe the stages involved in retailer selection of store brand suppliers.
- Understand the power dynamic between national brands, store brands, manufacturers and retailers.
- Outline the role of category management for retailers who wish to provide efficient customer response.

SUMMARY

This chapter examines the importance of retail brands and explores the role of brands in retailers' product portfolios. First, we focus on retailers as brands or the store as a brand. Selfridges and Harrods are retailer brands that attract and retain customers because of their store brand image and product range. As these brands have a specific and strong image, they attract manufacturer brands that benefit from their customer base. The second issue we examine in this chapter is the marketing of store brands. Manufacturer brands, or national brands, often carry high levels of customer brand loyalty but they frequently face the challenge of being positioned at a higher price than their store brand competitors. In the first section of this chapter, we discuss the retail brand and examine issues including store loyalty and brand image. In the second section, we take a detailed look at store brands and examine critical issues including shelf space allocation, customer loyalty and consumer response to store brands in specific product categories.

Creating Powerful Brands. DOI: 10.1016/B978-1-85617-849-5.10007-8

THE STORE AS A BRAND

Some of the strongest corporate brands in the world are retail brands (Martenson, 2007). For example, in 2009 the Swedish clothing retailer H&M was ranked 21 in the Interbrand Top 100 brands list, with a brand vale of $15,433 million. IKEA was ranked 28 with a brand value of $12,004 million. H&M differentiate though real time adaptability in products or 'nano fashion' and also through price, corporate social responsibility and designer collaborations. IKEA's popularity reflects increased consumer interest in spending time at home; and the retailer's website attracts over 450 million visits every day (http://www.interbrand.com/best_global_brands.aspx). When a retail organisation has a strong corporate brand, it is difficult to imitate it. Retail brands are different to other brands as their brand portfolio comprises houses of brands, which can include manufacturer brands, store brands and the image of the store itself (Martenson, 2007). For retail brands to be successful, customers need to perceive value, which occurs when needs, wants or desires are met relative to costs incurred.

In a study of 1000 shoppers aged between 30–50 years, Martenson (2007), found that customers rated stores based upon (in order of importance):

- the quality of relations with its customers;
- the neatness of the store and the shopping experience;
- the perception that the store understood its customers and offered a good assortment;
- the availability of low prices and value for money.

Superquinn, an Irish grocery retailer, is an example of a value-added retail brand with an emphasis on high quality in customer relations. Its founder, Senator Fergal Quinn, opened his first supermarket in 1960. He initiated a 'Boomerang Principle' ethos, i.e. 'keep the customer coming back'. To achieve customer retention, Superquinn employs a number of strategies. For example, the company ensures that managers spend time on shop floors listening to customers. Their 'sundown' policy means that all customer complaints receive a same-day response. Floor staff are encouraged to address regular customers by name. Customers also return to Superquinn because of their product quality. The company spends €2.5 million per annum on food safety and has developed traceability initiatives. It also provides wine specialists in-store to assist customers with product selection. The result of these initiatives has resulted in Superquinn opening 23 stores and employing over 3000 people in Ireland.

HOUSE OF FRASER

Retail brands often build their store image though the assortment of products offered. In this advert, House of Fraser promote Ray-Ban sunglasses which in turn lend their stylish, youthful image to the image of the House of Fraser brand

RETAINING CUSTOMER LOYALTY AND BUILDING ENGAGEMENT

While many retailers formerly engaged in price-based strategies, many leading brands are recognising the power of customer relationship management (CRM) in creating customer value and building lifetime brand loyalty. Loyalty schemes are often used to reward and retain customers. Buttle (2008, 267) defines a loyalty programme as *'a scheme that offers delayed or immediate incremental rewards to customers for their cumulative patronage'*. He dates the evolution of loyalty schemes back to 1844, when the Rochdale Pioneers developed a retailing cooperative that distributed surpluses to members as dividends, according to their spend. Those schemes were followed by S&H Pink Stamps and Green Shield Stamps in the 1950s and 1960s. These schemes provided stamps that customers collected in exchange for catalogue gifts. Buttle (2008) attributes many of the current loyalty card scheme structures to American Airlines who started their advantage programme in 1981. Similar scheme formats have proven popular with retailers including Tesco and Boots the Chemist.

Gilligan (2010) reviewed Tesco's CRM activities. The Tesco loyalty card was introduced in 1994, partly in response to similar international initiatives and also as a response to research findings. Analysis of customer data showed that in many stores, the top 100 customers were worth as much as the bottom 4000; and the top 5% of customers accounted for 20% of sales. The Tesco Clubcard offers points for each £ or € spent with the company. Points are redeemed as vouchers against future purchases or as reward tokens for partners including hotels, museums and restaurants. Gilligan (2010) emphasises the key benefit of the Clubcard to Tesco as the data provided by millions of weekly customer transactions. Data mining allows Tesco to assess customer purchases and trends or segment customers according to needs. Some uses of the Clubcard data are:

- managing the product range;
- developing new products based on customer trends;
- managing inventory;
- creating promotions with rewards for loyalty;
- enhancing inventory management;
- measuring promotional and media effectiveness.

Another loyalty scheme is that of Boots. Ninety percent of the UK's population visits a Boots store at least once each year (Buttle, 2008). The Boots' 'Advantage Card' was introduced in 1999 and is targeted primarily at the 83% of Boots' customers who are female, aged 20–45 years, who purchase non-essential, 'indulgence' items during 55% of their visits to the store. In 2008, Boots' 'Advantage Card' had more than 13 million cardholders (Buttle, 2008).

BROWN THOMAS

Brown Thomas launched a loyalty card 'Brown Thomas Black' which was designed to reflect the exclusive image of the store. This image is also reflected in the headline 'loyalty has its rewards & privileges'

As brand-building devices, loyalty schemes offer advantages to consumers. Buttle (2008) outlines that customers benefit from credit acquisition and redemption. When a customer acquires points, this may offer psychological value such as a sense of belonging and being valued, as well as anticipation of rewards. In addition, customers are motivated to acquire points for the material rewards at redemption. However, Buttle (2008) cautions that the consumer may be loyal to the scheme and not to the store. If a customer is already loyal to the store, the card will not change their patronage behaviour. However, if the customer was formerly a light buyer, they may increase their behavioural loyalty to benefit from buying from additional categories.

THE INTERNATIONALISATION OF RETAIL BRANDS

Strong brands are increasingly looking for international opportunities (Pappu and Quester, 2008). Coe and Hess (2005) observed the acceleration of internationalisation over the last 20 years, particularly among grocery retailers. They identify 'elite transnational retailers' including Wal-Mart, Carrefour, Metro, Aldi, Lidl and Schwarz, Intermarche and Tesco. These retailers have attained elite status by extending their core markets of North America and Western Europe to achieve dominant positions in markets of East Asia, Eastern Europe and South America. These retailers have also increased the scale and scope of their activities overseas. For example, Uncles (2010) explored retail change in China, where retail brands thrive due to converging retail structures and increased consumer demand. He cites the example of Tesco's growth. In 2004 Tesco established a 50:50 partnership with Ting Hsin International Group. The company has since increased its stake to 90% and in 2008 owned 58 hypermarkets in 22 Chinese cities. On the other side of the world, D'Andrea (2010) highlights Latin America as an opportunity for international expansion, due to trends such as increasing urbanisation and an increase in mean per capita income. These trends have facilitated retailers such as Carrefour to expand and increase store size. In 2010, Carrefour had 183 stores in Brazil and 152 in Argentina.

A strong brand offers leverage for retailers to develop an international presence (Coe and Hess, 2005). Such internationalisation also has implications for manufacturer brands. For example, the centralisation of procurement at an international level means that transnational retailers source suppliers at a global level, rather than the historical approach of importing brands from the home suppliers. In addition, expansion into overseas markets can pave the way for imports of local produce into the home market. When procurement is centralised, this can be an impetus for the supply base, favouring those suppliers who can provide the required quality and volume of goods to meet international market needs. We will explore the power dynamic between manufacturer and retailer in more detail later in this chapter.

Sometimes, firms look to expand overseas when domestic opportunities are saturated or when overseas markets become more attractive. Topshop, the UK clothing retailer, is renowned as a British high street fashion destination. It has subsequently exported its brand to over 100 overseas stores, including SoHo, New York. As Topshop has successfully shown, when retail brands export, they often stretch their brand equity into new territories under the same brand name, as exemplified by IKEA. In other instances, the parent brand may trade under a local brand name to gain market access or to retain a local image. For example, the Asda supermarket chain is the UK arm of Wal-Mart. Occasionally, internationalisation of retail brands incorporates co-branding to enhance the equity of the retail brand. An example of this is the collaboration between Karl Lagerfeld and H&M, the Swedish clothing retailer. When the designer, who creates haute couture and ready-to-wear collections for Chanel, launched a collection designed for H&M, more than 1000 people entered New York's Fifth Avenue store in the first hour to buy his pieces.

The internationalisation of retail brands often requires managers to think globally, yet act locally. Humby (2010) emphasises the importance of knowing the local customer and cites the example of adaptive strategies employed by Tesco in Asia. For example, Asian customers are more likely to visit the supermarket on a scooter than by car. Therefore, Tesco had to rethink their supermarket car parks. Understanding customers has also led to small changes in Tesco's loyalty programmes. For example, the South Korean version of the Tesco Clubcard is known as the 'Familycard'; and the familiar slogan *Every Little Helps* is translated into *We put our heart to serve you* in Thailand. In addition, Asian customers are more likely to receive Tesco promotions through SMS, compared with the mail-outs received by UK customers. Globalisation succeeds when brands ensure that their activities increase relevance for consumers at a local level.

STORE BRANDS

Store brands are products and services that bear the retailer's brand name. They play a critical role contributing to the value of the store as a brand. Store brands allow retailers to add value and they have evolved from 'repackaged labelling' options into 'a true product brand' characterised by product development and added value ranges (Johansson and Burt, 2004). Soberman and Parker (2006) describe the role of 'store', 'house' or 'private' brands as a form of 'upstream integration' by the retailer. These products are often of equal quality to manufacturer brands, but generally lower priced. Store brands offer advantages to the retailer. These include the attraction of price-sensitive customers, a competitive edge to negotiate better deals from manufacturers on wholesale pricing and enhanced store image (Rondán

Table 7.1 A Sample of UK and Irish Store Brands

Consumer Need	Tesco	M&S	Sainsburys	Superquinn
Indulgence	Finest	M&S Range	Taste the Difference	SQ Superior Quality
Low Fat	Healthy Living/Light Choices	Simply Fuller Longer	Be Good to Yourself	Eat Better
Organic	Organic	Organic	SO Organic	Organic
Low Price	Value		Basics/Economy	Euroshopper

Cataluña et al., 2006). Earlier we noted that a store image can be created in part by its store brands. This is a two-way relationship. For example, in the clothing sector, a study of Finnish consumers found that store image reduced perceived risk and therefore enhanced the quality perceptions of store brands (Liljander et al., 2009).

In the grocery sector, Nielsen (2009) estimates that store brands occupy between 25% and 33% of value sales in the major European markets — UK, Germany, France, Spain and Italy. In the UK, store brands account for almost half of all sales. Nielsen explains that store brands were successful after launch due to their 'good, better, best' tiers; for example, Tesco's Value and Tesco Finest ranges attract different market segments. Table 7.1 presents a sample of store brands offered to UK and Irish consumers, which illustrates the range of grocery products offered and the consumer needs met by each range.

This table shows that store brands can position themselves to target specific customer segments or to address specific customer needs. For example, appealing to environmentally conscious shoppers, each of these retailers offers an organic line within their own labels. Aware of consumer interest in healthy eating, these retailers have a range of own labels whose ingredients are deliberately focused around healthy eating. Many of the retail brands also build their marketing communications activities around meeting the needs of these customer segments. For example, Tesco offer a "Tesco Diets" online club, (www.tescodiets.com) where customers are provided with diet plans, with an option of home delivery of healthy food options. Sainsburys have a "Little Ones" baby and toddler club (www.sainsburys.co.uk/littleones/), where parents get advice on recipes and meal planning, as well as parenting tips.

SUCCESS FACTORS OF STORE BRANDS — CONSUMER BEHAVIOUR AND ECONOMIC CYCLES

Steenkamp et al. (2004) examined the success of store brands in Europe. They found that retailers were expanding their brand ranges and were developing

WAITROSE

A challenge for retailers is to offer competitive prices while retaining a quality image. In this advert for Waitrose, the company are emphasising cost savings for Easter, while presenting the slogan 'everyone deserves quality food'. (Reproduced by kind permission of Waitrose)

unique positioning for their stores as brands, enhancing customer loyalty. They identified factors within Fast Moving Consumer Goods (FMCG), which were contributing to the success of own labels:

- manufacturer brands were advertising less;
- manufacturer brands were less involved in innovations;
- there was little perceived difference between manufacturer and private label brands on quality, value or confidence;
- there were similarities in packaging between manufacturer and private label brands.

Store brands also benefit from increased marketing support as retailers give them prominent shelf space (Rondán Cataluña et al., 2006). Lamey et al. (2007) caution manufacturer brands against cutting brand support and product development during a recession; and highlight these two activities as a means to mitigate against the effects of economic contraction on manufacturer brands. Another issue challenging manufacturer brands is the growth of discounters (Steenkamp et al., 2004). In the retail sector, store brands are facing significant challenges from the 'hard discount' retailers, such as Aldi and Lidl.

Consumer needs change according to economic circumstances. Luijten and Nagtzaam (2009) mapped the behaviour of Dutch shoppers against market conditions. In 2002, consumers chose stores due to product assortment and service levels. From 2003, consumers focused on price and displayed less loyalty. When the economy improved in 2006 and 2007, customers were again less focused on price and more focused on service and accessibility. By 2009, consumers became focused on price. Lamey et al. (2007) also observed consumer response to economic conditions. Again, they identified that consumers switch to store brands when an economy is suffering. However, they also noted that consumer switching is asymmetric. In other words, the switch to private label during a recession is faster than the switch back to manufacturer brands after a recession ends. They also found that consumers did not drop private label brands once the economy picked up, which resulted in long-term 'scars' on manufacturer brand performance. A challenge for the manufacturer brand therefore is to offer relevant customer value to retain customer loyalty and avoid substitution during periods of poor economic growth. However manufacturer brands also face challenges from retailer support for own label, the rise of generics and the growing power of the retailer as a buyer. We examine these issues next.

RETAILER SUPPORT FOR OWN LABEL

Retailers' marketing activities are often designed to enhance store brand market share, which is the volume sales of store brands, relative to sales of all products in the product category. Rubio and Yagüe (2009) found four factors that influence store brand market share. The factors identified are:

1. *Competitive Strategy* – where there is a large gap between manufacturer brands and store brands, store brand market share is increased; where manufacturer brands advertise heavily, store brand market share decreases.
2. *Market Structure* – where retailers are concentrated, store brands have greater share due to enhanced negotiating power and communication economies.
3. *Economic Financial Results* – where retail profit margin or stock turnover for the product category is increased, store brand market share is larger.
4. *Basic Market Conditions* – where the product category is perceived to be high risk, store brand share is low. However, where demand is elastic and consumers are responsive to price, store brand market share is high.

In addition, Gómez and Okazaki (2009) identified factors that influenced store brand shelf space. They explained that store brand share was influenced by store differentiation, which in turn is created by the assortment of products offered, the number of in-store promotions, the store image and the store format. If manufacturer brands adopt strategies of differentiation and innovation, this improves customers' perceptions of these brands and reduces store brand shelf space. Finally, market structure, in terms of leading manufacturer brand rivalry and manufacturer brand concentration, reduces store brand space.

If retailers wish to optimise their store brand share, Rubio and Yagüe (2009) offer the following recommendations:

- invest in store brands of categories with high stock turnovers to compensate for stock and marketing costs incurred;
- apply wider differentials between manufacturer and store brands as an incentive to purchase store brands;
- take advantage of negotiating power to achieve lower list prices and better manufacturing agreements for store brands;
- take advantage over manufacturer brands on price where demand is elastic.

To some extent, the options available to the retailer are dependent on the positioning of the store brand. Chan Choi and Coughlan (2006) identified two options for positioning: quality differentiation or feature differentiation. Quality differentiation is the extent to which the differentiated characteristic is one which all customers value at the highest level when price is constant; for example, a product's formulation may create a unique taste, which is valued by all customers. On the other hand, feature differentiation is the level of variance in features such as size, or packaging; for example, a soup manufacturer may offer single serving cartons that are suitable for microwaving and would appeal to customers who use the product at work. Chan Choi and Coughlan (2006) advise that when national brands are differentiated, a high quality private label should be positioned close to a stronger national brand and a low quality private label should be positioned closer to a weaker national brand. If national brands are undifferentiated, the private label should differentiate itself from national brands, for example, by offering unique flavours or packaging.

The commitment behind a good value own label is apparent, both in the tight **product specifications** given to manufacturers and in the considerable effort invested in identifying and evaluating potential own label producers. Some retailers have significant R&D departments that undertake preliminary product development research prior to briefing potential producers. For example, at Marks & Spencer there are over 200 people working in its food R&D laboratories and in 2007, the brand announced Plan A, a €200 million 'eco-plan' that impacts on operations such as the proportion of foods that are 'healthy eating' products and the use of Fairtrade produce.

Grocery retailers have moved on from 'copy-cat' branding to developing a whole range of own label brands that are carefully targeted to meet the needs of different segments. Retailers were criticised for too often looking back and developing their own label range with the experience of manufacturers new product development programmes. Now they look forward, understanding their consumers' current needs, anticipating new needs and then developing innovative own labels.

Retailers have stringent standards about own label **pack designs**, insisting that they not only communicate the characteristics of the individual line but also reinforce their store's image. In the 1970s, when retailers fought on a low-price proposition, their own label packaging was functional and stark. Graphics did little more than identify the name of the store and convey an impression of low cost. Today, retailers recognise the 'silent persuader' value of own label packaging.

Quite clearly, retailers are using well-tried branding techniques to develop quality propositions for their own labels. They have devised well-conceived personalities for their own label ranges that support the image of their stores. They are using innovative advertising campaigns that memorably reinforce their positioning. For example, slogans such as Tesco's '*Every Little Helps*' and M&S '*Its not just food, its M&S food*' are synonymous with the retail brands and their brand images. In 2009, five retail brands were visible among Marketing magazine's top 100 advertisers in the UK. Tesco spent over £88 million on advertising, placing it at number 6. Asda, spending £76 million, was the ninth largest advertiser and Sainsburys, at number 12, spent £62 million. Marks & Spencer's spend of £51 million placed them at number 18, ahead of Boots at number 20 with a spend of £47 million. With the exception of Marks & Spencer, each brand had significant year on year increases in expenditure, with Asda increasing spend by 49%, predominantly in print media.

THE ARRIVAL AND DEMISE OF GENERICS

Generics are products distinguishable by their basic and plain packaging. Primary emphasis is given to the contents, rather than to any distinguishing retail chain name.

Carrefour in France pioneered this route in 1976 with its Produits Libres and was quickly followed by Promede's Produits Blancs, Paridoc's Produits Familiaux and Euromarche's Produits Orange. In Germany, Carrefour, Deutsche Super-markt and the Co-op encountered problems of poor quality perceptions, due to the low prices. In Switzerland, where there is a significant own label presence and consumers are primarily concerned about quality, generics had little success.

In the UK several retailers experimented with generics. Some were very committed to the concept, while others, notably Tesco, were less certain, particularly as this conflicted with their move to shift their store's positioning upmarket. The problem was that a cheaper range was being introduced with a quality level inferior to that of own labels. Consumers had previously been experiencing improvements in quality and service and were confused by the return of the 'pile it high, sell it cheap' era. Furthermore, what had happened was that retailers had not launched a unique, stand-alone range. Instead, they had created a secondary own label range, perceived by consumers as an alternative to the current own label range.

The true generic concept had not been implemented. By definition, generics have no promotional backing nor any distinctive labeling. Yet their launch was heralded by a considerable promotional spend. A significant effort was devoted to pack designs which, while less sophisticated than conventional own labels, still clearly linked each range with a specific store. In particular, Tesco's stark white packs with blue print resulted in a range that was very prominent on the shelves and was quickly recognised and associated with Tesco.

As consumers began to think of generics as an extension of own labels, they switched from the higher margin own labels to the lower margin generics. This was contrary to retailers' hopes that generics would take sales from manufac-turers' brands. Retailers saw generics weakening their overall profitability and tarnishing the quality image that they were striving for. They had misjudged consumers' needs. Low prices are no longer the prime reason driving choice. Instead consumers, as experienced buyers, were seeking value for money – rather than low prices – along with a cluster of added values. During the mid-1990s generics or, more correctly, budget own labels, reappeared in stores such as Sainsbury, Tesco and Safeway. To some extent this may have been a response to the move of aggressive marketers of discount stores, such as Aldi and Lidl.

THE INCREASINGLY POWERFUL RETAILER

There has been a notable swing in the balance of power from manufacturer to retailer, albeit as we explore later in this book, the advent of the internet is swinging the balance in favour of consumers. The early 1960s saw a boom in retailing, due to such issues as relaxation of building controls, the success of self-service and more professional management, that identified smaller retailers ripe for acquisition. The abolition of resale price maintenance in 1964 gave a further boost to retailers, who were freed from the pricing stipulations of brand suppliers. As evidence of concentrated retailer power, A. C. Nielsen's data shows that in 1971 multiple retailers accounted for 44% of grocery sales, yet by 2009 the top four — Tesco, Asda, Sainsburys and Morrissons accounted for 69% of the grocery spend.

Retailers became more efficient through centralised buying and centralised warehousing. They shut their smaller stores and opened a smaller number of stores, all of much larger selling areas. For example, according to A. C. Nielsen the average store size of a Sainsbury's store in 1992 was 24,040 sq. ft. In 2009, 34 of its stores were greater than 55,000 sq. ft. Part of their increased profitability was passed on to consumers in the form of cheaper prices, further enhancing their attractiveness to consumers.

The reasons for the increasing concentration of multiple retailers can be appreciated from Fig. 7.1. As certain groups became larger, many manufacturers found it difficult not to acquiesce to retailers' demands and they passed on more favourable volume discounts. They rationalised these above average discounts on the basis that they were justified by very large orders. Retailers used some of these discounts to fund their own label programmes and to present their range as being better value than those of independent or cooperative retailers. More consumers were attracted by this proposition, contributing to the profitability of multiples. Ever aware of the future, the multiples invested in better stores and further increased their share, forcing out less profitable smaller retailers.

Manufacturers were aware of this shift in power to the retailers but, surprisingly, took little action. Their response lacked any long-term strategic thinking. They failed to appreciate how retailers were becoming more innovative, more consumer driven, more concerned about developing strong images for their stores and increasingly committed to growing their own labels. Instead of communicating their brands' added values to consumers, they cut back on brand advertising in favour of buying shelf space.

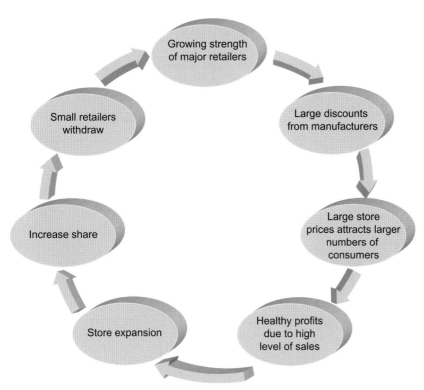

FIGURE 7.1
The wheel of increasing multiple retailer dominance

This shift in the balance of power resulted in retailers no longer being passive conduits for branded goods. Instead, they became highly involved instigators and coordinators of marketing activity. As one retailer said:

> We now see ourselves as the customer's manufacturing agent, rather than the manufacturer's selling agent.

THE RESPONSES OF WEAK AND STRONG MANUFACTURERS

Weaker brand manufacturers, particularly those lacking a long-term planning horizon, were unable to find a convincing argument to counter retailers' demands for extra discounts. They were worried about being de-listed and saw no other alternative but to agree to disproportionately large discounts. Many erroneously viewed this as part of their promotional budget and failed to appreciate the implication of biasing their promotion budget to the trade at the expense of consumers. Retailers' investment in own labels brought them up to the standard of manufacturers' brands. With increasing investment in own labels and less support behind manufacturers' brands, consumers began to perceive less differences between brands and own labels and choice began to be influenced more by availability, price and point of sale displays. As retailers had more control over these influencing factors, weaker brands lost market share and their profitability fell.

Weak manufacturers' brands were not generating sufficient returns to fund either maintenance programmes or investments in new products. At the next negotiating round with retailers, it was made clear that their sales were deteriorating and again they were forced to buy shelf space through even larger discounts. In the vicious circle shown in Fig. 7.2, they were soon on the spiral of rapid decline.

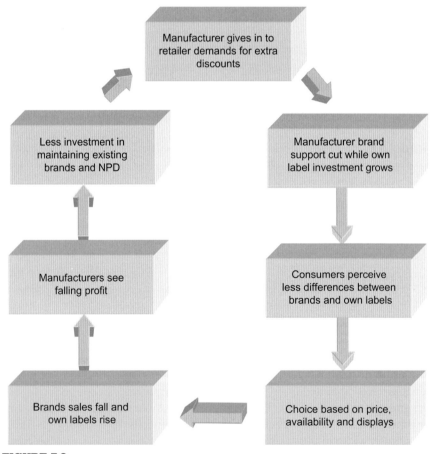

FIGURE 7.2

The weakening manufacturer's brand

From the vicious circle of deteriorating brand position, it can be appreciated that own labels are particularly strong in markets where:

- there is excess manufacturing capacity;
- products are perceived as commodities — inexpensive and low-risk purchases;
- products can be easily compared by consumers;
- low levels of manufacturer investment are common and the production processes employ low technology;
- there are high price gaps in the market and retailers have the resources to invest in high-quality, own label development;
- variability in quality is low and distribution is well developed;
- the credibility of a branded product is low because of frequent and deep price promotions as opposed to the increasing credibility of own labels;
- branded products are offered in few varieties and with rare innovations, enabling the own label producer to offer a clear alternative.

By contrast, strong manufacturers such as Unilever, Heinz and Nestlé, realised that the future of strong brands lay in a commitment to maintaining unique added values and communicating these to consumers. They 'bit the bullet', realising that to succeed they would have to support the trade, but not at the expense of the consumer. Instead, they invested both in production facilities for their current brands and in new brand development work. With strong manufacturers communicating their brands' values to consumers, these were recognised and choice in these product fields became more strongly influenced by quality and perceptions of brand personality. Retailers recognised these manufacturers' commitment behind their brands and wanted to stock them. Distribution increased through the right sorts of retailers, enabling brand sales and profits to grow. Healthy returns enabled further brand investment and, as Fig. 7.3 shows, strong brands thrived.

The confectionery market in the UK is a good example of a sector where strong manufacturers' brands dominate. The major players Mars, Nestlé and Cadbury are continually launching new brands and heavily advertising their presence. Interestingly, the power of multiple retailers is also dissipated by virtue of notable sales going through other channels, e.g. confectioners, tobacconists and news agents, along with vending machines and garage forecourts. The same circumstances apply to Coca-Cola and Pepsi-Cola, who are less dependent on the multiple retailers. However, these brands often engage in cooperative marketing with retailers — for example, Karray and Zaccour (2006) found that in cases where a strong manufacturer brand competes with the retailer private label, it can be profitable for manufacturer brands if they engage in cooperative advertising.

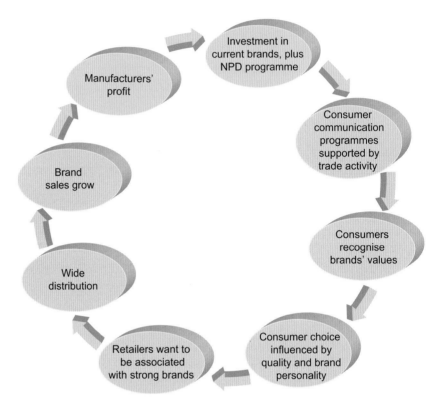

FIGURE 7.3
Strong brand's response

WHETHER TO BECOME AN OWN LABEL SUPPLIER

Increasingly, manufacturers are faced with the decision on whether or not to accept invitations from retailers to produce an own label version of their brand. The short-term attraction of extra sales needs to be weighed against long-term issues — not least of which is creating their own competitor. Firms like Coca-Cola are not prepared to produce an own label. They believe they have very successful brands whose formulation others find difficult to emulate. They argue that they have such strong brand assets that they have little to gain in the long-term from own label production. Mars is a particularly good example of this. Their experience of producing the world famous Mars Bar is such that no other company can emulate the quality of their brand at a cheaper or similar price. In fact, one manufacturer's trials for a retailer indicated that a poor quality own label could only be produced with a price 50% higher than the current Mars Bar!

One of the issues that needs considering is the economics of being a brand manufacturer versus an own label supplier. The firm needs to identify whether the payback in the long term from branding exceeds that from following an own label route. The analysis needs to identify all the activities involved in converting raw materials into final products and the costs put against these. The first series of costs are those based on supplying an own label. The premium that the firm must pay for undertaking the extra work involved in branding then needs to be identified. Finally, the extra margin, if any, attributable to marketing a brand rather than an own label, needs to be gauged. Providing the economics are sensible and any differential advantage can be sustained and as consumers recognise the quality difference over own labels, it would be wise to remain a brand manufacturer.

Production levels need to be evaluated. If there is 10% excess capacity in the factory and it is estimated that own label production will take 20% of normal production, the manufacturer is faced with the problem of deciding which lines to limit. Not only will the acceptance of the own label contract cut back the production of the manufacturer brand in the short term, but in the long term there is danger of even more detrimental production cuts. If the brand's differential advantage is difficult to sustain — for example, patents expire in a year's time — own label may be an attractive option. Before progressing down an own label route, however, the manufacturer needs to consider whether there is a lot of goodwill inherent in the brand's name, which in the short-term others may find difficult to overcome. If so, it is worth trading on this brand asset, rather than rushing into own label production.

It is necessary to question whether there is a commitment internally to investing in the future of brands. If a new director has recently been appointed, will they sway the board's views away from continually supporting brands? If the firm has recently been the subject of a takeover, will the new owners show the same concern about investment? To adopt a half-hearted approach may result in a secondary brand, which at best, can look forward to a short lifespan as retailers employ sophisticated financial evaluation systems such as the rationalisation of their range. As Fig. 7.4 indicates, in its early days a successful brand has well-differentiated benefits that consumers appreciate and for which they are prepared to pay a price premium. Without investment, the point of difference will fade as more 'me toos' appear, reducing the price premium once charged. As the brand slides deeper into the commodity domain, it is common for manufacturers to place more reliance on price cutting, further damaging the brand. In reality, this is just another representation of the process described in Fig. 7.4.

In some firms there is a motivational issue involved in deciding whether to supply own label. Some managers think that they will have less contact with their advertising agencies or that they will have to change their style of always looking for the best, to one of finding the cheapest. They worry that if

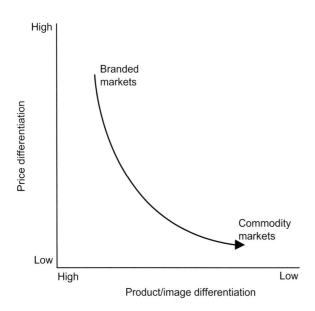

FIGURE 7.4

Sliding down the commodity curve

they subsequently stop supplying own label, the retailer will reveal their point of difference to one of their competitors. While previously they may have enjoyed negotiating from a position of strength, they may feel that they are becoming very dependent on the own label orders of a few, very large retailers.

Small manufacturers who underestimate sales potential may be particularly attracted by own label contracts. Taking a less than optimistic view of the future, they may short-sightedly see own label as their insurance. It would be far better if these firms evaluated the strength of their consumer franchise and assessed the potential for growing in a niche, selling their brand only through carefully selected retailers.

In some cases, own label can provide a basis for growth, particularly through expansionary retailers. However, against this must be weighed weak retailers who are ineffective in countering the encroachment of other retailers. They survive by adopting a policy of 'management by line of least resistance', giving prominence to those lines that sell best and putting little effort into slower moving lines. Producing own labels for these retailers is unlikely to lead to increasing profitability.

It can be misguided to take the view that own labels can be used to cover overheads. The marketer needs to question why overheads are becoming significant and take action to resolve this problem. Launching an own label range for this reason may be a short-term solution to a problem which will reappear. Retailers are likely to negotiate very low prices, probably giving the manufacturer a poorer rate of return for own labels than might otherwise be expected from brands.

Some authors argue that manufacturers gain from supplying store brands. Gomez-Arias and Bello-Acebron (2008) identify reasons often presented for manufacturer brands supplying private label. First, manufacturers are incentivised to supply private labels to fill idle capacity, which ultimately provides greater economies in production costs. Second, private labels buffer between leading and follower brands, which can offer manufacturers an additional competitive position. Finally, manufacturers may be required to respond to the increasing power of retailers. They explain that a retailer deciding to launch a store brand sometimes puts a 'squeeze' on manufacturers who have to compete and fight cannibalisation. If, for example, the manufacturer refused, the opportunity to provide a store brand may be taken by the competition.

The course of action taken by the manufacturer will depend on whether it is a high or low quality manufacturer. A high quality brand manufacturer's best option is to offer a store brand when it is a premium private label but not when it is a traditional private label or a generic. In contrast, the low quality brand manufacturer can offer either a low quality store brand or a generic. A possible risk for low-quality brand manufacturers is that the store brand will squeeze the low quality brand out of the market. Alternatively, if the retailer positions its store brand below the low-quality brand, the manufacturer would be forced to drop its price to force the generic out of the market. Gomez-Arias and Bello-Acebron (2008) suggest that low quality brand manufacturers may do best to drop their own brand and become private-label specialists. These strategic options are presented in Table 7.2:

Table 7.2 Strategic Options Facing Manufacturer Brands (*after Gomez-Arias and Bello-Acebron 2008*)

	Manufacturer Strategy	High Quality Manufacturer	Low Quality Manufacturer
Retailer Strategy	Premium Private Label	Offer Private Label	Drop manufacturer brand and become private label specialist.
	Traditional Private Label or Generic	Do not supply Private Label	Drop price to keep generic out of the market.

When managers argue for own labels because this helps build a closer working relationship with retailers, there are signs that they are not making the most of their current brands. There is no excuse for not maintaining regular contact with retailers through the existing portfolio of brands. This can be done through invitations to the manufacturer's sales conferences, to hospitality events to hear about new product developments, etc. If at all, there is an argument that broadening the portfolio with own labels may dilute the quality of discussions with retailers, since more lines need discussing at each meeting. Retailers are unlikely to be equally interested in selling manufacturers' brands and their own labels, as some managers incorrectly argue. They view some lines as being particularly good for reinforcing their store image or to generate store traffic or to boost profits. They place a different emphasis on different parts of their range and may give the own labels better shelf positioning, at the expense of the original brands.

However, Soberman and Parker (2006) observed an increase of manufacturer brands supplying store brands. They argue that store brands can increase profitability of a national brand manufacturer even when the national brand is suffering declining share. They identify quality equivalent store brands across the majority of consumer categories, including financial services (credit cards), disposable products (tissues, film), health and beauty aids (toothbrushes, vitamins) and cleaning products (liquid soaps, detergents). In the past, there was a perception that store brands were lower in price but also inferior in quality.

In 2010, 60% of American consumers surveyed believed that private labels were of the same quality as manufacturer brands. When manufacturer brands also supply store brands, this offers advantages to retailers. First, retailers can target different market segments; attracting those consumers who are 'advertising attracted' to manufacturer brands and those who are price oriented to store brands. Further, recent studies have shown that, while advertising allows the retailer to better discriminate pricing between store and manufacturer brand, it also allows the retailer to increase the price for both segments (Soberman and Parker, 2006). They also note that launching a store brand stimulates higher levels of manufacturer advertising and the manufacturer realises a significant gain when a retailer invests in launching a store brand.

HOW RETAILERS SELECT OWN LABEL SUPPLIERS

With increasing demands for higher quality own labels, retailers are becoming more selective when choosing potential own label suppliers. Johansson and Burt (2004) explain that close relationships need to be established with coordinated routines between manufacturer and retailer. The relationship between supplier and retailer takes on a complex dimension with the internationalisation of grocery markets. When a manufacturer is supplying a retailer who serves numerous international markets, this increases reliance on technology-based transactions and a strong relationship between partners is vital. Some of the considerations taken into account when assessing potential suppliers are:

- Can they produce the quality standards consistently?
- Do they meet compliance requirements such as hygiene, fair trade or work practices?
- Does their organisational culture fit with our brand values?
- Are they financially sound?
- Do they have adequate capacity to meet the current targets and sufficient spare capacity to cope with increasingly successful own labels?
- Do they have the logistics infrastructure to ensure reliable delivery?
- Is the production machinery up to date and well maintained?
- Do they have good labour relations?
- Will they be committed to the retailer?

- Are they already committed to supplying own labels to others or could they offer exclusivity?
- Do they have the flexibility to respond to short-term market fluctuations?
- Will they be able to hold adequate stock?
- Do they have a good marketing department that the retailer feels they can work with?
- Will the supplier maintain good communications with the retailer, regularly informing them of any relevant issues?
- Will the supplier agree to the retailer's payment terms?
- Would the retailer be happy to be associated with the supplier?

When retailers negotiate with manufacturers to supply store brands, a number of characteristics can influence how this process works (Johansson and Burt, 2004):

- The degree of vertical integration, which is the amount of coordination between central operations and store operations, will affect whether the buying is undertaken centrally and how many decision makers will be involved in the process.
- The degree of horizontal integration, which is the relationship between the retailer's outlets, will affect whether products are standardised. For example, if a chain is wholly owned, standardisation is generally higher than among a group of independent retailers.
- The degree of internal integration, which is the structure of the buying task, will vary depending on the level of vertical and horizontal integration. Among retailers with stronger vertical and horizontal integration, internal integration is simplified, information technologies are utilised and greater market orientation is emphasised.

In addition, manufacturers should consider the nature of the buying unit. How many participants will be involved in the buying centre? How many departments or functional areas within the firm will be represented? How many levels of the firm will be involved? Following a series of 50 interviews with experts in the UK, Sweden and Italy, Johansson and Burt (2004) set out the stages involved in the buying process, as is next considered.

An illustrative example is the buying process in the UK, as this market is at a high level of private brand development. Buyers are divided into trading units and adopt a category management approach. Units are broken down into teams, with sub-teams holding responsibility for product groups, such as biscuits or crisps. Teams incorporate product developers, buyers, marketers and merchandisers. For a new private brand product, the team will analyse the category and identify an opportunity gap. They will then put a product specification together, which would set out requirements including packaging, quality, price, size, colour and content. Buyers will then source suppliers to

FIGURE 7.5

Comparison of stages in the retailer buying process for Manufacturer and Private Label brands in the UK (*after Johansson and Burt 2004*)

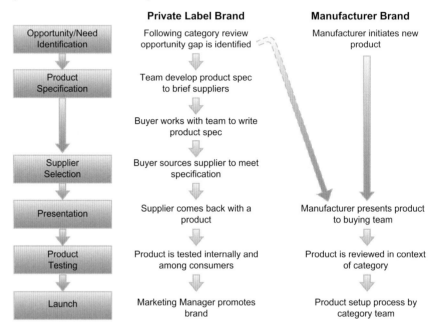

meet specifications and they will often start with their list of existing suppliers. The chosen supplier creates the product and when it meets requirements, it is subjected to consumer testing. Following trial, the product is developed. Johansson and Burt (2004) explain that this process can take six to nine months, which is lengthy in comparison with the launch of a manufacturer brand, which may take just three to four weeks because it is already a finished product. Following product development, the marketing manager develops communication and sales material and the product is launched into stores. Fig. 7.5 sets out the stages involved for the manufacturer following the private label route or the manufacturer brand route. Clearly, the level of complexity in developing a private label brand means there is a greater time requirement for this process.

PRIORITISING BRAND INVESTMENT THROUGH DIFFERENT RETAILERS

Even though retailer power is increasing, manufacturers, particularly those with strong brands, need to adopt an offensive rather than a defensive strategy when deciding where brand investment is needed. Ideally, manufacturers with strong brands always want to sell their brands through particularly attractive retailers. For some manufacturers, a retailer may be extremely attractive because of the high volume being sold, the retailer's image may be ideal for the brand and the retailer may have a policy of strongly supporting the brand through good in-store positioning and rarely being out of stock. Such retailers need identifying and nurturing.

It is rare, however, for a firm to have a portfolio of brands that are all very strong. Likewise, it is rare for all possible retailers to be classified as highly attractive retailers. To prioritise brand activity through different retailers, we have found a simple two-dimensional matrix to be particularly helpful. This is shown in Fig. 7.6.

Using the dimensions of brand strength and retailer attractiveness, it is possible to rank the order in which resources should be allocated.

The first requirement is to evaluate the **brand strength** through each current and potential retailer. A brand is strong if it scores well on the factors critical for brand success. This information could be obtained by talking directly with each retailer or by talking with the sales team who are in contact with retailers. The organisation therefore needs to identify retailers' views about the factors they consider when listing a brand. Their criteria could include issues such as profit, likely sales level, the extent to which the brand's image matches that desired by the retailer, the extent to which the brand has a sustainable functional advantage, etc. With knowledge of the critical success factors each retailer uses, the brand can then be evaluated against these and scores assigned, showing how the strength of the brand varies by individual retailer.

The next action is to evaluate the **attractiveness of each retailer**. In a workshop, the brand team needs to discuss and agree on the attributes that make a retailer attractive. These might include such issues as likely sales level, the extent to which they attract customers who fit the brand profile, geographical coverage, etc. Using these attributes, a score for the attractiveness of each retailer can be calculated.

The location of each retailer on the matrix can be identified by plotting their position, relative to the brand strength and the retailer attractiveness dimensions. By then examining the quadrants in which each retailer is located, this part of the audit enables managers to prioritise their retailer investment strategy.

FIGURE 7.6

Brand strength—retailer attractiveness matrix

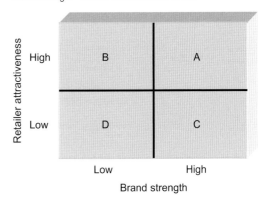

Those retailers who are in quadrant A of the matrix in Fig. 7.6 should be considered primary retailers for the brand. In this quadrant the brand and retailers are ideally matched and in the (probably unlikely) event of these retailers demanding better terms, the brand owner is in a strong position to counter such demands by clarifying the value of the brand. These retailers are ideal partners for joint marketing campaigns to reap the benefits of the brand. In particular, if these retailers introduce store brands, Karray and Zaccour (2006) advocate that manufacturer brands should engage in cooperative advertising, to counter the harmful effects of the private label brand.

The next priority retailers are those in quadrant B where the brand is not particularly strong, but the retailer is attractive. This represents an opportunity for the brand, since the changes to make the brand strong are under the control of the brand owner and the attributes needing attention would be known from the scoring procedure. Once these changes are addressed, the strengthened brand can capitalise on the synergistic effects from trading with highly attractive retailers. The danger in this quadrant is that the retailer also regularly reviews their brand mix portfolio and this brand would be regarded as underperforming.

The third priority retailers are those in quadrant C where the brand is strong but the retailers are not particularly attractive. The audit needs to identify whether there are any joint initiatives that could make these retailers become attractive (e.g. joint advertising, better training for the retailers' merchandisers, etc.). If it is felt that investment in this type of retailer is unlikely to improve its attractiveness, questions need to be asked as to why the retailer is being employed.

Finally, retailers in quadrant D need considering. Where the brand is weak and the retailers are not attractive, serious consideration needs to be given to cutting any ties (if they exist). Where the brand is of mediocre strength and the retailers are of mediocre attractiveness, thought needs to be given as to whether any investment might transform the situation over the longer horizon.

By undertaking a brand strength and retailer attractiveness audit on a regular basis, an organisation is able to prioritise its investment and working relationship with individual retailers. This can also stop the 'knee-jerk' response of many brand owners when faced with a demand from a retailer for a larger discount. By identifying in which quadrant the retailer lies, an appropriate response can be formulated.

It is not uncommon to hear about managers feeling forced into a particular position because the retailer had the 'upper hand'. Another way to appreciate the balance of power between a brand owner and retailer is to analyse the proportion of a brand's sales through each retailer and then for each individual retailer to make an assessment of how important the brand is for that retailer.

Table 7.3 Power Analysis

Deskpro's Sales Through Brightspace:	%	Share of Office Equipment Brands Through Brightspace:	%
Officerite	27	Alpha brand	36
Brightspace	25	Companion brand	33
Ranger	17	Deskpro brand	16
Workspace	16	Other brands	15
Designer	15		
Total	100	Total	100

For example, Table 7.3 shows a hypothetical analysis for an office equipment manufacturer, Deskpro.

In this particular example Deskpro is more reliant on Brightspace than Brightspace is on Deskpro. A quarter of Deskpro's sales depend on maintaining distribution through Brightspace, yet Brightspace only achieves 16% of its office equipment sales through Deskpro. This type of analysis better enables brand owners to appreciate which retailer might exert pressure on their brand.

To stop selling its brand through Brightspace would hurt Deskpro more than Brightspace. Therefore it needs to find ways of growing business for its brand through those retailers other than Brightspace, at a *faster rate* than is envisaged through Brightspace.

WINNING WITH BRANDS RATHER THAN OWN LABELS

Own labels have become a strong force and look likely to pose a serious threat to brands because of their improved quality, the development of premium own labels and their expansion into new categories. Despite this trend, brands remain strong and healthy and are likely to remain so in the future. This is due to three reasons. First, the very nature of the purchase process favours manufacturers' brands, as consumers need the quality assurance and the choice simplification they offer. Secondly, manufacturers' brands have built a solid foundation through years of investments in advertising and consistent quality. Retailers need the traffic building power of these brands and cannot afford to rely solely on their own brands. Finally, retailers may overstretch their names by moving into too many categories, resulting in customers doubting their quality across the entire product range.

Brand manufacturers should carefully weigh the strengths and weaknesses of becoming an own label producer. They usually enter into own labels production to fill occasional excess capacity, to increase production in categories where the brand is weak or to lower overall manufacturing and distribution costs. Although these may appear valid reasons at the time, in the long run companies run the risk of cannibalising their own branded products.

Analysis of the financial viability of own label production is usually undertaken on an incremental marginal cost basis. This exaggerates the benefits of selling own labels by neglecting the fixed overheads associated with excess capacity. A more precise evaluation based on the full cost, shows that in many cases production of own labels is less profitable, particularly when compared with branded counterparts.

Some companies move into own label production for fear that a competitor might do so. It is sometimes argued that dual production enables them to influence the category, the shelf-space allocation, the price gaps and the promotion timings. It also allows them to learn more about consumers, thus improving their ability to protect their brands. In reality, however, few companies have used own label production as a strategy to achieve competitive advantage. In the majority of cases, own label production appears to increase a company's dependence on a few large retailers, forcing it to disclose its cost structure and share its latest innovations with them.

From these considerations, companies should be more cautious about producing own labels. If they already manufacture own labels, they should evaluate whether this decision is still beneficial. They should undertake an own label audit and calculate the own label profitability on both a full cost and a marginal cost basis. They also need to evaluate the impact that own labels have had on the market share of their brands. This will enable them to consider whether it would be wise to withdraw from own label production.

Quelch and Harding (1996) suggest further strategies that brand owners can use to defend themselves against own labels. To reinforce their brands' positioning, companies need to invest consistently in product improvements. They should beware of introducing brands positioned to compete between their traditional brands and the own label equivalent. Few companies have been successful in doing this. Often, these 'middle' brands prove unprofitable and end up competing with the company's traditional brands.

By building trade relationships with retailers, brand owners may be able to demonstrate that own labels are usually less profitable than expected because of additional promotion, warehousing and distribution costs. They may also find that consumers of branded products spend more in a store compared to purchasers of own labels.

Manufacturers need to monitor the price gap between their own brands and own labels and to adjust prices continually to maximise their brand's profitability.

Brand producers would be wise to invest in separate management of each product category and strive for greater efficiency in promotions and merchandising. Performance should not be measured by market share and volume but by using category profit pools, in which the profit is calculated as a percentage of the total profit generated by all companies competing in the category. Historically this measure has shown own labels scoring weakly, due to low volume and low profit; and it highlights the danger of becoming an own label producer.

Hoch (1996), proposed a series of strategies that brand manufacturers can consider in their attempt to hold own labels at bay. Some of these include:

(i) innovate through either radical innovation or product improvements;
(ii) offer 'more for the money', i.e. maintain prices but provide additional value. This could be achieved through actions such as enhanced pack design or strengthening the brand;
(iii) reduce the price gap through price reductions;
(iv) offer a lower priced, possibly lower quality offering;
(v) wait and do nothing – albeit this, in effect, lets the environment dictate strategy!

Verhoef, Nijssen and Sloot (2002) tested whether these strategies were being used by Dutch grocery manufacturers. Their conclusions were that companies generally do not directly attack own labels using price reductions (strategy iii) or by offering new lower priced, possibly lower quality offerings (strategy iv). Rather, organisations sought to increase the perceptual space between their brands and own labels. Some firms focused on technological innovation and reinforcing their brand's strength (strategies i and ii), while others just concentrated on their brand's strength (e.g. awareness, preference and relationships in strategy ii).

UNDERSTANDING THE BALANCE OF POWER

The previous section showed some of the challenges brand manufacturers face due to increasing retailer power. To be able to cope with stronger retailers, manufacturers need to understand their bases for power. The following framework by French and Raven (1959) explains how one party can exert power over another and helps to identify the most powerful party.

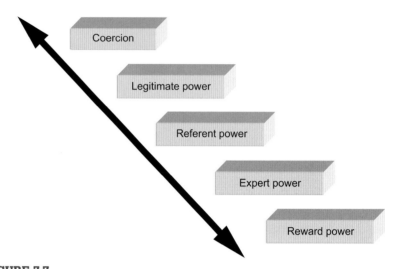

FIGURE 7.7
The continuum of power (*adapted from French and Raven 1959.*)

Power is the ability of one party to control the actions of another and can be conceived in terms of the continuum in Fig. 7.7.

The **coercive basis for power** is the expectation of one party that the second will be able to punish them if they fail to fulfill the other's wishes. For example, a retailer may delist a supplier who does not agree to their demands or a brand manufacturer may refuse to supply a retailer who sells their premium brand at a substantial discount. The magnitude of the power base depends on the effect that carrying out the threat can produce. Thus in markets where products are undifferentiated, where retailers carry broad product ranges or where a few retailers account for a large proportion of sales, most suppliers are too small to pose a credible threat.

One party may exert **legitimate power** when the other accepts its legitimate right to influence them. In franchising, for example, a franchisee agrees to run the business according to the standards specified by the franchisor.

Referent power occurs when members of the distribution channel are willing to defer to a highly-regarded channel member. For example, the strict controls required by Marks & Spencer ensure that their suppliers gain an excellent reputation for quality in their own circle and thus enables these suppliers to win further business.

Expert power exists when a retailer recognises that a supplier has a certain expertise, for example, in sales training or in stock control and they are likely to defer to the manufacturer and follow their requests if this expertise leads to better and more effective decisions.

A manufacturer can exert **reward power** when a retailer expects financial reward by deferring to them — a discount for an increase in the display area, for example.

Another helpful way of appreciating where the power lies in manufacturer–retailer relations is provided by Davies (1993). His model, shown in Fig. 7.8, illustrates the balance of power between suppliers and retailers and highlights possible strategies that either party might implement to respond to possible threats. The logarithmic scales on both axes indicate that power becomes an issue once either party accounts for about 10% of the other's business.

A car dealer with a Ford franchise would be an example of **supplier power**, as the manufacturer can maintain exclusive control over the dealership. The dealer is wholly dependent on Ford, while Ford with its numerous dealers sees only a small proportion of sales going through each dealer. By contrast, Marks & Spencer's dictation of standards to most of its suppliers is an example of **retailer power**. A large proportion of each supplier's sales come from Marks & Spencer, who have a considerable number of suppliers. The close relationship that exists between some white goods suppliers and electrical retailers represents a clear case of **mutual dependency**, in which a retailer distributes most of the manufacturer's products and the manufacturer accounts for most of the retailer's turnover. Finally, many retailers and manufacturers fall in the bottom left quadrant, as they operate at arm's length, in a relationship of **interdependence**.

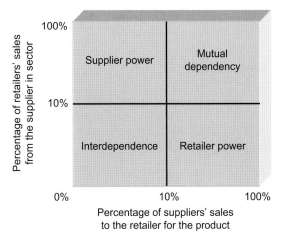

FIGURE 7.8

Sources of dependency (*after Davies 1993*)

Although the degree of power is a function of the level of concentration in the sector, the actual extent to which power is exerted depends on individual firms. For example, as we discussed earlier, the grocery sector is marked by high retailer power but in the case of strong brands such as Coca-Cola and Kellogg's, power still remains in the suppliers' hands. However, a brand is only a source of power for the manufacturer when consumers leave the store that does not stock it, refusing to accept any of the alternative brands. To ensure that the retailer needs the manufacturer, the brand needs to guarantee and communicate quality and innovation.

In the absence of either a dominant retailer or supplier, the relationship between the two may evolve as a strategic alliance with clearly defined roles and goals. The retailer sells and provides sales data and forecasts, the manufacturer supplies promptly and both are responsible for brand promotions. During the

FIGURE 7.9

Possible strategies on how to serve different retailers (*after Davies 1993*)

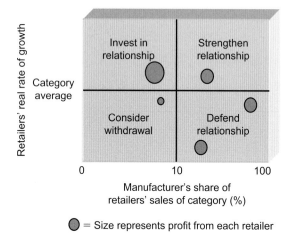

= Size represents profit from each retailer

relationship, conflicts may arise even though responsibilities and objectives have been set. Companies that have developed a clear trade marketing strategy based on both long-term and short-term planning are more likely to resolve conflicts than those relying entirely on negotiating skills.

A further way for manufacturers to decide upon their retailer strategy is to consider the use of the matrix in Fig. 7.9. Different retailers will be growing at rates above or below the average for the category sector. Those growing faster than the average category growth are the more attractive retailers. Where the manufacturer has a high share of the retailer's sales of the category, particularly where the retailer is growing fast, this represents a very attractive business partnership. By considering where each of the manufacturer's retailers lies on this matrix, along with some indication of the brand's profit through each of these retailers, strategies are shown for each of the quadrants.

CATEGORY MANAGEMENT

The previous section considered the implications of an uneven balance of power between retailers and manufacturers and suggested appropriate strategies. However, the issue of whether retailers or manufacturers have more power is less relevant nowadays, as we are moving from adversarial to cooperative relationships.

Retailers and manufacturers recognise that they are more likely to achieve long-term success if they are collaborative rather than confrontational. One way in which manufacturers can support retailers is to recognise that retailers do not sell individual brands, such as Pantene, Salon Selectives, Wella or L'Oreal, but rather categories, such as shampoos. Manufacturers should also be aware that retailers do not think only about brand leaders per se, indifferent to manufacturers trying to increase the sales of their brands at the expense of all the others. Retailers are more interested in gaining the collaboration of manufacturers to increase overall sales and profits of the category. In the USA, companies such as Procter & Gamble began to promote this type of relationship to cope with the concentration of retailer power and changes in consumer purchasing behaviour. They placed greater emphasis on collaborating with retailers and developing marketing programmes that grow retailers' total category sales. Since then, more manufacturers have sought to build their brands by working in collaboration with retailers to grow the total category, following a category management route.

The Institute of Grocery Retailing (2010) defines category management as:

'The strategic management of product groups through trade partnerships which aims to maximise sales and profit by satisfying consumer and shopper needs'

Categories can be based around product groups, such as confectionery, or based around occasions, such as 'food on the move'. Table 7.4 provides more insight comparing the traditional relationship between retailers and suppliers with the category management relationship.

In 2007, the IDG group conducted a category management survey. Their findings revealed that all the UK firms surveyed practised category management and that manufacturers and retailers were adopting a strategic, rather than tactical approach. While the tactical approach focused on operational aspects of category management, such as ranging and promotions, a more strategic approach focuses on cross-functional collaboration, customised research and strategic alignment.

Manufacturers often have information that retailers do not have. In addition, some brands, such as Coca-Cola and Kellogg's have sufficient reputation to 'pull' consumers to retail stores to seek out those brands (Bandyopadhyay et al, 2009). For manufacturers managing families of brands and competing for limited shelf space, category management offers the opportunity to increase sales and build the brand share by helping retailers to identify the best category mix for a given store or area.

Table 7.4 Appreciating the Essence of Category Management (*after Harlow 1995)*

Traditional approach	Category management approach	
Adversarial	*Joint*	Cooperative
12-month fiscal horizon	*Strategic*	Long-term perspective (3–5 years)
Pushing supplier's agenda	*Planning with retailers*	Retailer committed to the strategy by being involved in its development
Promoting supplier's brand	*Total category*	Focus on total category rather than individual brands
Increase offtake of own products	*Sales and profit*	Increasing total consumer offtake
Concern is only for supplier's benefits	*For mutual benefits*	The retailer benefits and the supplier must also be able to derive benefits through increased sales of their products

When retailers ask the most dominant brand in a category to set the strategy for all brands in that category, the dominant brand is known as the 'Category Captain' (Bandyopadhyay et al, 2009). The Category Captain is a manufacturer of a leading brand that uses its own data and detailed information about all of the brands within a category to establish a plan for the optimum product/brand mix and planograms depicting product placement and the amount of shelf space allocated to each brand. Often, these firms have multiple brands in a category and a key objective is to avoid brand cannibalisation. For example, Bandyopadhyay et al. (2009) identify Procter and Gamble's positioning of Head and Shoulders as 'strong dandruff fighting' shampoo, whereas Pantene is positioned as a 'healthy' shampoo. Therefore, category management by the company can help prevent competition between its own brands. There may be pitfalls associated with the allocation of a Category Captain. For example, they may reduce the number of suppliers, they may ensure a gradual price increase or they may use information to manipulate the retailer in ways that benefit their brands. Therefore, retailers often appoint another manufacturer as 'validator' to verify the Category Captain's recommendations (Bandyopadhyay et al., 2009).

Consumers have also 'learned' to view products in terms of category. For example, some debate exists about the advertising of certain food and drink brands which explicitly target children. However, a Scottish research study (Preston, 2010) showed that at age 12 years, children base habits on parent's involvement with certain categories and product category has a greater influence on behaviour than individual brand advertising.

Traditionally, manufacturers put the emphasis predominantly on their brands. Category management has shifted this focus. The brand still plays a fundamental role but it is no longer the focal point: instead, managers consider how the brand fits with the category and how the manufacturer could be better structured to deliver better brand benefits. The company needs to be structured so that the emphasis is placed on the long-term building of the brand through innovations within its category. Traditional structures have chopped firms up into different skills, very much akin to triangles with boards of directors at the apex and functional departments beneath them. Category management helped companies to recognise that business processes are critical and that these run across the functional silos. By restructuring to concentrate on business processes, firms are better able to capitalise on category management opportunities. For example, in some companies, the Marketing Director is responsible for the processes of brand development. Day-to-day questions from retailers about such issues as price changes and short-term promotions are dealt with by the Customer Development Director. Innovation Managers take responsibility for developing completely new propositions for each brand. One of the results of category management is that retailers are better able to identify weaker brands and this put more pressure on the poorer performing suppliers to improve.

Category management is regarded as a very helpful strategic tool to improve customer satisfaction and profitability. In particular, manufacturers see it as a route to strengthen their brands to become category leaders. Retailers see it as enhancing their expertise in pricing and promotion, better satisfying consumers and improving efficiency and profitability. Its implementation requires a major cultural change for manufacturers as they allow categories to take precedence over brands, and for retailers the category has precedence over the traditional department. Companies have restructured their organisations into more adaptable structures and invested in better communication systems. Greater collaboration builds more trust and achieves better performance.

Howleg et al. (2009) examined the drivers of customer value offered by category management, which were traditionally identified as optimising assortments, optimising promotions and optimising new product introductions. Reviewing the findings of their study of Austrian consumers, they advocate the inclusion of two new components of category management. These are personnel/image and quality of point of sale (POS) and they are additional influencers of consumer value. To optimise personnel and POS, for example, new areas of cooperation between retailers and manufacturers are required. These new areas of cooperation would require joint training and exchange of category knowledge. To enhance POS, retailers and manufacturers could align activities to place a new emphasis on cleanliness and atmosphere.

USING INFORMATION IN CATEGORY MANAGEMENT

Retailers and manufacturers achieve most from category management when they work together sharing information and areas of expertise. One of the dangers is being drowned by a sea of data and so they need to identify efficient ways to share market research data about aspects such as merchandising, shopper studies and scanning. Better software gives retailers instant and more user-friendly access to information on issues such as space allocation, consumer trends and promotion tracking. Retailers are gaining more knowledge of the category and market situation through their loyalty cards. A key requirement for effective category management is high-quality databases to link consumer and retail data. Furthermore, adaptable organisational structures and the full involvement of suppliers' and retailers' senior management ensures greater likelihood of greater benefits.

FIGURE 7.10

The process of category management (*after Nielsen 1993*)

Category managers have a rich set of data at their disposal, but to use this efficiently they need a coherent and flexible framework. The model in Fig. 7.10 was developed by Nielsen (1993) to help managers achieve better category profits.

The following case provides an illustration as to how categories can be managed. By tracking their database, a retailer became aware of the poor performance of their health and beauty sales, in particular, shampoos. They collaborated with one of their leading suppliers to address this problem. They jointly **reviewed the category** and discovered that the manufacturer's leading brand was underperforming in the retailer's stores because of less competitive prices, infrequent promotions and the manufacturer having too narrow a range of shampoo brands. The two companies examined the consumer panel data and the geo-demographic databases. These helped identify the most appealing shampoo range for the retailer's consumers. The analysis revealed that the consumer profiles varied significantly depending on the location of the stores. Three distinct consumer segments with different purchasing needs were identified: 'upscale affluent', 'middle- class families' and 'inner city'. To improve sales, it was decided that each location needed to **target their consumers** by providing a product range appropriate for that specific group of shoppers.

The retailer and the manufacturer then worked to optimise the product range, the pricing and the promotion. This **merchandise planning** involved adding new brands as well as delisting some of the manufacturer's weaker brands.

Both companies worked on the **implementation of the strategy** and developed detailed price plans, promotion schedules and shelf allocation fixtures. Sales information and consumer databases were used to track the effectiveness of the strategy and to assess the response of key target groups to promotions. These **results were evaluated** and they demonstrated the success of the new relationship: the retailer's overall sales and profits increased and the manufacturer's leading brand achieved a larger market share.

One of the challenges arising in category management is the optimum allocation of shelf space. Firms often use planograms to determine where products should be stored, the amount of space to be allocated to each brand and the assortment of products required to maximise customer value. Landa-Silva et al. (2009) reviewed the factors presented in previous shelf allocation models that firms might take into account when allocating space. These include:

- assortment — the desired range of products on display;
- cross elasticity — the impact of sales of one product on sales of another product;
- facings — the number of units of a product facing straight out to the consumer;
- groupings — the effect of placing products from the same category together or apart;
- location — fixture location, product category location and item location;
- manufacturer demands — the shelf space demands of specific manufacturers;
- marketing variables — including price, promotion and advertising;
- space elasticity — the ratio of change in unit sales of a product in relation to the change in it's allocated shelf space;
- stockouts — the extent to which a product is missing from shelf;
- target audience — optimising location for accessibility to the target audience.

They developed a heuristic approach to automate shelf space allocation, which included checking to ensure that enough shelf space is available, making iterative changes such as swapping products between shelves and using planograms to allow shop managers to visualise shelf layouts. Category management should be regarded as a programme for continual improvement. Analysis of the results should prompt regular reviews of the category so that retailers and manufacturers can jointly work out the best strategies and ensure long-term success.

EFFICIENT CONSUMER RESPONSE (ECR)

After adopting the principles of category management, many retailers and manufacturers started to consider how they could further enhance the benefits

FIGURE 7.11

The relationships in ECR

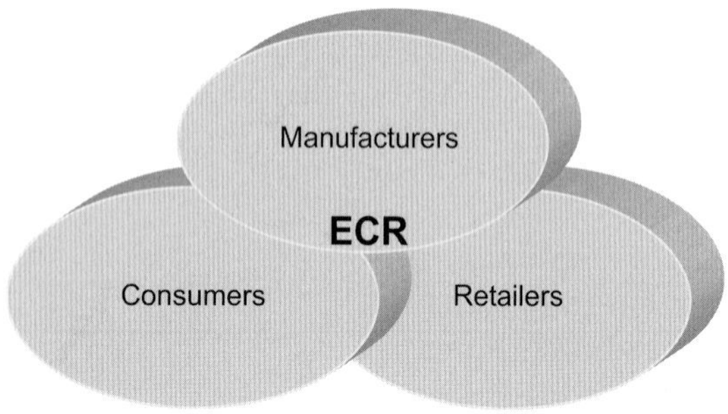

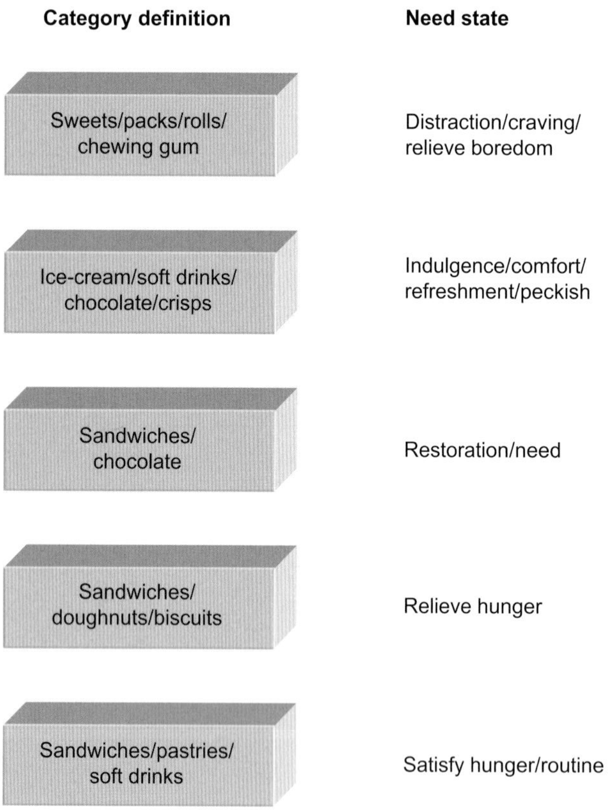

Category definition	Need state
Sweets/packs/rolls/chewing gum	Distraction/craving/relieve boredom
Ice-cream/soft drinks/chocolate/crisps	Indulgence/comfort/refreshment/peckish
Sandwiches/chocolate	Restoration/need
Sandwiches/doughnuts/biscuits	Relieve hunger
Sandwiches/pastries/soft drinks	Satisfy hunger/routine

FIGURE 7.12

How the category impulse snacking products for immediate consumption varies by need-state (*Johnson 1999*)

of working closely together and examined tasks that they had historically undertaken individually. They identified new opportunities to integrate their resources and activities and the era of 'efficient consumer response' (ECR) evolved. Beside the core issue of category management, the key elements of ECR are: new product development, consumer and trade promotions, store assortment and product replenishment. Their integration, as shown in Fig. 7.12, is the challenge managers are addressing to achieve better performance and higher consumer satisfaction. At the core of ECR lies category management.

In the early days, categories were developed through managers' assumptions about the most appropriate brand groupings. Increasingly, categories are being defined that better reflect consumers' usage and purchasing behaviour, by placing more emphasis on consumers' values. Thus ECR is at the heart of the interactions between manufacturers, retailers and consumers, as shown in Fig. 7.13. Lindblom et al. (2009) explain that these relationships vary in strength along a 'relationship continuum' with 'purely transactional' relationships at one end, and 'long-term cooperative partnerships' at the other. Those suppliers who are involved in category management are at the 'cooperative' end of the spectrum.

THE EVOLUTION OF ECR

Like category management, ECR was developed in the USA. In the mid-1990s, Wal-Mart and Procter & Gamble reassessed their relationship and recognised that their individual attempts to maximise their gains were at the expense of the other party. For example, Procter & Gamble often used money-off coupons that generated sudden increases in demand, which necessitated Wal-Mart having to increase stock and manage short bursts of shelf dynamics. On their side, Wal-Mart changed its buying patterns, preferring to purchase very large quantities of P&G brands only when on special trade discount promotions. In turn this caused Procter & Gamble significant production and logistics problems. They recognised that these promotions confused consumers and were detrimental to Wal-

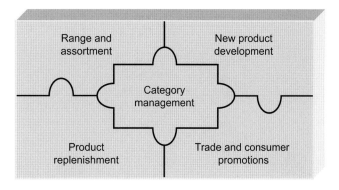

FIGURE 7.13
The key elements of ECR

Mart and P&G. They therefore decided to switch to a policy of every day low pricing, which delivered value to consumers and reduced promotion, production and storage costs. The implementation of ECR allowed them to improve consumer satisfaction by delivering more desirable brands at lower prices and with higher quality. It also improved business effectiveness by eliminating non-value-added activities and saving costs. In brief, for manufacturers, retailers and consumers the value of ECR could be summarised as follows:

> ECR is a supply chain strategy to satisfy customers with a better range, faster and at a lower cost.

These two companies were soon followed by others who recognised that ECR is not just about improving logistics: it completely challenges the way companies do business. It is not so much about a new set of techniques but rather a new way of bringing previous techniques together in a coherent and coordinated way. Companies have to learn to re-evaluate the way they do business. They need to collaborate to design promotions, product ranges and new products, which often requires a change in the company's culture.

Amongst other benefits, the implementation of ECR offers manufacturers and retailers the following advantages:

* business-related criteria: improvements in sales, market share and turnover rates;
* consumer-related criteria: buying frequency and consumer loyalty;
* product-related criteria: product visibility, shelf space, product location and launch (Lindblom et al., 2009).

Moreover, when companies closely collaborate to implement this programme of continuous improvement, costs are not reallocated but removed. The increased integration of activities between manufacturers and retailers leads to further benefits, especially during the process of brand development including:

- activity-based cost accounting enables manufacturers to show retailers the true costs of meeting their needs;
- electronic data interchange (EDI) improves stock holding;
- continuous replenishment programmes enable manufacturers to replenish brands on the basis of actual store demand.

HOW FIRMS IMPLEMENT ECR

To understand why companies have adopted ECR and how it is implemented, it is useful to examine the following principles of ECR:

1. consumer/shopper focused;
2. total category consideration;
3. data based;
4. multifunctional/total system perspective;
5. collaborative/trust based.

Procter & Gamble used these principles to reassess their business processes and the way they support their relationships with retailers. First, they realised that they understood the *consumer at home* but were less sure about them as *shoppers in-store*. They knew when their brands were used and how satisfied consumers were with them but were less confident about who bought them in the stores, why they bought them or how they felt during the purchase. In order to better understand shoppers, they gathered data on their purchasing behaviour and their response to promotions. In the laundry category, for instance, it emerged that promotions attracted consumers who were neither loyal nor likely to increase their spending. This meant that the company was spending money but did not encourage loyalty or market growth. Furthermore, they discovered that most consumers were not interested in promotions and would have preferred constant low prices to special offers, for which they knew they had to pay in some other way. This newly gained knowledge of consumers as shoppers allowed the company to make better decisions about the whole category, covering, for example, promotions, range, replenishment and new brands.

This holistic approach on the whole **category** instead of only on single brands required a shift in the focus of Procter & Gamble. Their analysis revealed significant differences about the impact of promotions in different categories. Investigating audit data showed that promotions were not necessarily the best way to grow their market share and that their impact varied by category. Previously they had been analysing the impact of promotions on individual brands. Looking at the impact of promotions on whole categories gave insight about the varying effects of promotions and led to ideas about different ways of rewarding loyalty.

Suppliers and retailers already have a vast amount of data on their markets, consumers and shoppers. However, to maximise category development and profit, they need to share it more between each other. The mutual benefits to be gained are enormous. By implementing ECR with their key customers, for example, Procter & Gamble reduced inventory by one-third and increased the volume of one category by one-fifth.

Sharing a wide amount of data with partners is not sufficient if the company itself is unable to work *multifunctionally* in their strategy development. At Procter & Gamble, a multifunctional team was set up to look at the range of brands being supplied. This team asked different departments to identify the fundamental factors for the most efficient product range. Building on these findings, they discovered that in the laundry category, 40% of the existing stock-keeping units could be eliminated and yet 95% of consumer needs could still be met.

If these principles are followed by both the manufacturer and the retailer in a relationship based on **collaboration** and **trust,**they can meet their customer needs and mutually achieve long-term success. Companies are still reluctant to share information and trust their partners.

One of the problems organisations face is the danger of forcing consumers to fit the definition of the category rather than defining the category based on an understanding of consumer behaviour. Unless retailers know why consumers use their stores, their merchandising strategies may be ineffective. For example, should the 'snacks' category include food and drink, should it include savoury foods and confectionery? One way to help identify the category definition is through group discussions. In this context, consumers are invited to create category shelves in a store environment, using packs of products. As they build these arrays, they are asked to explain their thinking process, showing why they arrived at particular arrangements. By deliberately probing to assess their motivating need-states, further insights are provided about brand marketing strategies. For example, a team initially started with a category 'impulse purchased snacking products for immediate consumption'. By probing their need-states, several different category definitions were identified, as shown in Fig. 7.13.

In this example, the brand's team needed to assess the extent to which their stores and their suppliers best fitted the different need-state groups. For example, some of their stores close to offices may have been ideally located to capitalise on the 'relieve hunger' group particularly at lunchtime, while their out of town stores may have been well positioned to target the 'indulgence/comfort/refreshment/peckish' group. While each store may have argued that they stocked and merchandised impulse snacking products, the category compositions needed to be different between the 'close to offices' and the 'out of town' stores. Furthermore, communicating to consumers in these different types of stores what the compositions of the categories were, facilitated consumers' decision making.

HOW FIRMS MEASURE THE SUCCESS OF ECR

Manufacturers and retailers are keen to assess the contribution their partners or potential partners are making to enhancing ECR. With the aid of the ECR scorecard, companies can measure the strengths and weaknesses of their current and potential partners. The ECR scorecard, which has been developed by the Institute of Grocery Distribution, is shown in Fig. 7.14.

The ECR scorecard allows marketers to challenge every activity a company undertakes within the supply chain by asking whether this activity delivers value to the end consumers. With such a radical, and some might say ruthless, assessment of each activity, marketers can determine their real contributions. For example, they may wonder whether the benefits derived from a line

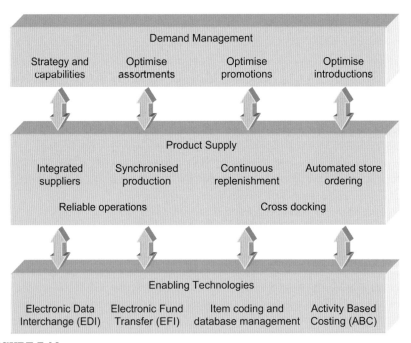

FIGURE 7.14
ECR scorecard (*after IGD 1997*)

extension compensate for the extra costs and increased operational complexity this introduced.

Some companies have already adopted the ECR logic to distinguish between strong and weak brands and seriously question brand proliferation. As a result, the range of brands within their portfolio has decreased. In these companies the focus has shifted from 'consumer loyalty to brands' to 'shopper loyalty to both store and brands'. This change brings a new perspective to the core concept of consumer satisfaction because extra brands do not necessarily generate extra sales. For example, one analysis found that at one stage in the confectionery category, the top 10 brands in the UK market accounted for £900 million sales and the following 40 brands for £1100 million sales. Are these extra brands enhancing the choice experience or just confusing consumers?

CONCLUSIONS

Issues in retailer branding have evolved to encapsulate the branding of the retailer itself, as well as the strategic activities involved in developing strong manufacturer and store brands. Retailers require strong brands to compete in a dynamic market, where inerternationalisation and collaboration have changed the manufacturer-retailer dynamic and fostered new challenges for creating customer loyalty. On their home ground, retailers face challenges from price elasticity and new competitors in the form of discounters. A challenge for retailers is to create a differentiated brand that offers value but at the same time presents a quality offering.

Store brands pose opportunities and challenges for retailers and manufacturers. For retailers, store brands offer a means of store differentiation, and a competitive response to powerful manufacturers. For manufacturers, store brands represent competition for shelf space and market share. Whether a manufacturer chooses to supply private label depends in part on the quality perception of its brand. If it is a high quality manufacturer, it may prefer to offer a private label brand to target a new segment or gain additional shelf space. If it is a lower quality manufacturer, it may question whether it should solely produce store brands.

The balance of power affects mutual interaction. A powerful manufacturer such as Kellogg's can dictate terms to retailers. On the other hand, retailers can 'squeeze' weaker manufacturers, who may be forced to cut prices or reduce terms to remain listed.

Strategically, it is best for manufacturers and retailers to cooperate. Category management is an approach which allows both parties to share information and areas of expertise. Outcomes of efficient category management include optimal range and assortment, appropriate new product development and targeted consumer promotions. When category captains are appointed, they can advise retailers of optimal shelf layouts. Adopting an ECR approach, retailers and manufacturers consider products in terms of consumer needs. By adopting a perspective from 'consumer loyalty to brands' to 'shopper loyalty to both store and brands' customer satisfaction is enhanced and retail brands are strengthened.

Own labels are now high quality retailer brands, backed by significant corporate promotional campaigns, reinforcing clear personalities. Retailers are increasingly attentive to changing environmental circumstances, launching innovative own labels to capitalise on new consumer trends. Pricing policies position own labels as good value lines, rather than cheap alternatives to manufacturers' brands. In view of these developments, we feel the term 'own label' is no longer appropriate and should be replaced by 'retailers' brands'. This recognises that the full repertoire of branding techniques are being employed by retailers, who are sophisticated in their strategic marketing.

Retailers are developing alternative categories of own labels, such as 'lifestyle own labels'. An essential ingredient for their success in such cases must be consumer-relevant added values — not just lower prices. As the generics experience has shown, only a small proportion of consumers are prepared to trade off added values for low prices.

Retailer dominance is likely to be a feature common to more markets. More professional retail management will be better served by information technology, enabling them to make more rapid adjustments to their portfolio of brands as new opportunities evolve. Manufacturers will succeed with their brands if they recognise the basis of their consumer franchise and continue to invest in this — rather than diverting their marketing budgets to buying shelf space. Manufacturers can adopt a more proactive approach to brand marketing in a retailer-dominated environment. They can assess the strengths and weaknesses of being an own label supplier by considering issues such as the economics of branding, the strength of their consumer franchise and the more effective use of production facilities. They can evaluate how well equipped they are to meet the needs of own label contracts. Furthermore, they can use the brand strength–retailer attractiveness matrix to prioritise brand investment programmes through different retailers.

Despite the increasing strength of own labels, brands have managed to remain strong due to their ability to offer consumers quality assurance and choice simplification. Despite the threat of premium own labels, brand owners can successfully maintain their position and strengthen their brands. Own label production should not be regarded as a short-term solution for excess capacity and brand owners should consider implementing some of the strategies reviewed earlier to protect themselves against own labels.

The relationship between retailers and manufacturers is often characterised by an uneven balance of power. Both parties need to understand the degree to which they depend on each other so that they can identify appropriate strategies to optimise the relationship. In the absence of a clear leader, a suitable strategy is to regard the relationship as a strategic alliance and clearly define roles and goals.

In the new era, manufacturers and retailers can improve both consumer satisfaction and profitability by adopting a 'category management' approach that involves managing product categories as business units and developing marketing programmes for the total category. The experience of many retailers and manufacturers has shown that by collaborating and exchanging information, both parties benefit.

Further integration between retailers and manufacturers can be achieved through ECR. This strategy enables companies to improve consumer satisfaction by delivering good value brands and improves business effectiveness by eliminating non-value-added activities. For its successful implementation, companies need to gather and store data on consumer purchasing behaviour for the whole category. They also need to ensure that each firm understands and is fully committed to ECR as a multifunctional approach. Additionally, there needs to be a relationship of trust and collaboration.

The ECR scorecard helps assess the strengths and weaknesses of a potential partner and examines the real value to consumers of supply chain activity. This scorecard is no doubt likely to make firms more cautious about widening their brand portfolios.

MARKETING ACTION CHECKLIST

It is recommended that, after reading this chapter, marketers undertake the following exercises to test and refine their brand strategies.

1. On a two-dimensional map for a specific market, with the axes representing price and quality, plot the position of your brand(s), your other branded competitors' brands and the own labels from the main retailers. Repeat the exercise, but do so thinking back two years. As a management team, consider why any changes have occurred and assess which factors have had a particularly strong impact on your brand.

2. Two years ago, how did the leading retailers in your product field manage to sell their own labels more cheaply than popular brands? Today, how are these own labels achieving their price advantage? What are the implications of this for your brands?

3. In the market where your brands compete, what propositions do each of the major retailers' own labels offer? Thinking back two years, how have

these propositions changed, if at all? What are the implications of these changes for your brands?

4. For any one of your markets where your brands compete against own labels, what are the differences between the leading retailer's own labels and your brand(s)? Where the retailer's own label has an advantage over your brand(s), evaluate how this advantage has been achieved and consider how you could better your brand on this attribute, if consumers would value such a change. Where your brand(s) has an advantage over the leading retailer's own labels, consider what is required to sustain this.

5. What proportion of your brand sales in a particular market go through multiple retailers? How does this compare against the situation two, five and ten years ago? What factors are giving rise to multiple retailer dominance? Will these factors continue to aid the growth of their power? How do you plan to compete in a market dominated by increasingly powerful retailers?

6. What personalities do consumers associate with your brands and the own labels against which they compete? How clear are these personalities? Has your brand's personality become less distinctive over time, while retailers' own label personalities have become sharper? Do the personalities of the leading retailers' own labels mirror their stores' brand personalities? If there is any difference between the personality of a retailer's store and its own labels, evaluate why such a difference has occurred and what the implications are for your brands.

7. Are your brands sold through convenience or non-convenience outlets? Does your promotional strategy take into account the need for simple on-pack information through convenience outlets and the need to educate sales assistants in non-convenience outlets?

8. Should you be debating whether to supply an own label version of one of your current brands, focus your decision by scoring the advantages against the issues:
   - whether others will find it difficult to emulate your brand;
   - whether there is a lot of goodwill tied up in the brand name;
   - the economic implications of own labels and brands;
   - whether patents are soon to expire;
   - the production implications in terms of existing capacity;
   - whether there will be continuing internal support for brand investment;
   - what internal morale will be like if own label contracts are accepted;
   - whether own label production will help block competitors;
   - the accuracy of sales forecasts for brands and own labels.

9. Should you wish to work on own label contracts, assess whether you will be in a strong or a weak negotiating position when pitching for this

business by applying the audit questions in the section 'How retailers select own label suppliers'.

10. Evaluate the appropriateness of your priorities for brand support through different major accounts by using the brand strength–retailer attractiveness matrix.

11. Is your company currently manufacturing own labels or considering this option for the future? What analysis have you undertaken to assess the level of profitability resulting from own label production? What will be the long-term consequences of your own label production for your brands? Are you running the risk of cannibalising them? What competitive advantage will you gain from the label production? If the answer to this last question is none, you should carefully consider whether the problem of excess capacity could be solved in another way.

12. (For a manufacturer) Are your brands threatened by own labels? As defensive strategies to cope with these threats:
    − are you investing consistently in product improvements?
    − do you have any plans to introduce new brands between your own traditional brands and own labels?
    − what type of trade relationship are you building with retailers?
    − do you monitor the price gap between your brands and own labels?
    − do you measure performance by using category profit pools?

13. Consider the relationship you have with your trade partner using the matrix shown in Fig. 7.8. How would you describe the balance of power between your trade partner and yourselves? Which strategy emerges from this position? How do you plan to develop this strategy in the short and long term?

14. What type of relationship exists with your trade partners to encourage the long-term growth of product categories as well as the success of specific brands? Are you satisfied with the degree of trust and collaboration between your trade partners and yourselves? Are the senior managers of both parties committed to category growth? Do you freely share your sources of information and areas of expertise?

15. How complete is your knowledge of your consumers as shoppers? What do you know about their purchasing behaviour and their response to promotions in your category? Could this knowledge be improved by undertaking further market research?

16. What kind of information systems do you currently use to exchange the information with your trade partners? Is the information exchange efficient and effective? Do you assess whether the data and sources used are really necessary? Have you established high-quality databases to link consumer and retail data?

17. Do all the activities within your supply chain really deliver value to consumers? Do your consumers need the new brands and line extensions

offered or are they confused by the ever-increasing range? What information do you have on their reactions to promotions?

STUDENT BASED ENQUIRY

1. Identify a retailer that you believe has a strong brand presence. Discuss the methods used by this retailer to position itself relative to its competitors.
2. Select a retail loyalty scheme you are familiar with. Identify three sources of data that the loyalty card provides to the retailer and suggest three ways a retailer might use this information to enhance its overall brand offering.
3. Visit a grocery store in your neighbourhood. Select i) a store brand and ii) a manufacturer brand in the same product category. For the two brands, evaluate the differences and similarities in shelf position, packaging, and pricing. Which brand would you purchase? Discuss your reasons for your choice.
4. For each of the following groups identify two sources of power and two ways these groups can strengthen their strategic position: discounters, manufacturers of leading national brands and manufacturers of private label brands.
5. Select a product category in a nearby grocery store. Evaluate the shelf space allocated to each brand in that category. Identify, in your own opinion, who the market leader is. Evaluate the role of shelf layout on your interpretation of brands in the product category.

References

Bandyopadhyay, S., Rominger, A., & Basaviah, S. (2009). Developing a framework to improve retail category management through category captain arrangements. *Journal of Retailing and Consumer Services, 16,* 315–319.

Butttle. (2008). *Customer Relationship Management: Concepts and Technologies* (2nd ed.). Oxford: Butterworth-Heinemann.

Chan Choi, S., & Coughlan, Anne T (2006). Private label positioning: quality versus feature differentiation from the national brand. *Journal of Retailing, 82*(2), 79–93.

Coe, N. M., & Hess, M. (2005). The internationalization of retailing: implications for supply network restructuring in East Asia and Eastern Europe. *Journal of Economic Geography, 5,* 449–473.

D'Andrea, G. (2010). Latin American retail: where modernity blends with tradition. *The International Review of Retail, Distribution and Consumer Research, 20*(1), 85–101.

Davies, G. (1993). *Trade Marketing Strategy.* London: Paul Chapman.

French, J., & Raven, B. (1959). The bases of social power. In D. Cartwright (Ed.), *Studies in Social Power.* East Lansing, Michigan: University of Michigan Press.

Gilligan, C. (2010). CRM at Tesco: from understanding to engaging customers. In D. Jobber (Ed.), *Principles and Practice of Marketing* (6th ed.). Maidenhead: Mc-Graw Hill.

Gomez-Arias, J. T., & Bello-Acebron, L. (2008). Why do leading brand manufacturers supply private labels? *Journal of Business and Industrial Marketing, 23/4,* 273–278.

Gómez, M., & Okazaki, S. (2009). Estimating store brand shelf space: a new framework using neural networks and partial least squares. *International Journal of Market Research, 51*(2), 243–266.

Harlow, P. (1995). Category management: a new era in FMCG buyer-supplier relationships. *Journal of Brand Management, 2*(5), 289–295.

Hoch, S. (1996). How should national brands think about private labels? *Sloan Management Review, 37*(2), 89–102.

Howleg, C., Schnedlitz, P., & Teller, C. (2009). The drivers of consumer value in the ECR Category Management model. *The International Review of Retail, Distribution and Consumer Research, 19*(3), 199–218.

Humby, Clive (2010). International differences. *MarketingWeek*. Available at: http://www.marketingweek.co.uk/opinion/international-differences/3010621.Article. Accessed 3 March 2010.

Johansson, U., & Burt, S. (2004). The buying of private brands and manufacturer brands in grocery retailing: a comparative study of buying processes in the UK, Sweden and Italy. *Journal of Marketing Management, 20*, 799–824.

Johnson, M. (1999). From understanding consumer behaviour to testing category strategies. *Journal of Market Research Society, 41*(3), 259–288.

Karray, S., & Zaccour, G. (2006). Could co-op advertising be a manufacturer's counter strategy to store brands? *Journal of Business Research, 59*, 1008–1015.

Lamey, L., Deleersnyder, B., Dekimpe, M. G., & Steenkamp., J. B. E. M. (January 2007). How business cycles contribute to private-label success: evidence from the United States and Europe. *Journal of Marketing, 71*, 1–15.

Landa-Silva, D., Marikar, F., & Le, K. (2009). Heuristic approach for automated shelf space allocation. *Proceedings of the 2009 Symposium on Applied Computing*, 922–928.

Lindblom, A., Olkkonen, R., Ollila, P., & Hyvönen, S. (2009). Suppliers' roles in category management: a study of supplier-retailer relationships in Finland and Sweden. *Industrial Marketing Management, 38*, 1006–1013.

Marks & Spencer. (2007). *Marks and Spencer Launches 'Plan A' - €200m "Eco Plan"*. Available at: http://corporate.marksandspencer.com/media/press_releases/RNS/15012007_MarksSpencerlaunchesPlanA200mecoplan. Accessed 12 February 2010.

Nielsen. (1993). *Category Management in Europe: A Quiet Revolution*. Oxford: Nielsen.

Nielsen. (2009). *Nielsen Retail Performance Survey*. Available at: http://uknielsen.com/site/index.shtml. Accessed 1 March 2010.

Pappu, R., & Quester, P. G. (2008). Does brand equity vary between department stores and clothing stores? Results of an empirical investigation. *Journal of Product and Brand Management, 17*(7), 425–435.

Preston, C. (2010). Parental influence upon children's diet: the issue of category. *International Journal of Consumer Studies, 34*, 179–182.

Quelch, J. A., & Harding, D. (Jan–Feb 1996). Brands versus private labels: fighting to win. *Harvard Business Review*, 99–109.

Rondán Cataluña, F. J., Navarro García, A., & Phau, I. (2006). The influence of price and brand loyalty on store brands versus national brands. *International Review of Retail Distribution and Consumer Research, 16*(4), 433–452.

Rubio, N., & Yagü, M. J. (2009). The determinants of store brand market share: a temporal and cross-sectional analysis. *International Journal of Market Research, 51*(4), 501–519.

Soberman, D. A., & Parker, P. M. (2006). The economics of quality-equivalent store brands. *International Journal of Research in Marketing, 23*, 125–139.

Steenkamp, J. B. E. M., Koll, O., & Geyskens, I. (2004). *Understanding the drivers of private label success: a western European perspective 2004*. Tilburg University, Europanel and GfK Panel Services Benelux, Dongen, The Netherlands: AiMark.

Uncles, M. D. (2010). Retail change in China: retrospect and prospects, The International Review of Retail. *Distribution and Consumer Research, 20*(1), 69–84.

Verhoef, P., Nijssen, E., & Sloot, L. (2002). Strategic reactions of national brand manufacturers towards private labels. *European Journal of Marketing, 36*(11/12), 1309–1326.

Further Reading

Anon. (18 March 1991). Value of own label at Sainsbury. The Grocer, 6 July, 18.

Ashley, S. (1998). How to effectively compete against private-label brands. *Journal of Advertising Research, 38*(1), 75–82.

Broadbent, T. (2000). *Advertising Works II*. Henley on Thames: World Advertising Research Centre.

Burk, S. (1997). The continuing grocery revolution. *Journal of Brand Management, 4*(4), 227–238.

Burt, S. (2000). The strategic role of retail brands in British grocery retailing. *European Journal of Marketing, 34*(8), 875–890.

Burnside, A. (15 February 1990). Packaging and design. *Marketing,* 29–30.

Caulkin, S. (7 May 1987). Retailers flex their muscles. *Marketing,* 37–40.

Davidson, H. (1987). *Offensive Marketing*. Harmondsworth: Penguin Books.

Davis, I. (1986). Does branding pay? *ADMAP, 22*(12), 44–48.

Day, J. (5 October 2000). Private labels on parade. *Marketing Week,* 46–47.

de Chernatony, L. (1987). *Consumers' perceptions of the competitive tiers in six grocery markets. Unpublished PhD thesis*. London: City University Business School.

de Chernatony, L., & Knox, S. (1991). Consumers' abilities to correctly recall grocery prices, 151–69. In N. Piercy et al. (Eds.), *Proceedings of Marketing Education Group 1991 Conference*. MEG: Cardiff Business School.

de Kare-Silver, M. (Nov. 1990). Brandflakes. *Management Today,* 19–22.

Dunne, D., & Narasimhan, C. (May-June 1999). The new appeal of private labels. *Harvard Business Review,* 41–52.

Economist Intelligence Unit. (Dec.1971). The development of own brands in the grocery market. *Retail Business, 166,* 27–35.

EPM. (2009). *Women's Retail Brand Awareness*. New York: EPM Communications Inc.

Euromonitor. (1989). *UK Own Brands (1989)*. London: Euromonitor.

Furness, V. (24 October 2002). Straying beyond the realms of own label. *Marketing Week,* 23–24.

Henley Centre for Forecasting. (1982). *Manufacturing and Retailing in the 80s: A Zero Sum Game? Henley*. Henley Centre for Forecasting.

IGD. (2010). Category Management. Available at: http://wwwigd.com/index.asp?id=1&fid=1&sid=6&tid=38&folid=0&cid=125. Accessed 18 February 2010.

Laaksonen, H., & Reynolds, J. (1994). Own brands in food retailing across Europe. *Journal of Brand Management, 2*(1), 37–46.

Liebling, H. (7 November 1985). Wrapped up in themselves. *Marketing,* 41–42.

Lijander, V., Polsa, P., & van Riel, A. (2009). Modeling consumer responses to an apparel store brand: store image as a risk reducer. *Journal of Retailing and Consumer Services, 16*(4), 281–290.

Luijten, T., & Natzaam, R. (2005). Consumer confidence and purchasing behaviour. In *Key information for marketing and policy plans* (pp. 34—37), GFK 2005 Annual Guide.

Macrae, C. (1991). *World Class Brands*. Wokingham: Addison Wesley.

MacNeary, T., & Shriver, D. (1991). *Food Retailing Alliances: Strategic Implications*. London: The Corporate Intelligence Group.

Marketing. (2009). *Marketing's Top 100 Advertisers 2009*. Available at: http://www.marketingmagazine. co.uk/news/wide/893106/. Accessed 12 February 2010.

McGoldrick, P. (1990). *Retail Marketing*. London: McGraw Hill.

McGoldrick, P. (2003). Retailing. In M. Baker (Ed.), *The Marketing Book*. Oxford: Butterworth-Heinemann.

Mitchell, A. (25 October 2001). Assessing the scope of grocery super-categories. *Marketing Week*, 28.

Moss, S. (1989). Own-goals. *Marketing, 16 February*, 45—46.

Nedungadi, P., Chattopadhyay, A., & Muthukrishan, A. V. (2001). Category structure, brand recall and choice. *International Journal of Research in Marketing*, 34(8), 191—202.

New York Times. (2010). Times Topics: Karl Lagerfeld. Available at: http://topics.nytimes.com/top/reference/timestopics/people/l/karl_lagerfeld/index.html?scp=3&sq=H&M& st=cse. Accessed 17 February 2010.

Nielsen. (1989). *The Retail Pocket Book*. Oxford: Nielsen.

Nielsen. (2005). The Power of Private Label. Available at: it.nielsen.com/trends/documents/2005_ privatelabel.pdf. Accessed 11 February 2011.

Nielsenwire. (2009). Store brand success around the world. Available at: http://blog.nielsen.com/nielsenwire/consumer/store-brand-success-around-the-world/. Accessed 11 February 2010.

Ohmae, K. (1982). *The Mind of the Strategist*. Harmondsworth: Penguin.

Porter, M. (1976). *Interbrand Choice, Strategy and Bilateral Market Power*. Cambridge: Harvard University Press.

Quinn, F. (2006). *Crowning the Customer: How to Become Customer Driven*. Dublin: The O'Brien Press.

Rapoport, C. (16 February 1985). Brand leaders go to war. *Financial Times*, 24.

Sainsbury plc, J. (2010). *Company Profile*. Available at: http://www.j-sainsbury.co.uk/index.asp? pageid=327 (accessed 12 February 2010).

Sambrook, C. (7 March 1991). The top 500 brands. *Marketing*, 27—33.

Sciullo, M. (2010). UK-based Topshop is top of list for fashionistas, bargain hunters. Pittsburgh Post-Gazette. Available at: http://www.post-gazette.com/pg/10038/1033570-314.stm. Accessed 17 February 2010.

Segal-Horn, S., & McGee, J. (1989). Strategies to cope with retailer buying power. In L. Pellegrini, & S. Reddy (Eds.), *Retail and Marketing Channels*. London: Routledge.

Sheath, K., & McGoldrick, P. (1981). Generics: their development in grocery retailing and the reactions of consumers. *A report from the Department of Management Sciences, the University of Manchester Institute of Science and Technology*.

The Grocer (1988). *Multiple price pressure is blamed for the 'debasement' of ice cream*. (p. 4) 23 April.

The Grocer (1991). *Why the squeeze was eased*. (p. 10) 10 August.

The Grocer (1991). *The ultimate proposition? Money back and replacement pack*. (p. 10) 16 March.

Thompson-Noel, M. (9 April 1981). Big time grocery brands — the beginning of the end ? *Financial Times*, 11.

Walters, D., & White, D. (1987). *Retail Marketing Management*. Basingstoke: Macmillan.

Walford, J., & Edwards, T. (1997). Where own label is heading: a recommendation. *Journal of Brand Management, 4*(5), 320–326.

Whitaker, J. (1990). Single market – multiple opportunities. *Paper presented at Private Label Manufacturers Association Conference.*

Wilkinson, A. (2001). Retailers are doing it for themselves. *Marketing Week, 27,* 19–20.

www.ikea.com. Accessed 11 February 2010.

www.safeway.com/ifl/Grocery/TopCategoriesDisplay?identifier=OOOOrganics. Accessed 11 February 2010.

www.superquinn.ie/aspx/Content.aspx?id=747. Accessed 11 February 2010.

CHAPTER 8

Brands on the Internet

OBJECTIVES

After reading this chapter you will be able to:
- Identify the challenges and opportunities presented by the virtual environment.
- Understand the factors that influence the online brand experience.
- Explain the significance of the online brand experience and outline the factors critical to creating this experience.
- Evaluate the role of Web 2.0 for brand building and understand in particular the role of brand communities in shaping the brand experience.
- Describe the main ways firms can build brand equity online.

CONTENTS

SUMMARY

This chapter considers how branding principles can be used to help grow brands on the internet. It opens by considering the new mindset that managers need when developing internet-based brand strategies. It argues that the brand concept is the same as that in any other medium, even though the enactment is different. A framework is presented to appreciate the elements of the brand's online experience. It traces the evolution of online branding from e-commerce to the role of Web 2.0 in shaping the customer's experience of the brand. It also addresses the role of brand communities as a force for brand building or brand damage. Challenges resulting from the internet such as shifting power, pricing and online service delivery are addressed. The chapter finishes by exploring the nature of online brand equity and explores how firms can harness aspects of their website to create customer trust, satisfaction and equity.

MOVING INTO A VIRTUAL ENVIRONMENT

Electronic commerce and e-marketing are giving rise to new brand opportunities, along with new challenges. Electronic commerce occurs where goods or

293

Creating Powerful Brands. DOI: 10.1016/B978-1-85617-849-5.10008-X services are sold through a website. Electronic marketing on the other hand is any marketing activity that is conducted through the medium of the internet. It is rather myopic to regard the opportunities presented by the internet as being solely enhanced sales and profit since it is important to appreciate how these are achieved. As this chapter will explore, increased sales and profit comes through new forms of customer behaviour and revised business models. For example, improved performance emerges from issues such as:

- listening to communities rather than telling customers;
- enabling customers to have direct input into product development;
- recognising that communities co-create brand value, rather than passively consuming brands;
- becoming more community focused and rapidly responding to new suggestions;
- relaxing control over brands.

Brands that have succeeded using the internet — be they traditionally based bricks and mortar brands or "pure" internet brands, for example, Amazon, Google and Yahoo! — have thrived because they view the internet as a force that levels imbalanced playing fields in favour of customers. These brands are based on revised business models, significantly improving the value equation through greater customer interactivity, enhanced consumer experiences, tailoring and rapid responses. Furthermore, successful brands on the internet show commitment to web-based branding and they take a long-term view about brand building.

Some organisations are typified as having incremental thinking about web-based marketing, even though a new mindset is needed. As such, a perusal of their web pages shows remarkable similarities to their company brochures, resulting in brochureware. These brands are based on the questionable classical premise of *telling* customers and the erroneous assumption that the organisation is in control of brand information. The internet has spawned new ways for customers to share information with considerable frankness, bypassing brand owners. Negative Double Jeopardy (Kucuk, 2007) is the term used for the unpleasant scenario that more attractive brands have more anti-brand sites. For example, sites such as www.mcspotlight.org, chasebanksucks.com or starbucked.com facilitate networked customer contact, deflating brands' advertising claims. Kucuk (2007) explains that all internet nodes are equal, therefore internet users have 'speech equality'. As customers co-create brand value, this extends to the websites that customers create about brands. Four types of anti-brand sites exist (Kucuk, 2007):

- Experts — experts use external information to discuss issues. While the consumer might be a customer of the brand, they recognise issues and present extensive and detailed information based on general media sources.

- Symbolic haters – negative messages are based around word of mouth and the brands targeted may simply be fashionable to 'hate'.
- Complainers – these sites are created by customers who have experienced a problem with the brand and tend to complain about the process or product, rather than the brand's philosophy. The company may have ignored initial complaints, and the site provides a way for aggrieved customers to get their point across.
- Opportunists – these sites are fed by media reports of service failures. Opportunists are less involved with the brand and more interested in maximising visibility.

Clearly, managers of successful brands must always be vigilant that they are not receiving unwelcome internet coverage!

A cause for concern is the brand marketer who primarily looks at internet-based brand opportunities in terms of being mechanisms to save costs. Cost savings arise for brands on the internet for a variety of reasons. For example, customers taking over some of the work traditionally done by staff, reduced salaries bills from fewer staff, savings from bypassing intermediaries and from fewer offices. It was once estimated that cashing a cheque through a banking employee in a bank costs around $1.10, while withdrawing money over the internet only costs about $0.15 per transaction. The problem is that with intelligent search engines that compare competing brands, unless the organisation is able to respond rapidly and is geared to continually driving down any operating costs, a short term advantage can become a threatening challenge. Zettelmeyer et al. (2006) explored how the Internet lowers prices. In their study on the US auto retailing industry, they found that the internet provides information, especially about the invoice prices of dealers; and the process of online buying services enables customers to take advantage of lower prices. In addition, they found that customers differ, as those who dislike bargaining benefit more from internet information.

An interesting example of internet pricing can be observed in the music industry. Typically, customers can buy an album on iTunes for £7.99 or cherry-pick their favourite songs for 99p. The UK group Radiohead launched their album 'In Rainbows' solely as a digital download through their website. Radiohead took the radical approach of inviting their fans to put their own price on the download, plus a 45p transaction charge. Sherwin (2007) explains that customers clicking on the price box received the message 'it's up to you' and a subsequent refresh of the screen revealed 'no really, it's up to you'! Using an 'honesty box' allowed customers to pay nothing or the full price for the album. This approach was a success for the group. The NME reported that most fans opted to pay nothing for the download. The download facility was removed after three months and the album was physically released. It generated more money before its physical release than the band's previous album and went to number one in the UK and USA once it was physically released (NME, 2008).

The strategic brand winners recognise that, while prices initially attract customers, added value brings them repeatedly back. Thus Amazon's added value of a one-click system for revisiting customers, its new titles alerting system and suggested companion books help it to not have to compete on an ever-decreasing vicious spiral.

RECONCEIVING THE CONCEPT OF 'BRAND'?

Some might wonder, from the points in the previous section, as to whether there is a need for a new theory of brands when they are in an online, rather than offline, environment. We firmly believe that a brand is a brand, regardless of its environment. What is different is the way the brand's promise is executed. Consider the brand evolution spectrum in Fig. 8.1.

In the early days of internet-based branding, an opportunistic marketer may have regarded branding as being about providing copious brand details, i.e. brand as a data centre. In their over zealous attempts to provide brand data, they regarded branding as something they did to a visitor. If their brand is to thrive, customers need a better customer experience. This is achieved through a customer-centric approach allowing customers to visit the site and get more tailored information. (e.g. train times from http://www.tgv.co.uk/). If the focus is primarily upon information, albeit tailored, this is still characterised by marketers wanting to be in control. Eventually organisations recognised that the internet is not a passive medium that acts as a conduit for information. Rather it is a dynamic environment that celebrates customer involvement in an experience, with both a brand and a community interested in the brand. Disney grasped the opportunity when it launched www.family.com. Its family values attract people who respect the entertainment origins of this brand but then want to reach out to other like-minded individuals to discuss family-related matters, for example, education, parenting and health.

The point of this evolutionary spectrum is that the early emphasis on brands in an internet environment

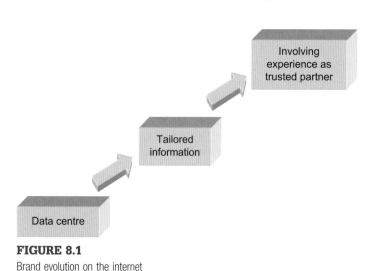

FIGURE 8.1

Brand evolution on the internet

was at the rational level, rather than progressing to an experiential state. The characteristics of a brand can be understood from recognising that a brand is a cluster of rational (functional) and emotional values that enable a stakeholder to recognise a promise about a unique and welcomed experience. This perspective of a brand is depicted in Fig. 8.2.

In the bricks and mortar environment, the Co-op Bank has the rational value of being responsibly fair through its policy of ethical investments. This links to the emotional value of caring. These values enable stakeholders to perceive the promised experience of being in control of their savings for their own good and for the good of others. The migration of this bank to the internet resulted in the brand undergoing changes. For example, it is branded Smile.co.uk and uses atypical pink colouring to enable it to stand out from the dull associations of other banks. The easily navigated site and the animated smile, amongst other aspects of the site, reinforce the brand values of responsibility and caring, enabling a similar promise to be perceived between the offline and online brand.

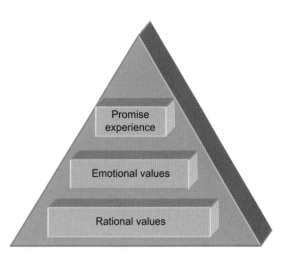

FIGURE 8.2

Understanding the nature of a brand

In the case of financial services, people are reticent about sharing experiences and tend to be wary of branding experiences that involve participation with others. However, investors feel empowered through the richness of information to express disparaging views about unacceptable corporate practices and therefore signal their opinions about which organisations should not receive investments from an ethical bank. If the ethical bank is not attentive to the growing views of their community, an ever-amplifying message of distrust will damage the brand.

The promised brand experience through the internet needs to take account of the way that consumers are co-creators of the brand experience. For example, Weight Watchers, through their brand promise of aiding slimming, would appear to have support and accountability as two of its brand values. In the bricks and mortar environment, this brand promise is delivered, amongst other ways, by having regular supportive group meetings with 'weigh-ins'. When this brand migrated to the internet (www.weightwatchers.co.uk) the brand values and brand promise remained similar but the brand enactment changed. In terms of guided eating, subscribers have access to a database which assigns points against different types of food to facilitate diet tracking. Online recipes and the encouragement for individuals to have their own progress charts are provided. A menu planner is available and to make it easier for the subscribers to keep to their diets, a shopping list tool is available which prints the items that need purchasing for a week's menu. There is a "panic button" which provides support if a member is seriously tempted to eat forbidden food. Each individual can personalise the site through their inspirational icon and a trophy cabinet stores stars for weight loss. If it is difficult to attend the regular group weigh-in meetings, support is provided through a dedicated chat room replicating the community of weekly meetings. Moderators watch out for any wrong advice that is offered in the chat room.

THE IMPORTANCE OF THE ONLINE BRAND EXPERIENCE

The way a visitor draws an inference about a brand online is different from the offline environment where conventional factors such as staff and the store environment have a notable impact. The brand triangle in Fig. 8.2 enables the brand's team to define the characteristics of their brand promise and its enactment can be appreciated from the elements shown in Fig. 8.3. This model enables the coherence of the brand to be assessed. For example, if one of the rational values underpinning the brand promise is convenience, yet the speed with which the brand's site is downloaded is slow and navigation of the site is difficult, there is a need for further work on the brand's site. It is important to recognise that the brand experience is not just assessed on the content of a site but rather through an amalgam of the elements in Fig. 8.3. Each of these elements will be addressed.

(i) Locating the brand and speed of download

The first challenge facing a potential visitor is that of locating the brand on the internet. Their task might have been facilitated as there could be a link between the site they are currently using and the new brand, referred to as affiliate marketing. This is a strategy employed by iTunes to get more visitors to their site. For example, the Apple iPhone offers an app. called 'Shazam'. When a customer hears a piece of music but does not recognise the song name or artist, activating Shazam allows the iPhone to 'tag' the song. The name of the song and artist are provided to the customer, who can use their phone to access iTunes and download the song directly to their iPhone.

In the case of linked brand sites, visitors may draw some inference about site similarities, however trivial. It is therefore wise for brand marketers to have a sound rationale for linkages, rather than being just driven by the opportunism of further sales. Furthermore, some thought about the bridging between the two sites is needed. For example, a customer seeking premium flights may appreciate a weblink to car rental sites and luxury hotel accommodation.

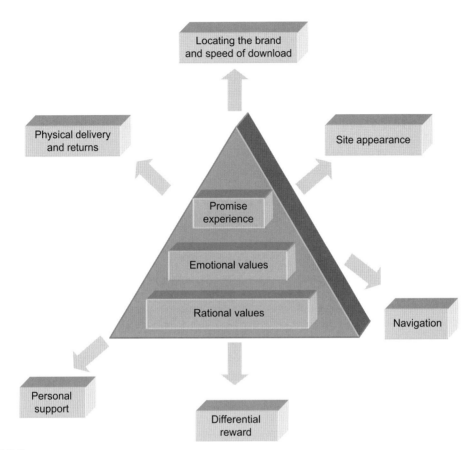

FIGURE 8.3
Enacting the brand promise online

Another way for a visitor to find the brand is through well-targeted banner advertisements. However, managers must be mindful of the media used and target audience. For example, many customers carry mobile phones, which incorporate web browsers. A study in Ireland by Mediaedge:cia (Adworld, 2009) found that 60% of customers would not watch banner ads on their mobile phones unless they carried vouchers, discounted apps. or offered entertainment. Therefore, managers must consider the benefits offered to consumers and build these positive messages into the creation of their banner ads.

An alternative location strategy is for an individual to either search directly for a specified brand or for them to search details about a particular category. Ward and Lee (2000) showed that as users gain more experience of the internet, they become less reliant on brand names. This does not imply that brands no longer act as a rapid route to brands' sites. Rather it implies that brand names can be useful mechanisms short cutting the search to a brand's website. Provided users then perceive the site to be credible, they have confidence to better appreciate the brand's characteristics through searching out more information from the site. The challenge becomes one of creating positive associations and awareness offline, then engendering trust online to enable the user to better understand the brand.

Seeking to move a brand name to the top of the names published in a search is one way of raising a brand's profile. As each of the search engines uses different rules for the order in which brand names emerge, consultants offer search engine optimisation campaigns to increase a brand's information market share. To successfully implement search engine optimisation, firms must understand the key words consumers use when searching for their site or product categories. The company must ensure that its website carries relevant keywords and phrases in its 'tags', to ensure that the site appears in a prominent position on the customer's search.

Migrating a brand to the internet evokes the automatic assumption that there will be a consistency of brand names. This assumption may in some cases be erroneous. Using "&" on the internet is problematic, so B&Q registered their online brand as www.diy.com. A good brand name on the internet, if a pure internet brand is being created, should be intuitive, memorable, specific and short. Creating a brand offline, with intent to subsequently capitalise on the internet, should involve a search as to whether the name is already in use.

Speed of download becomes the next issue after locating the brand's website. A home page which has animations and sound may be impactful but this could cause irritation waiting for the page to form. To then be greeted by a need to download special software to realise the full benefits of the brand's site can further irritate and encourage visitors to rapidly depart. Consumers' image of a brand's site is influenced by the speed of download. If the organisation is insistent on having a website that will take time to download, having a countdown mechanism will at the least make consumers aware of this issue and possibly reduce some of the irritation.

(ii) Site appearance

The site's appearance engenders a view about the brand's promised experience. First appearances are important as visitors are rapidly scanning to decide whether to continue. At this point, it is worth considering the extent to which the home page extends a warm welcome, with a notable emphasis on greeting the visitor or whether it jumps to being company centred and is biased to transmit information. As the mouse moves around the screen, chance movements over particular areas that prompt new pictures can prove engaging. The style of writing, the 'tone of voice' and the initial colour all give clues about the brand's character. For example, Zucker (2009) observes how luxury fashion brands have migrated to the internet. Brands moved online to follow their customers who now spend more time online on Facebook or Twittering. Fashion brands are offering bloggers front row seats at shows as their influence on other consumers is immediate. In addition, top designers, such as Dolce and Gabbana and Louis Vuitton, streamed their fashion shows online for the first time in 2009 to capture the online audience. Burberry offers an excellent example of using their website appearance to build their brand image online. The brand is synonymous with the trench coat, which was originally created for use by soldiers. Burberry's 'Art of the Trench' website (http://artofthetrench.com/) uses social networking to present photographs of different people in Burberry coats. This provides a point of difference for the brand, engages consumers and allows customer participation in research, as interested customers can state which coats they like and dislike (Zucker, 2009).

(iii) Navigation

Navigating the site provides a further array of clues about the brand. For example, university websites meet the needs of a wide range of stakeholders including current and prospective students, alumni, employers and staff. Students have different information needs from day to day. They may wish to locate a lecturer's contact details, find out about past exam papers or view library articles. New students may wish to find out about course programmes or fee structures. Employers may wish to find out about programmes or student exemptions. Therefore navigation must be logical to the user. When the National University of Ireland Galway relaunched its brand, it also revitalised its website (www.nuigalway.ie). Previously, information was presented according to its categorisation within university structures. In the past, students often had simple queries such as 'How do I apply for a postgraduate course?' or 'How do I register for my course?'; yet they were required to 'drill down' within sections to find the information they needed. Altering the layout of the site to address consumer needs simplified its appearance and significantly improved its navigation. This supported the university's brand and its message of 'real learning'. An understanding of the website's customers through research can prove invaluable here to understand the informational pathways that visitors expect to follow.

(iv) Differential reward

Differential rewards should alert the brand's team to consider why someone would want to use their web-based brand, rather than have an offline interaction. Is it because managers think it is much easier for a user to gain their personalised information, while a consumer might find the offline staff behaviour unhelpful — and the web negates the need for any personal contact? For example, online banking may offer privacy and convenience to some customers, while others who prefer the personal interaction at the branch may feel 'pushed away' into an online environment. Some organisations take advantage of the internet to augment their brands. For example, enabling the online brand to become co-created by visitors is an increasingly important aspect of the enactment of the brand's promise. Dell's online brand (Dell.co.uk) enables a visitor to "build" their own computer and at each stage to learn about the price. Dell has extended this feature into a key selling message in their TV advertising. Consumers can personalise their laptops and decide on trade offs between enhanced features and increased prices.

(v) Personal support

The personal support component of the brand recognises that visitors use several channels to make a decision. There are frequent instances where the presence of a helpline biases visitors' preferences in favour of using a brand. For example, several hotels' online brand sites provide details about availability and prices for one-, two- and four-bedroom accommodation. However, for the three-person family seeking a room, this is not particularly helpful. By having a central telephone line that can be contacted 24 hours a day, a visitor can rapidly make a decision. For example, Dell provides access to a person in technical support if a potential consumer is unsure about how best to configure their potential computer. When an e-mail contact helpline is available rather than a customer service employee, it should be recognised that it takes a visitor a few minutes to pose an electronic question but a reply can sometimes take hours or even days. Organisations therefore need to build a prompt response into their systems.

(vi) Physical delivery and returns

Physical delivery is another way the brand's promise is tested. Some online brands have had considerable investment in their sites, with less thought about the logistic system to facilitate physical distribution. In addition, organisations need to consider how the delivery representative interacts with the consumer. For example, Amazon.com allows customers to view their orders and track delivery via their online shopping basket. As another moment of truth, which enables users to assess a brand, there is scope for competitive advantage by ensuring the delivery staff is aware of the values of the brand being delivered, in order to act in a way that reinforces the brand. In case of problems with the brand, there needs to be consideration of the most suitable way for purchasers to return the brand and protection provided to both buyers and sellers. Customers and sellers who use online sites such as eBay are offered protection through the use of PayPal (www.paypal.com). Customers benefit as their account details are not provided to the seller and sellers have greater protection as payment is lodged directly to the seller's account.

THE EVOLVING ROLE OF THE INTERNET IN BRAND BUILDING

The internet is both an opportunity and also a threat for brands. This section will focus on the golden opportunities for brands, before moving on to consider some of the threats. While there are numerous opportunities, these will be addressed under the headings of Web 2.0, social communities, customer participation in brand building and the future role of the web in brand building.

i) Web evolution and Web 2.0

So far, we have discussed examples of websites where information is provided or e-commerce occurs. This form of online activity is classified by Kambil (2008) as Web 1.0. Since the late 1990s, Web 2.0 has received significant attention. The web has become a collaborative platform with collective users sharing power. This challenges managers to tap into user-generated content and cope with the issue of transparency (Kambil, 2008). Examples of Web 2.0 technologies include social networking sites, wikis and shared workspaces, blogs and podcasts, polling and social bookmarking (Chui et al., 2009). Such technologies present opportunities for new internal management practices but also for customer interaction with brands and with each other. Online communities have created new opportunities for brands and we discuss examples of these next.

ii) Building social communities

Carlson et al. (2008) explain the concept of brand communities as networks of individuals joining groups of like-minded brand admirers. Where there is no social interaction, they explain that these communities can be classified as psychological communities; and those where customers interact together can be classified as social communities. Where the community is psychological, the customer is most interested in the brand, but not necessarily with other parties. Muniz and O'Guinn (2001) describe the characteristics of social communities. They:

■   Have a consciousness of some kind — there is a sense of belonging, that members are different to non-members;
■   Have rituals and traditions — members share values and members reproduce the community meaning through their actions online and offline;

- Have a sense of moral responsibility — community members feel morally committed to the group, they integrate members by sharing experiences and they support the correct use of the brand.

Social communities therefore are powerful ways for marketers to harness individuals who are loyal to the brand, to create stronger customer bonds and a sense of belonging to a group. Sicilia and Palazón (2007) explain the social community relationship as a triangle since members have two types of relationship — one with the brand and the other with other community members.

Weber (2009) argues that marketers will become 'aggregators of social communities'. In other words, marketers will begin to speak *with* customers, instead of *to* them through traditional media. Weber (2009) explains that 90% of people avoid TV ads by using TiVo, DVD recording or their remote controls. Therefore the web offers companies a way to communicate with customers who share a common interest and can express their thoughts and opinions about brands. He provides several examples, including Microsoft, who use Xbox Live to create a community entertainment experience (http://www.xbox.com/en-US/LIVE).

Traditionally, many of the virtual communities were developed for high-involvement brands. A good example of a community is Harley Davidson's HOG (the Harley Owners Group), which helps to make 'the Harley Davidson dream a way of life'. Social communities can also be effective for more consumable brands in the fast moving consumer goods (FMCG) category. When social communities use an online platform to communicate and where this platform is provided by the company, this is known as a virtual brand community (Sicilia and Palazón, 2007). Thus the brand itself can create a social space within its website. An example of an FMCG brand that successfully implements a social space online is Coca-Cola. Sicilia and Palazón (2007) examined its operation in Spain (http://www.cocacola.es/home). Coca-Cola chose young people as a key segment to build their virtual community. For Coca-Cola, the space allows customers to interact through an internet platform. The site invites users to join by providing details such as email address and demographics, as well as customer information about product usage and purchasing habits. This provides invaluable information to Coca-Cola. On the other hand, customers participate in a loyalty programme where they are rewarded for repeat buying. Customers get their loyalty points online, which, along with lively content, turns visitors into community members. The community has been very successful for Coca-Cola, with 50,000 daily hits. Consumers wishing to have a 'space' in which to describe their brand experiences can join an online community. For example Coca-Cola's Spanish website (www.coca-cola.es/home) offers space for social networking.

Another example of a strong FMCG community is the Italian Nutella brand community. Nutella is a hazelnut spread owned by the Ferrero company. Cova and Pace (2006) explain that Nutella 'is a metaphor for desire, softness, sin, idle transgression, sometimes followed by immediate doubt and a sense of guilt'. They also explain that Nutella is a 'she', with the female suffix 'ella' as part of the brand name. With this insight, Ferrero set out to create a community of Nutella fans in Italy, the brand's birthplace, where it has 90% of the cream spread market (Cova and Pace, 2006). The purpose of the site was to spark dialogue with the tribe of Nutella fans, create feedback for marketing and inform customers about offline events. As Nutella is a cult brand, users are passionate and share their thoughts about the role of Nutella in their lives. Unlike Coca-Cola who provide an interactive space to create social interaction without necessarily discussing the brand, the Nutella site, www.mynutella.com, allows fans to share their thoughts about Nutella, discuss their first experience of using Nutella, publish a diary of their life with or without Nutella and create sub-groups of fans within the brand's tribe.

Finally, Casaló et al (2009) set out methods of encouraging customer participation in brand communities. In their study of Spanish speaking members of virtual communities, they found that trust in a community increases participation in that community; and participation with an online brand community increases the customer's affective commitment, or intention to remain with the brand. They recommend that managers promote virtual meetings among members to promote communication and cohesion, offer benefits to satisfy members' needs and evolve to meet changing consumer needs.

iii) Customer participation in brand development

In addition to membership of brand communities, consumers can spread the brand message through viral marketing. Viral marketing refers to the process where customers pass a company's marketing message on to friends and colleagues (Dobele et al., 2007). Many companies have used viral marketing, often through YouTube videos. The accessibility of YouTube is evident in the popularity of its videos. For example, the official trailer for the Twilight movie generated over two million views. When videos create an emotional connection, consumers are inspired to forward them to friends. The Blair Witch Project is an example of successful viral marketing. Before the film opened, people were accessing and talking about the website (www.blairwitch.com/), which suggested that the events depicted in the film were based on reality and frightened viewers. Dobele et al. (2007) explain that viral marketing is particularly effective for Generations X and Y, as they are 'jaded' with conventional media. The use of emotions including surprise, joy, sadness, anger, fear or disgust can provoke customers into action or communication with others. What is most important for managers is to achieve a fit between the emotion and the message, to target their message appropriately and to capture the imagination (Dobele et al., 2007). An example of viral marketing was T-Mobile's Trafalgar

Square Sing-Along campaign on 30th April 2009 (Gidman, 2009). A sudden gathering, known as a 'flash mob' was generated by an email which contained detailed instructions. The company attracted nearly 14,000 people to Trafalgar Square to sing along to the Beatle's 'Hey Jude'. The resulting content was used by T-Mobile across media on TV advertising and on YouTube, where the video generated almost two million views (http://www.t-mobile.co.uk/sing/).

Consumer advertising can also be created independent of the brand's input. Berthon et al. (2008) explain that advertisements have been liberated from company control and 'now expresses a myriad of heterogonous voices'. Consumers create advertisements about brands for enjoyment; for example, consumers who place videos of themselves using products on YouTube. Berthon et al (2008) advise mangers to facilitate this kind of advertisement development. Consumers also create advertisements for self-promotion, perhaps to piggyback on the success or positive image of the brand. Berthon et al. (2008) give an example of an iPhone advertisement created in New York by Alec Sutherland. Finally, consumers create advertisements to change perceptions by promoting or disrupting the brand. Managers can choose to repel such activities but run a risk of appearing like a 'Goliath' fighting a weak customer in a market, which does not see the 'attack' as a major issue (Berthon et al., 2008). However, managers must be aware that the internet offers challenges and we address these in the next section.

In addition to communication, consumers can be actively engaged in co-creating their product or service. For example, iTunes customers can post play lists of their chosen songs which other music fans can listen to, rate and purchase. Nike offer a sub-brand, iDNation, which allows customers to customise footwear or clothing. For example, a customer can select 'running shoe' from the footwear category and choose colours, materials and technology to customise the shoe to their needs. Mini also offer a customisation option for customers, who can select colours, bonnet stripes and wheel types to create a bespoke car.

It should be recognised that when people visit a brand's website they don't just want to interact with the brand, but also with like-minded people. The brand is the host who needs to welcome new visitors, facilitate introductions to other visitors and ensure that in a relaxed and open environment, visitors can form their own community. The role for the brand needs to be made explicit. There are instances where the brand needs to be unobtrusive and it gains respect as the community members decide when they want to ask questions about the brand. By contrast, there are other instances where the brand as a host plays a more obtrusive role through actively moderating cyber conversations. One way of conceptualising the extent and nature of the brand's intervention in community conversations is to consider how the brand's values lead to particular types of relationships. For example, building on some research once presented by Research International, one could characterise a brand's relationship with its community in terms of two dimensions, as shown in Fig. 8.4. The first dimension relates to the extent to which the brand is dominant ('let me tell you about....') or submissive ('what would you like to know about....'). The second dimension describes whether the relationship is close ('hello Peter, welcome back') or distant (e.g. treating a returning visitor as if they were a first time visitor).

Using the matrix in Fig. 8.4, it is possible to identify four roles for the brand as it acts as a host to a community. Where the brand values translate into a relationship best described as 'critical expert', there are 'visible' moderators who seem to the community to be controlling the conversations, in sofar as there appear to be guidelines about the nature of dialogues, which are strictly enforced. The tone of voice is akin to being stern and impersonal. The role of 'cold servant' is seen when the brand appears distant and submissive. No attempt is made to intervene in dialogues and the brand only responds when requested, but even then without any hint of personalisation. In the friendly guide role, the brand is welcoming, warm, personable but possibly seen as being 'fussy', taking the lead role in conversations and wanting to always show its knowledge. By contrast in the final quadrant, where the brand is a cooperative question answerer, it is more relaxed and it doesn't come across as being akin to a nervous individual, who seeks to control conversations.

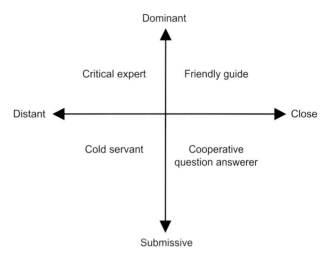

FIGURE 8.4

The brand's hosting role with its community

THE CHALLENGES FACING BRANDS ON THE INTERNET

There are numerous challenges that brands face on the internet and this section will consider some of these under the headings of shifting power, comparison-shopping, pricing and service.

(i) Shifting power

As authors such as Dussart (2001) and Mitchell (2000) have shown, the balance of power is swinging away from distributors and manufacturers to customers. Through the wider dissemination of information and reduced search costs, consumers are more able to signal their discontent about the way organisations deal with them and new responses can be seen. Informediaries are emerging, who act as spokespersons for consumers. Their new line of marketing is akin to 'I have a very large number of people interested in buying brands in the category. What price will your organisation bid for their business?'. Forward thinking organisations are responding to the growth of consumer participation by devising channels that facilitate brand selling and purchasing. For example, auction sites are emerging that enable consumers to exchange brands with each other; for example, eBay (http://www.ebay.co.uk/) allows customers to auction brands across a diverse range of product categories.

Through being far more attentive to evolving needs and responding faster, brands will continue to thrive on the internet. The shift for brand owners will be away from *talking to* customers to *listening to* customers. An example of this is the site WeddingGuideUK.com. (http://www.weddingguideuk.com/). This has a notable number of message boards that revolve around marriage and the subsequent lifestyle changes, for example, grooms forum, wedding forum, fashion and beauty forum and weight loss forum. The organisation takes a back seat and listens to the conversations. It analyses emergent trends and frequently raises issues, then provides contact details about organisations who offer goods and services relevant to the discussion topics. Examples of these might include florists and photographers. The host site makes its money through changing the service providers. Another example of a site targeted towards a specific market segment is www.rollercoaster.ie. This Irish site provides pregnant women and new parents with advice and allows users to share discussions across a range of issues, such as advice for mums to be, baby names and nutrition. Such sites are invaluable for brands, as loyal customers can share positive 'word of mouse'. In addition, brand managers can gain invaluable market insights by reading about customers' product concerns and their issues with the competitors' products.

While this shift of power towards consumers will necessitate new ways of marketing brands, it does not signal the death of brands. Customers are more cautious on the internet and brands that have built trust will be more warmly considered by consumers, as they represent guarantees about issues such as quality, reliability or consistency.

ii) Comparison shopping

Brands are evolving that, as the consumer's guardian, are able to search out brand information and undertake comparison shopping. Examples of these include Kelkoo.co.uk, Halifaxshoppingextra.co.uk and Travelsupermarket.com. These comparison brands gain consumers' trust when they establish themselves as being independent and reputable. A powerful route to winning trust is to have high quality editorial content, preferably written by independent teams of writers. Blocking requests for information could damage goodwill, since a reluctance to supply information may be noted by the inquiring brand and the position of this response may lead potential purchasers to wonder whether they wish to deal with a secretive organisation.

It should also be recognised that consumers' buying criteria are numerous and price is but one factor. Trust and convenience are some of the reasons why consumers will not solely depend upon low prices. As such, comparison shopping brands should not be treated as a threat but rather recognised as another implication of greater transparency. Their evolution should encourage organisations to ensure they have fully identified their brands' benefits. They should be regarded as potentially powerful allies communicating the cluster of advantages inherent in each competing brand.

(iii) Pricing

As previous sections have explained, the internet makes price comparisons easier. Price comparison sites, such as Pricescan.com, rapidly compare the prices and brand characteristics of a wide variety of categories. With the advent of sites such as Pricerunner.co.uk and ebay.co.uk, consumers have begun to consider brands as being open to negotiation. In addition, as buyers had to state the price they are willing to pay for an item, when they are granted this price they may try for a lower price next time.

When brands offer their products to different international markets, this can also create pricing problems. For example, customers in the UK will expect to pay Sterling prices, while other EU customers may expect to pay in Euro. Currency fluctuations and local taxes can result in different prices in different markets. While this price difference is not as evident in a physical retail environment where a price tag bears the local price, when customers shop online they may feel they are getting better value by shopping in another currency. To offset the impact that this may have on sales at a local level, many retailers are now launching country-specific websites to prevent customers ordering from an overseas site.

There is a perception that when brands are marketed on the internet they will be cheaper. However as we have already noted, price reductions may be made more to those who would be reluctant to ask for discount in a face-to-face situation (Zettelmeyer et al., 2006). This study showed that customers can be segmented according to their beliefs about pricing. Providing prices online enables customers to be better informed than other buyers and where buyers dislike bargaining, they can expect to pay less online (Zettelmeyer et al., 2006).

Ratchford (2009) found that features of online shopping affect pricing. These include switching costs, reputation, heterogeneity in search costs, heterogeneity in demand for services and offline pricing. Therefore, some sites will be perceived by customers to offer better value than others.

(iv) Service

It is not infrequent to find organisations developing websites for their brands that assume consumers' questions can all be adequately answered by just the internet. Many customers want to talk through questions, yet there is not always a telephone number to call. In such instances many consumers find it easier to click through to a competitor's brand. By looking at a brand's website through the consumers' eyes, the limitations of a single channel for customer service can be appreciated. Many sites now offer Frequently Asked Questions (FAQ) sections, which present typical queries that consumers may have. For example, the US Government provides a questions and answers section on their IRS site for customers unsure about taxation issues (http://www.irs.gov/faqs/index.html). An important aspect of e-service is its quality. Studies have found that e-service quality is multi-dimensional. Among customers of online book and record stores, Ribbink et al. (2004) found that variations in e-service quality resulted in e-satisfaction; and that satisfaction was influenced by an attractive, easy to use interface. Another study was conducted on online banking customers in Australia (Herrington and Weaven, 2007). In this sector, trust is important and this is created through security and privacy of information. Herrington and Weaven (2007) had some interesting findings. They discovered that 'efficiency' did not affect customer satisfaction, as this was a feature that customers expected automatically with online banking. They also found that 'functional, safe and reliable' were insufficient in generating customer satisfaction. A key problem for customers of online banking was the absence of human interaction. Herrington and Weaven (2007) recommend that bank e-retailers should also consider the role of traditional services in relationship building. Therefore, in some sectors it may be best to combine the convenience of e-service with the human contact of physical premises. However, for banks who offer an online service only, managers must find other ways of creating relationships; for example, through online communities.

BRAND EQUITY OF ONLINE BRANDS

Argyriou et al (2006) explain that brand equity models have traditionally assumed advertising as a core brand building activity. However, they explain 'elements of brand equity (i.e. awareness) have been linked to online marketing activities'. Further, in the online context, building equity with customers extends to building relationships with all stakeholders including employees, buyers, suppliers, as well as customers (Argyriou, 2006). Jevons and Gabot (2000) suggest that online brand equity is more akin to services brand equity because the customer is interacting with an intangible object and trust is critical. For example, a customer who is buying a new book through Amazon.com is reliant on other customer reviews and the online imagery to make their choice. Jevons and Gabot (2000) explain that the customer's first experience of the brand is about creating trust and creating a positive experience for the customer, rather than creating prior expectations. However, to create equity, it is important that the consumer perceives value in visiting the brand online and receives a positive brand message. Stuart and Jones (2004) explain that the purpose for some websites is to allow customers to book or check and pay their bills, which reduces costs to the firm. Stuart and Jones (2004) explain that this is often evident in service firms, for example, in banking, air transport and telecommunications. In building brand equity, Stuart and Jones (2004) caution that the brand message arising from directing customers online should not be 'Do it yourself so that we wont have to provide any service'!

Christodoulides et al. (2006) created a scale, which managers can use to measure brand equity among customers. They suggest that customers evaluate online brands across five aspects:

— Emotional Connection — the affinity between the customer and the brand. For example, someone may only use Skype for talking with a friend abroad and the association of Skype bringing a sense of togetherness may result in a strong emotional connection with this brand.
— Online Experience — how the customer experiences the brand in real time. A customer who enters their financial details online to apply for a mortgage, only to discover that the site has crashed, will have a poor online experience and this will negatively affect the brand.
— Responsive Service Nature — the response and level of the customer service interaction provided. For example, a customer ordering a Dell computer has an opportunity to have a Dell expert install their system, which offers reassurance to the first time buyer who may be overwhelmed by technical specifications.
— Trust — the level of confidence customers have in the brand's reliability and their intentions where they perceive risk. For example, when booking a flight online, customers may look for a 'lock' icon to reassure them that their details are safe.
— Order Fulfilment — provides a connection between online and offline. Customers using ASOS.com, the online fashion store, are instructed to follow five simple steps to select their chosen items. This ensures that the goods they receive match their order and that the customer is satisfied.

CONCLUSIONS

This chapter has explored how to help grow brands on the internet. It has shown that brand building in a cyber environment necessitates managers having a new mindset. For example recognising that there are communities rather than customers, that there is need to listen rather than talk to internet visitors and that communities co-create value rather than passively consume brands. In addition, managers must recognise that customers seek a positive experience from online brands; and this can be created by interaction with the brand or with other like-minded customers. We believe a brand is a brand regardless of its environment — what is different is the way the brand's promise is executed. A brand that exists in a bricks and mortar environment can be migrated to the internet using the same brand promise, provided it remains true to its original values. If not, a schizophrenic brand results, giving rise to confused consumers who will leave.

As organisations become more confident with branding on the internet, their strategies are less likely to be just about providing considerable information for visitors. They will, rather, involve them in desired experiences and treat them as trusted partners. The rational, functional characteristics of the brand will have formed solid foundations, enabling management to develop more tailored experiences that reflect unique, emotional and rational values.

The coherence of the brand can be assessed from the extent to which its brand triangle is reflected in the experience that visitors have. This involves assessing factors such as the ease of locating the brand and speed of download, site appearance, navigation, differential reward, personal support and physical delivery/returns. In certain sectors that carry additional perceived risk, we explain how the brand image created by the internet can be supplemented by the relationship created by human interaction with front line staff.

Successful internet branding will generate strong brand equity. The intangible nature of online brands means that creating brand equity on the internet is similar to creating equity for services brands, thus managers are challenged to create tangible cues about brand value. To create brand equity, managers should be less concerned with creating prior expectations and should focus instead on the creation of a positive experience online. To enhance equity, the internet should also create an emotional connection with the customer, provide responsive service, establish trust and ensure that online and offline services are consistent.

MARKETING ACTIONS CHECKLIST

It is recommended that after reading this chapter, marketers undertake some of the following exercises to test and refine their brand strategies.

1. The first section of this chapter considered some of the differences that require new thinking when marketing brands on the internet. It may be helpful to convene the brand's team and to get them to consider how

appropriate the current business model is for your brand, given the differences identified. For example, given that communities want to be more involved in the co-production of value, how are you encouraging and responding to this? Has your organisation adopted an incremental model to branding on the internet or is there any radical thinking?

2. After talking with your colleagues in the brand marketing team, summarise why you are marketing your brands on the internet. If you have a few reasons that primarily centre around cost saving and increasing sales without any customer based reasons, it is worth reconsidering your internet branding. While it is important to increase sales profitably, it is important to recognise that this comes about because of customers appreciating the value in your brand. If this scenario applies to you, how could you refine your approach so there is also a benefit for customers?

3. If you have not done so, it is worth undertaking interviews with some of your customers to understand why they use the internet when interacting with your brand. This should enable you to appreciate the importance of different motivating factors and where price is ranked. Given these findings, it may be helpful to explore with customers how closely your internet branding meets their motivational criteria and therefore what changes may be suited to better satisfying their needs.

4. Given that brands evolve on the internet from emphasising copious brand details to tailoring information based on better knowledge of the customer, to an involving experience as a trusted partner, which of these categories best describes your brand? Where do your competitors' brands reside on this typology? How can you uniquely develop your brand so that its trajectory encompasses the involving experience category?

5. One way of characterising a brand is through the brand triangle, which portrays a brand as a cluster of rational and emotional values and enables a unique and welcomed promise about an experience. Taking your offline brand, what are its three rational (functional) values, its three emotional values and what is the promised experience? Repeat this exercise for your online brand. Where there are any inconsistencies, consider how changes could be instigated to ensure homogeneity.

6. Undertake interviews with key stakeholders about your brand on the internet. Firstly, explore with them the experience they perceive from your brand. By assessing this against your intended experience, you can assess amongst which stakeholder groups there is greatest need for actions to better align the intended and perceived brand promise. With respondents close to their internet connected pcs, ask them to locate, then explore your brand's website. Ask them to specify the brand that most closely competes with your brand, then get them to locate and explore this brand's website.

Having undertaken this online exploration, probe their views about your brand's website regarding:

— locating your brand and speed of download;
— site appearance;
— ease of navigating around the site;
— differential rewards;
— extent to which they perceived there is personal support for the brand;
— physical delivery and returns (if they have experience).

With this insight, consider how your online brand can be refined to better match its desired experience.

7. What are you doing to build genuine relationships between your brand and its communities on the internet? Whenever you ask internet visitors for information, are you giving them something in return? How could you incorporate new technologies such as blogs and wikis into your website to deliver an enhanced customer experience?

8. From what you know of your internet users, consider how you could further segment these users. The intention is that you might be able to offer these finer segments better personalised variants of the brand. For each newly devised segment, are you still likely to generate satisfactory returns on your investment?

9. By talking with your stakeholders, appreciate their views about the limitations which they perceive as restricting them from having greater interactivity with your brand. Consider how, as a result of their comments, you could enhance the interactivity of your brand's website. While talking with your stakeholders, also assess the extent to which they perceive the online relationship as being dominant versus submissive and distant versus close. Using these two dimensions, then assess where your brand is positioned, referring to Fig. 8.4. If the stakeholder perceptions disagree with your intentions, consider what changes would help.

10. Undertake some interviews with customers who have recently visited your brand's website as they were actively looking to purchase a brand in this category. Thinking about the service they experienced, what are their views about:

- the extent to which your site met their expectations;
- whether they were pleased with any options to personalise the brand;
- the time it took to reply to their enquiries;
- the extent to which the information on the site was seen as useful.

11. Examine the websites of your competitors' brands and then assess the extent to which your brand and the competing brands are tightly controlled by the organisations. How appropriate would it be and how could you begin to relax control so your brand moves closer into its community's domain?

12. Examine your website against the five cues customers use to determine brand equity. Are there areas you could develop further? How are you creating trust and an emotional connection with your customers? To what extent do the actions of your front line staff support your web activities?

STUDENT BASED ENQUIRY

- Select two websites, one which uses Web 1.0 and one which uses Web 2.0 technologies to interact with its customers. Evaluate how each site's approach helps to build its brand. Which approach do you think is most effective?
- Conduct an online search for brand communities, select one that is initiated by a brand to encourage its customers to interact and one that is initiated by consumers as an anti-brand site. Explore the tone and content of the dialogue used in each site and discuss how customer interaction can help or harm a strong brand.
- Select a social networking site you are a member of, or are familiar with. Outline three brands that, in your opinion, would benefit from using the site and for each brand, outline two ways in which the site could be used to build a strong customer experience.
- Search for the online site of a physical product or service you regularly use. Comment on the role of: i) search engine optimisation, ii) the domain name, and iii) site navigation, in supporting or detracting from your perception of the brand.
- Choose i) a skiing holiday, ii) a current CD, or iii) a fashion item and search for the best price online. How do price comparison sites influence your decision? Comment on the role of brands in your choice.

References

Adworld (2009). *Irish consumers don't like mobile advertising.* Available at: http://www.adworld.ie/news/read/?id=1a2ce80f-7ff4-4a4e-a6ad-14075320ddef. Accessed 7 December 2009.

Argyriou, E., Kitchen, P. J., & Melewar, T. C. (2006). The relationship between corporate websites and brand equity: a conceptual framework and research agenda. *International Journal of Market Research, 48*(5), 575–599.

Berthon, P., Pitt, L., & Campbell, C. (2008). Ad lib: when customers create the ad. *California Management Review, 50*(4), 6–30.

Christodoulides, G., de Chernatony, L., Furrer, O., Shiu, E., & Abimbola, T. (2006). Conceptualising and measuring the equity of online brands. *Journal of Marketing Management, 22*, 799–825.

Chui, M., Miller, A., & Roberts, A. P. (February 2009). Six ways to make Web 2.0 work. *The McKinsey Quarterly.*

Cova, B., & Pace, S. (2006). Brand community of convenience products: new forms of customer empowerment — the case 'my Nutella the community'. *European Journal of Marketing, 40*(9/10), 1087–1105.

Dobele, A., Lindgreen, A., Beverland, M., Vanhamme, J., & van Wijk, R. (2007). Why pass on viral messages? Because they connect emotionally. *Business Horizons, 50,* 291–304.

Dussart, C. (2001). Transformative power of e-business over consumer brands. *European Management Journal, 19*(6), 629–637.

Gidman, J. (2009). *Guerrilla event marketing: a mob in a flash.* Available at: http://www.brandchannel.com/features_effect.asp?pf_id=493. Accessed 8 December 2009.

Jevons, C., & Gabbott, M. (2000). Trust, brand equity and brand reality in internet business relationships: an interdisciplinary approach. *Journal of Marketing Management, 16,* 619–634.

Kambil, A. (2008). What is your Web 5.0 strategy? *Journal of Business Strategy, 29*(6), 56–58.

Muniz, A. M., Jr., & O'Guinn, T. (2001). Brand Community. *Journal of Consumer Research, 27,* 412–432.

NME. (2008). *Radiohead reveals how successful 'In Rainbows' download really was.* Available at: http://www.nme.com/news/radiohead/40444. Accessed 7 December 2009.

Ratchford, & Brian, T. (2009). Online pricing: review and directions for research. *Journal of Interactive Marketing, 23*(1), 82–90.

Ribbink, D., Allard, C. R., van Riel, V. L., & Streukens, S. (2004). Comfort your online customer: quality, trust and loyalty on the internet. *Managing Service Quarterly, 14*(6), 446–456.

Sherwin, A. (2007). *Name your price for the latest Radiohead album.* Available at: http://entertainment.timesonline.co.uk/tol/arts_and_entertainment/music/article2569511.ece. Accessed 7 December 2009.

Sicilia, M., & Palazón, M. (2007). Brand communities on the internet: a case study of Coca-Cola's Spanish virtual community. *Corporate Communications, 13*(3), 255–270.

Stuart, H., & Jones, C. (2004). Corporate branding in marketspace. *Corporate Reputation Review, 7*(1), 84–93.

Ward, M., & Lee, M. (2000). Internet shopping, consumer search and product branding. *Journal of Product & Brand Management, 9*(1), 6–20.

Weber, L. (2009). *Marketing to the social web.* London: Wiley & Sons. www.shazam.com/. Accessed 7 December 2009.

Zettelmeyer, F., Scott Morton, F., & Silva-Risso, J. (2006). How the internet lowers prices: evidence from matched survey and auto transaction data. *Journal of Marketing Research, 43*(2), 168–181.

Zucker, S. (2009). *Luxury fashion houses woo consumers online.* Available at: http://www.brandchannel.com/home/post/2009/12/04/Luxury-Fashion-Houses-Consumers-Online.aspx. Accessed 8 December 2009.

Further Reading

Aaker, D., & Joachimsthaler, E. (2000). *Brand Leadership.* New York: The Free Press.

Ancarani, F. (2002). Pricing and the internet: frictionless commerce or pricer's paradise? *European Management Journal, 20*(6), 680–687.

Annon. (September 2000). How internet booking cuts costs. *Internet Business,* 59.

Baker, W., Marn, M., & Zawada, C. (February 2001). Price smarter on the net. *Harvard Business Review,* 122–127.

Bezjian-Avery, A., & Calder, B. (1998). New media interactive advertising vs traditional advertising. *Journal of Advertising Research, 38*(4), 23–32.

Carlson, B. D., Suter, T., & Brown, T. (2007). Social versus psychological brand community: the role of psychological sense of brand community. *Journal of Business Research, 61*, 284–291.

Casaló, Luis, V., Flavián, C., & Guinalíu, M. (2008). Promoting customer's participation in virtual brand communities: a new paradigm in branding strategy. *Journal of Marketing Communications, 14*(1), 19–36.

Chaffey, D., Mayer, R., Johnston, K., & Ellis-Chadwick., F. (2000). *Internet Marketing*. Harlow: FT Prentice-Hall.

Curtis, J. (19 April 2000). Next weekend offers some down-to-earth hedonism. *Revolution*, 5.

Dallaert, B., & Kahn, B. (Winter 1999). How tolerable is delay? Consumers' evaluations of internet web sites after waiting. *Journal of Interactive Marketing, 13*, 41–54.

Dayal, S., Landesberg, H., & Zeisser, M. (2000). Building digital brands. *McKinsey Quarterly, 2*, 42–51.

Dignam, C. (14 March 2002). Prosumer power. *Marketing*, 24–25.

Dutta, S., & Biren, B. (2001). Business transformation on the internet: results from the 2000 study. *European Management Journal, 19*(5), 449–462.

Ghouse, S., & Dou, W. (1998). Interactive functions and their impacts on the appeal of internet presence sites. *Journal of Advertising Research, 38*(2), 29–43.

Green, J. S. (Ed.). (2000). *e-media*. Henley on Thames: Admap Publications.

Herington, C., & Weaven, S. (2007). E-retailing by banks: e-service quality and its importance to customer satisfaction. *European Journal of Marketing, 43*(9/10), 1220–1231.

Hoffman, D., & Novak, T. (May-June 2000). How to acquire customers on the web. *Harvard Business Review*, 179–188.

Ind, N., & Riondino, M. (2001). Branding on the web: a real revolution? *Journal of Brand Management, 9*(1), 8–19.

Johns, T. (October-December 2002). e-service strategy. *Marketing Insights*, 4–7.

Kraft, J. (2000). The electronic magaine. In J. Swinfen Green (Ed.), *e-media* (pp. 183–190). Henley on Thames: Admap Publications.

Kucuk, S. U. (2008). Negative double jeopardy: the role of anti-brand sites on the internet. *Brand Management, 15*(3), 209–222.

Kunde, J. (2000). *Corporate Religion*. Harlow: FT Prentice Hall.

Lawes, R. (2002). *Demystifying semiotics: some key questions answered*. Brighton: CD-ROM of The 2002 Market Research Society Conference.

Locke, C., Levine, R., Searle, D., & Weinberger, W. (2000). *The Cluetrain Manefesto*. Harlow: FT Prentice Hall.

Macrae, C. (2000). 21st century brand knowledge – towards the ADEP*T standard for brands' promise and trust. *Journal of Brand Management, 7*(4), 220–232.

Marshak, R. (March 2000). Ten customer-costly mistakes that internet merchants keep making. *Internet Retailer, 2*, 60–61.

McWilliam, G. (Spring 2000). Building strong brands through online communities. *Sloan Management Review*, 43–54.

Mitchell, A. (2001). *Right Side Up*. London: HarperCollins Business.

Page, C., & Lepkowska-White, E. (2002). Web equity: a framework for building consumer value in online companies. *Journal of Consumer Marketing, 19*(3), 231–248.

Reichheld, F., & Schefter, P. (July-Aug 2000). E-loyalty. Your secret weapon on the web. *Harvard Business Review*, 105—113.

Rubinstein, H. (2002). Branding on the internet — moving from communications to a relationship approach to branding. *Interactive Marketing*, 4(1), 33—40.

Schmitt, B. (2000). Creating and managing brand experiences on the internet. *Design management Journal*, 11(4), 53—58.

Sinha, I. (March-April 2000). Cost-transparency: the net's real threat to prices and brands. *Harvard Business Review*, 3—8.

Straw, J. (September, 30 2002). Serendipity online. *Brand Strategy*.

Torrance, S. (2000). Bright — a story from the future. *Journal of Brand Management*, 7(4), 296—302.

Zeithaml, V., & Bitner, M. (2003). *Services Marketing*. Boston: McGraw Hill.

http://www.harley-davidson.com/wcm/Content/Pages/HOG/HOG.jsp?locale=en_US. Accessed2 8 December 2009.

Winning the Brands Battle

CHAPTER 9

How Powerful Brands Beat Competitors

OBJECTIVES

After reading this chapter you will be able to:

- Understand the role of the brand as a strategic device for the firm.
- Distinguish between the strategies of cost-driven brands and value-added brands.
- Apply the value chain concept to illustrate a brand's competitive advantage.
- Discuss the characteristics of successful brands.
- Evaluate the contribution and limitation of brand extensions for the parent brand.

CONTENTS

SUMMARY

The purpose of this chapter is to review the diverse ways of positioning and sustaining brands against competitors. It explores the two broad types of brand competitive advantage — being cost-driven or value-added — and considers how value chain analysis can help identify the sources of competitive advantage. In considering the competitive scope of brands, strategies to develop different brands are reviewed. Methods of sustaining competitive advantage are described within a context of clarifying who the competitors are and how their responses can be anticipated. The core competencies of the brand and the extent to which the brand generates momentum are discussed as sources of sustainable competitive advantage. Strategic approaches to competitive activities are presented and factors influencing response time and competitive signaling are explored The strategic implications from a knowledge of brand share are presented and the contribution of share of voice to brand strength is examined. Characteristics of winning brands are presented. Issues about building or buying brands are raised. A structured approach to brand extensions is described and the influence of factors including perceived extension fit, cultural congruence and the nature of the industry on the success of extensions is examined.

Creating Powerful Brands. DOI: 10.1016/B978-1-85617-849-5.10009-1

BRANDS AS STRATEGIC DEVICES

In Chapter 2, we showed that firms interpret brands in different ways and as a consequence, place different emphasis on the resources they use to support their brands. Some firms believe that brands are primarily differentiating devices and as such they put a lot of emphasis on finding and promoting a prominent name. Others view brands as being functional devices and their marketing programmes emphasise excellence of performance. Our research has shown, however, that the really successful companies adopt a holistic perspective by regarding their brands as strategic devices. In other words, they analyse the forces that can influence the well being of their brand, identify a position for their brand that majors on the brand's unique advantages and defend this position against competitors. By adopting this perspective, the marketer does not just emphasise design or advertising but instead coherently employs all the company's resources to sustain the brand's advantage over competitors.

Once confident about the unique advantages of the brand and their relevance to purchasers, brand plans will be developed and followed through to ensure that the brand's differential advantage is sustained. At this point, it is worthwhile repeating our definition of a brand from Chapter 2, which encapsulates these points:

A brand is a cluster of functional and emotional values that enables organisations to make a promise about a unique and welcomed experience.

The strategist subscribing to this perspective of branding recognises that the key to success lies in delivering benefits, which satisfy buyers' rational and emotional needs. Successful brands make a unique and welcome promise and they do this through their functional and emotional values. The customer's brand experience creates a competitive advantage that others find difficult to copy. Unless the brand has a sustainable competitive advantage, it will rarely succeed in the long run. Ewing et al. (2009) explain that satisfying material needs, which they call 'constitutive utility', and the satisfaction of self-respect, self-image and self-confirmation needs, which they call 'symbolic utility', are both required for brand sustenance, as consumption is a vital force behind the life and death of brands. Brands such as TWA (airlines), Sinclair Spectrum (computers) and Triumph (cars) disappeared because they were unable to sustain their added values against more innovative competitors who were attentive and responsive to changing consumer needs. To some extent, brand death is natural and inevitable. Ewing et al. (2009) explore brand senescence and death and posit that because brands show human characteristics such as birth, growth and reproduction (through brand extension), we should also accept that they can die. In the long run, they suggest, the demand side of brands, market dynamics and consumer behaviour, will determine the length of brand life. In the process of sustaining life, *'the executioners call can be deftly delayed through clever revitalisation and repositioning strategies'* (Ewing et al., 2009, 62).

In the market for overnight delivery of letters and parcels there are many competent brands to choose from — yet Federal Express proudly boasts their positioning in their advertising as 'When it absolutely, positively has to be there overnight'.

Federal Express's ability to sustain their competitive advantage of rapid delivery is due to their efficiency in integrating a variety of supporting issues. All of their employees are carefully selected and trained to deliver superior customer service. Everyone is committed to the dictum, 'satisfied customers stay loyal to my firm' and individuals are encouraged to be independent, resourceful and creative in helping to meet customers' requests. Logistics planning and investment in physical distribution enable the firm to have the infrastructure to provide timely delivery. Supporting systems have been carefully designed and installed to ensure total customer satisfaction. For example, information technology enables any enquirer to be rapidly informed of the location of their parcel. By integrating all of these issues within a clearly communicated strategy, Federal Express has a powerful set of competitive advantages acting as a strong barrier to competitors. A consequence of these activities saw 'FedEx' ranked 13 among Fortune magazines' 'World's Most Admired Companies' in 2009.

In the industrial sector, Snap-on Tools' competitive advantage of superior service enables them to sell their tools successfully at a higher price than competitors — and sustain this position. Dealers regularly call on customers, mainly garages, with well-stocked vans. In the UK and USA, the markets are similar as it is the mechanic, not the garage proprietor, who buys the tool. Call schedules are designed so that their customers know when to expect visits. Typically, on each visit, the dealer goes in with new products, allowing customers to try them out. Part of their motivation for buying Snap-on Tools is the sense of pride of ownership. Well-trained salespeople learn to understand each of their customers. With a first class product warranty, good dealer training and the reputation of the sales people, the firm stays away from competing on price. In 2009 the company celebrated its 45th anniversary by achieving record UK sales, which exceeded £100 million for the first time.

Both of these examples show brands succeeding through having a competitive advantage and a strategy to ensure that competitors cannot easily share this position.

The first stage in developing a competitive advantage is to analyse the environment in which the brand will compete. One of the most helpful ways of doing this is to use the framework shown in Fig. 2.4 in Chapter 2. This logically enables marketers to consider the opportunities and threats facing the brand from within their own organisation, and from distributors, consumers, competitors and the wider marketing environment. Fully aware of the forces that the brand must face, the strategist can then start to find the most appropriate positioning for the brand.

COST-DRIVEN OR VALUE-ADDED BRANDS?

Brands succeed because they are positioned to capitalise on their unique characteristics, which others find difficult to emulate − their competitive advantages. This positioning is a coherent, total positioning, since it is backed by all departments in the firm. Everyone should be aware of what the brand stands for and they all need to be committed to contributing to its success.

A brand's competitive advantage gives it a basis for out-performing competitors because of the value that the firm is able to create. Consumers perceive value in brands when:

■ it costs less to buy them than competing brands offering similar benefits, i.e. 'cost-driven brands';
■ when they have unique benefits which offset their premium prices, i.e. 'value-added brands'.

Cost-driven brands

The website for McDonalds' 'Eurosaver mmmmenu' is a good example of a cost-driven brand. The website makes the point that the consumer can benefit from reduced price, without sacrificing quality. For example, the company asserts that their coffee is 'made from 100% Arabica beans sourced from Rainforest Alliance certified farms', and that their double cheeseburger is '100% pure Irish beef'. Those consumers to whom this advertisement is targeted are likely to perceive value in the McDonalds brand as it offers quality, traceability and fair trade at a low price.

McDonalds is positioned as a cost-driven brand through their Eurosaver menu.

Cost-driven brands thrive because action is continually being taken to curtail costs. As Fig. 9.1 shows, compared against the industry average, their total costs are always lower. A profit margin at least equivalent to that of other competitors can then be added and the selling price will still remain lower than that of the average competitor.

FIGURE 9.1

The economics of cost-driven brands

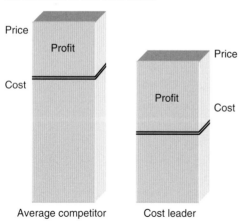

Average competitor   Cost leader

Some marketers shun the idea of cost-driven brands since they equate low cost with low quality. In some cases, of course, they are justified. Reducing quality standards, however, is only one way of reducing costs but is not the recommended route to follow since it can cause consumer dissatisfaction. Some of the other ways of cutting costs are through economies of scale, gaining more experience faster than competitors, more selective raw material sourcing, dealing only with large order customers, introducing new technology in production, streamlining the product range or reducing the service level offered. By understanding precisely what the target market wants, unnecessary frills can be eliminated and an attractively priced proposition developed. In grocery retailing, for example, Aldi, Netto and Lidl strive for low cost position not by cutting back on quality but by eliminating frills and services not relevant to the target market. Their low-cost competitive advantage is achieved by:

- not price-marking individual items;
- not accepting credit cards and their handling fees;
- having a small number of lines;
- not stocking own labels and thus being able to negotiate better discounts from manufacturers;
- having small stores that are cheaper and quicker to open than superstores;
- taking advantage of information technology;

Likewise in the motel sector, some of the ways Travelodge and Premier Inn are able to offer budget prices for comfortable, clean rooms to travellers are:

- making the most of spare land;
- not having telephones or drinks cabinets in bedrooms;
- being well located, close to main routes;
- not having to provide dining facilities in the motels since they are close to restaurants;

In the airline market easyJet and Ryanair are good examples of cost driven brands. They have reinvented the business delivery model in Europe, eradicating products and activities that failed to satisfy the primary need of providing safe, reliable air transportation. Some of the ways they drive their costs down are:

- encourage customers to book fares and check in online, eliminating staff costs;
- charge passengers for check-in luggage to reduce weight and plane fuel consumption;
- use secondary airports close to major cities;
- only offer in-flight snacks to those passengers prepared to pay for these;

- standardise their fleet on one aircraft enabling better negotiating strength when acquiring these and facilitating economies of scale when servicing the aircraft;
- turn flights around fast, so they spend longer flying than being on the ground.

Following a cost-driven branding route should not mean demotivating employees. If anything, it can act as a motivator through presenting challenges that stretch the organisation and allow individual creativity. For example, to succeed against Xerox in the personal copier business, Canon set its engineers the task of designing a home copier to sell profitably at under half the price of Xerox's current range. By literally reinventing the copier, substituting a disposable cartridge for the more complicated image-transfer mechanism, they were able to meet this challenge. Without the resources of Xerox, Canon had to become more creative in cost-effectively selling its range. This was achieved by:

- distributing through office-product dealers rather than meeting head-on Xerox's massive sales force;
- designing reliability and serviceability into its range and delegating servicing to its dealers rather than setting up service networks;
- selling rather than leasing their machines, thus not having to administer leasing facilities.

Cost-driven brands succeed when everyone in the firm knows that each day they have to become more independent and more creative in curtailing the costs of good quality brands. Newly launched competitor brands are subjected to careful scrutiny to see if further cost improvements can be made. Cost advantages are not just sought from one source, but from many areas.

In common with all strategies, there are risks in developing cost-driven brands. R&D activity may result in lower cost, superior brands, which some firms may short-sightedly consider a sustainable edge. However, technological advances need to be continually made, as others can soon emulate technology. Another danger is that marketers fail to foresee marketing changes because of their blinkered attention to costs. A classic example of this was during the 1920s when Ford was busy improving production efficiency and achieving lower costs on models whose sales were falling; whereas General Motors had introduced a wider range of cars, better enabling drivers to express their individuality. The days of 'Any colour, as long as it's black' were not to last that long.

Before following a cost-driven brand strategy, the marketer needs to consider how appropriate it is. For example:

- Is the price-sensitive segment sufficiently large and likely to grow?
- How will buyers respond when competitors launch low price alternatives?

- How fast can experience be gained to reduce costs?
- Is the culture of the firm geared to reducing costs?

If there are doubts about this route, the alternative of value-added brands needs to be assessed.

Value-added brands

Value-added brands are those that offer more benefits than competitors' brands and for which a premium price is charged. Apple Mac computers have the competitive advantages of multimedia processing capabilities and innovative design, as well as a range of software applications such as photo libraries, music recording and video. The company emphasises their 'out of the box' ease of use and simple connectivity to other devices such as digital cameras and external drives. They are not susceptible to viruses and they are secure from first installation. This brand represents a lifestyle for today's family that has diverse computing needs. As a consequence of their features and benefits, Apple Macs are sold at a premium. In 2009, a basic Macbook was priced at £816, compared with Dell's basic model, the Inspiron Mini 10v which retailed at £308.

Value-added brands do not succeed just through functional excellence; for companies like Apple, a strong image can also be a powerful competitive advantage. Another example is found in the car industry. For example, both the Lotus 2-Eleven and the Porsche 911 Turbo have similar accelerations from 0–60 mph (3.8 and 3.5 seconds, respectively) yet the Porsche image, created by years of advertising, helps contribute to this car's price premium over the Lotus.

To produce and market a value-added brand, the firm usually incurs greater costs than the average competitor in that sector. There is something special about the brand that necessitates more work to make it stand out. Because the brand is differentiated, it is likely that consumers will notice this; and for relevant added values, they will be prepared to pay a higher price. In such a situation the marketer is able to anticipate a higher margin than his competitors and set a price which fully reflects the benefits being sold. The resulting economics of value-added brands are shown in Fig. 9.2

In the grocery retailing sector, Harrods' Food Hall is a good example of a value-added brand. Consumers recognise that they are always certain of high quality produce, backed by a no-quibble guarantee. They enjoy the tastefully designed environment and appreciate being served by knowledgeable staff. Furthermore, the image surrounding the well-known brand of Harrods adds further value to the grocery shopping experience and justifies the price premium.

In the potato market, Marks & Spencer transformed a cheap commodity into a value-added brand through innovative marketing. They developed this by only accepting a certain size of potato, washing its skin, cutting it, placing

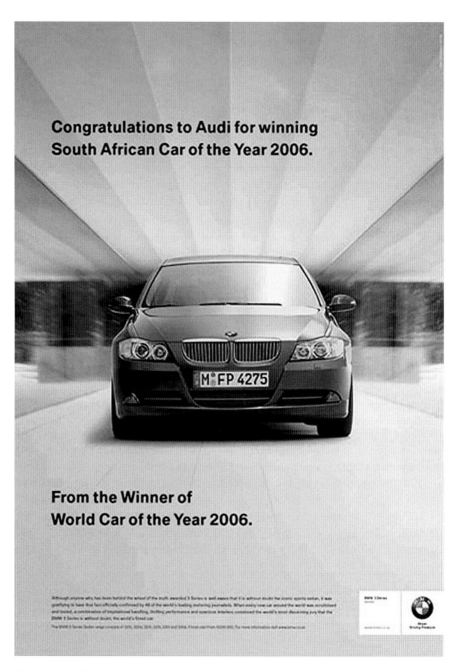

AUDI AND BMW

Audi and BMW have indulged in competitive 'sparring' in their advertising. For example, in a 2006 advert, BMW congratulate Audi for winning South African Car of the Year, adding 'From the Winner of World Car of the Year 2006'. In 2009, Audi challenged BMW to participate in a chess game with outdoor advertising containing the incitement 'Your move BMW'. Overnight, on Santa Monica Boulevard, BMW added their advertisement stating 'Checkmate'

cheese in it, aesthetically wrapping it and proudly giving it a prominent shelf position in their stores. Through this value-adding process, they were able to charge a price considerably higher than the cost of the raw materials.

Ideally, value-added brands differentiate themselves using a variety of attributes, as can be appreciated from the way that Citibank once managed to increase its business in the hostile Japanese retail banking sector. Its original six branches operated in a cartelised sector, where all banks were required by the Ministry of Finance to pay the same interest on yen investments. After undertaking a situation analysis, it transformed its retail banking operation into a value-added brand through a series of actions.

FIGURE 9.2

The economics of value-added brands

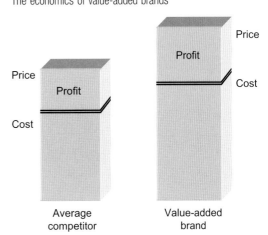

1. *It competed on the basis of foreign yield.* The regulations forbade offering competitive rates on yen investments but said little about foreign currency deposits. By offering a higher interest rate on deposits in American dollars, they attracted a significant new number of customers.
2. *It advertised aggressively.* Breaking the traditional Japanese promotional approach of not having comparative advertising, they aggressively advertised the higher rates on American dollar deposits.
3. *It redefined service.* It was common for customers to have to wait to be served in banks. As such, the Japanese made their banks comfortable places to wait in. By launching a telephone banking service, Citibank changed the rules of engagement by not making customers wait.
4. *It defined an image and targeted customers.* Citibank stressed its international, sophisticated image and targeted itself at clearly defined groups who travelled abroad a lot and were likely to have considerable liquid assets. These included expatriates, business executives and, following their service orientation, professionals who had little time to queue.
5. *It widened distribution.* An agreement was made with a bank in Tokyo, which had a significant number of automated teller machines (ATMs), to allow Citibank customers to use their ATMs to make cash withdrawals.
6. *It expanded the product portfolio.* High interest deposit accounts were developed for other foreign currencies, as well as gold deposits.

Developing value-added brands, however, also has its risks. The price differential between the value-added brand and its lower-cost competitors may widen to such an extent that consumers may no longer be prepared to pay the extra cost, particularly when there is little promotional activity justifying the brand's more exclusive positioning. Another threat is that of competitive imitations. If the technology supporting the value-added benefits is not that difficult to copy, competitors will soon appear at a lower price. Alternatively, buyers become more sophisticated as they repeatedly buy brands in the same product field and they start to take for granted the most recent value-added change, expecting more from the brand.

VALUE-ADDED BRANDS WITH COST-DRIVEN CHARACTERISTICS

It should not be thought that brands can only have the competitive advantage of being *either* cost-driven *or* value-added. These two scenarios represent extreme cases. Instead, it is more realistic to think of the *extent* to which brands have a cost-driven component as well as a value-added component. By considering which of these two components is more dominant, the marketer is able to think in terms of having a brand that is predominantly cost-driven or predominantly value-added.

By considering the contribution of the cost-driven and value-added elements of branding, it is possible to classify brand types and identify appropriate strategies. In Fig. 9.3, we have presented a matrix that shows the classification of brands on a strategic basis. This model of strategic brands can best be appreciated through examples of different brands of hotels.

An example of a **commodity brand** would be a hotel in a small market town, probably without any other competitors in that town and dangerously unaware of much more competitive and customer orientated hotels in the nearby towns. It is likely to be a small, independently run business with few rooms and little modernisation. The hotel would be open during summer months only. It would offer a limited selection of food in its restaurant, which would close at 7 p.m., with no room service. Staff in the hotel are mainly local teenagers hired during school holidays. They are not interested in offering customer service and do not extend their work beyond what is absolutely necessary. With limited training, they are unable to assist with unusual queries such as families requesting facilities for young children or guests with dietary requirements. Online booking is not facilitated and there are no special package deals available. This commodity brand offers consumers a poor service compared with other hotels and its chances of surviving are not that good.

A **benefits brand** is one which is not able to offer significant cost savings but instead offers particularly good service. Typical benefits brands are hotels specialising in business conferences. They have very good computerised systems to cope with requests, often at very short notice, to book, organise flexible room

Liverpool to
**Lanzarote &
Fuerteventura**

from
£59.99
single
inc. taxes

easyJet.com
Flights · Hotels · Cars · Holidays

Price correct as at 2 August 2010. Flights available to book now for travel between
2 August and 30 October 2010. Variable charges for hold baggage apply and some
payment methods attract a handling fee. See website for details.

EASYJET

EasyJet are an example of a cost-driven brand. As a result of driving costs down, they offer cheap fares to
customers. (Reproduced by kind permission of EasyJet)

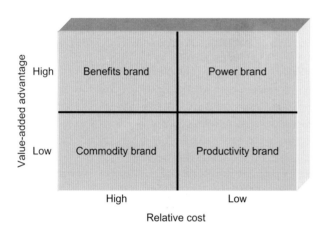

FIGURE 9.3

Classifying brands on a strategic basis

layouts, accommodate large numbers for banqueting and provide audiovisual support. The staff are well-trained and adapt their service according to company needs. They spend a lot of time ensuring that they have as wide a portfolio of contacts as possible with local and national companies. These hotels tend to target their services at organisations where there is a continual demand for 'stress free' conference space. They tend not to offer any discounts.

A good example of **productivity brands** are those hotels specialising in room-only offers. They offer a frill-free environment in return for a heavily discounted price. Generally, breakfast is not included in room prices. Very few additional services are provided by this type of hotel.

Power brands are those successful brands that offer consumers many relevant extra benefits. As a consequence of high consumer satisfaction, they have a high relative share of the market, enabling them to take advantage of economies of scale and the experience effect. By passing on some of these cost savings to consumers in the form of lower prices, they are able to give their brands a price advantage over competitors and maintain a virtuous brand marketing circle. They are likely to offer consumers a wide range of additional services such as dining menus and beauty treatments. Also, rooms may be equipped with modern equipment such as music systems and flat screen TVs. Also, well trained staff are able to address customer queries and are trained in cross-selling. For example, a couple on a weekend break may be encouraged to book an evening meal at the hotel and visit the hotel spa.

Commodity brands offer no advantages over any other brand and they are not particularly good value for money. History has shown that the commodity brand domain is an area to be avoided.

Benefits brands succeed because they are targeted at a specific segment, with a company-wide commitment for a process which delivers extra benefits that consumers particularly appreciate. It is essential for these brands that the firm maintains regular contact with its market, to continually assess satisfaction with the brand and identify ways in which it could be improved. The firm will continually invest in R&D, new services, production, logistics and marketing to ensure that the brand remains the best. Any cost-saving programme that might have an adverse impact on the brand's quality must be resisted.

Productivity brands need to be supported by a company-wide mission that stresses the need for each individual to be continually questioning 'Why do we have to do it this way?'. These brands' cost advantages can only be sustained if all aspects of the business system are continually subjected to tight cost controls. Wherever possible, standardisation and narrowing of the product mix need to be encouraged. Potential segments which fall below a critical size must be ignored.

Power brands like Coca-Cola, The Four Seasons, Sony and Nescafé thrive through being very responsive to changing market needs and continually trying to improve their brands, while at the same time looking for cost advantages. When North America shrugged off the appearance of small Japanese motorcycles as an event unlikely to succeed, they failed to recognise the significance of much larger Japanese motorcycles being raced on European circuits. The racing experience provided valuable learning about designing and producing larger motorcycles. Building up volume production and selling capability in small motorcycles quickly resulted in cost advantage, enabling the returns to be used to invest in widening the portfolio and adding value through attractively priced larger motorcycles.

IDENTIFYING BRANDS' SOURCES OF COMPETITIVE ADVANTAGE

When managers are faced with the problem of identifying their brands' competitive advantages, Porter's Value Chain (Porter, 1985) can be a very useful tool. An example of this is shown in Fig. 9.4. A flow chart first needs to be constructed showing all those actions involved in transforming raw material into profitable sales — the value-creating processes. This is divided into the stages:

- in-bound logistics, e.g. materials handling, stock control, receiving goods;
- operations, e.g. production, quality control, packing;
- out-bound logistics, e.g. storing finished goods, delivery, order processing;
- marketing and sales, e.g. pricing, promotion;
- service, e.g. installing, training, repairing.

The services supporting these activities are categorised into purchasing (procurement), technical development, human resource management and infrastructure. These are presented in the format shown in Fig. 9.4, since each of these services can support many of the value-creating processes. For example, different departments within the firm may be buying raw materials, the skills of industrious employees, delivery lorries, creative advertisements and an after-sales support unit.

Benchmarking itself against its competitors, the firm should then identify those activities that its managers believe they do better or cheaper. As competitive

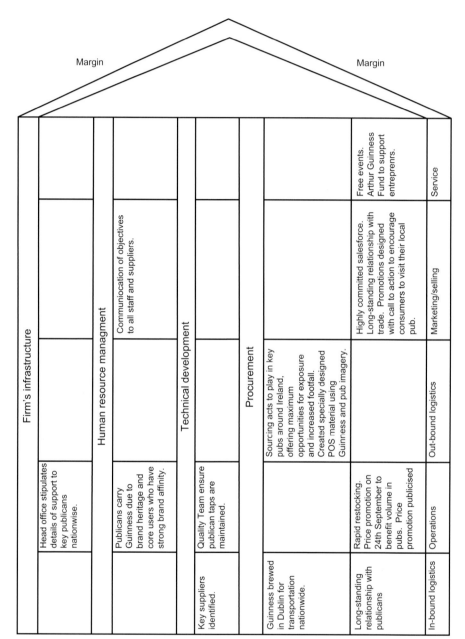

FIGURE 9.4

Guinness' 250 celebrations value chain, using Porter's (1985) analysis

advantage is a relative concept, it is important that the key competitors are identified for that particular segment. If a firm has a brand competing in several distinct segments, it should produce value chains for each segment.

Within the template of the value chain, managers can start to identify their brands' competitive advantages. This may be better appreciated using the example of Guinness. This was a case of excellent management of a heritage brand that retained its heritage while creating a relevant and contemporary image in a changing market. Using the strengths within its value chain, the brand capitalises on its strong relationship with its key suppliers, the Irish publicans, who are supported by Guinness' fleet of sales reps. Reps are further supported by a Quality Team that has responsibility for cleaning and maintaining taps in outlets such as pubs and hotels, to ensure that the product is supplied at the desired specifications. Among consumers, the popularity of Guinness is well known; one in every three pints ordered in Ireland is a Guinness pint. A challenge has been the attraction of new customers through a modern brand image, without alienating the affinity long-term customers held with the brand's heritage. In particular, the 21−25-year-old male market is a desirable segment. Within this segment, Guinness faces the challenge of changing tastes, as many younger consumers are used to sweeter tasting alcohols products, such as alcopops. In 2009, Guinness built its marketing operations around a special birthday − the 250th anniversary of the signing of the lease on their Dublin premises. Guinness wanted to build on an already strong relationship with Irish publicans and attract new users to the brand. To achieve this, they sought to create a strategy to build on the Guinness legacy and yet recruit consumers through a position of modernity. The marketing campaign had to therefore consider the timelessness of the brand. In order to emphasis their legacy, the company embarked on a series of 'Time Machine' style advertisements, which featured popular Guinness advertising from the 1970s onwards. These advertisements captured the public's nostalgia and gained substantial PR coverage. To build further on the heritage of the brand, the company identified competitive advantage to the internal culture of the organisation and its philanthropic nature, initiated by Arthur Guinness. In the 1800s, Guinness workers were provided with housing and excellent working conditions. To continue in this spirit, Guinness established the 'Arthur Guinness Fund' to support social entrepreneurs. As a component of the brand elements is the Arthur Guinness signature − the public were invited to add their signature to Arthur's. For each signature collected, the brand donated €2.50 to the fund. Finally, the brand used a call to action to support their distributors. The company considered actions that were intrinsic to the brand and decided to use the toast as a means of celebration. As the pint of Guinness dates to 1759, one minute to six o clock in the evening (17:59) on 24th September 2009, was set aside as the time for 'A toast to Arthur'. In addition, Guinness hosted concerts by well known acts in small local pubs, to enhance the relationship with publicans and to strengthen the association between visiting the pub and ordering Guinness. The campaign was highly effective. Publicans were supported as the footfall in pubs increased by 500,000, compared with a typical Thursday evening. In addition, the campaign was active in 2000 pubs across Ireland, and attracted new, younger users to try the brand for the first time.

Besides acting as a guiding framework to help managers identify *how* brands achieve a competitive advantage, the value chain also helps managers check *whether* they are reinforcing and capitalising on their competitive advantages. This is done by considering how well-linked each of the activities are. For example, if the managers believe that the brand achieves its competitive advantage because of its unique taste, this may be protected by buying higher priced but better quality raw materials. As a consequence, less wastage and lower costs could result from the smoother running of the distribution channel and from less wastage at the finished goods stage. Coordinating internal activities should ensure that all the processes in the value chain are optimised to give the brand the best chance of capitalising on its competitive advantages. At Disney, for example, which prides itself on its excellence of customer service, telephones are discreetly located in its theme parks so that employees can quickly have access to advice and, if need be, extra resources when they spot a problem occurring.

When considering linkages between activities, firms should not take a myopic perspective and solely consider their internal linkages. Instead, they should also identify advantages by linking their value chain back to their suppliers and forward to their customers. Clearly, the better suppliers understand how their customers will use their products, the more scope they have for adding more value by designing their value chains to integrate better with that of their buyers.

The value chain, however, has one major disadvantage. To be used effectively, managers need to have a good database describing the processes and economics of each aspect of every competitor's value chain. It is rare to have such a rich database. Trade journals, industry reports, sales peoples' reports, distributors' comments and Competition Commission reports can help to build a database, but any remaining gaps will have to be filled by management judgment. Clearly, if there are a lot of gaps in competitor analysis and if the wrong assumptions are made, any comparative analysis may well be flawed.

FOCUSING BRANDS' COMPETITIVE ADVANTAGES

So far we have concentrated on strategic brands in terms of the two broad competitive advantages of 'value-added' and 'cost-driven'. Beside the *type* of competitive advantage, marketers also have a choice about the *scope* of the market that they wish their brand to appeal to. Again, drawing on the work of Porter (1985) marketers are able to refine their brand strategies further by considering whether a cost-driven or value-added competitive advantage should support their brand for either a narrow or a broad target market. The four possible generic strategies are shown in Fig. 9.5.

Samsung is a good example of a **focus cost** brand in the television market. With their almost utilitarian style approach, they focus on first time purchasers or budget conscious consumers looking for good value for money. By contrast, Panasonic is an example of a **brand cost** player. They are interested in broader target groups and with these larger markets are able to capitalise on economies of scale with higher specification televisions that again represent good value for money. Bang & Olufsen represent a **focus differentiation** brand, offering a narrowly targeted group of design conscious householders aesthetically startling home cinema systems, which carry much higher prices. Part of their integrating branding approach involves carefully selecting a smaller group of distributors dedicated to providing a unique retailing environment that tastefully echoes their eye catching designs. Sony's continual emphasis on innovation and rapidly launching new televisions globally, exemplifies this as a **brand differentiation** brand.

FIGURE 9.5

Generic strategies for brands (adapted from Porter 1985)

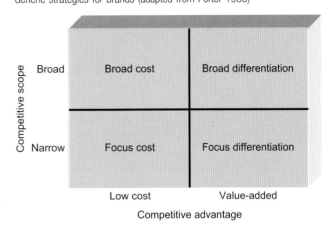

SUSTAINING A BRAND'S COMPETITIVE ADVANTAGE

Having identified the sources of the brand's competitive advantages and having positioned the brand with the most appropriate generic strategy, the marketer is then faced with the problem of sustaining the brand's uniqueness. If the brand is successful, competitors usually work hard to understand the basis of this success and then rapidly develop and launch their own version – often with an improvement. The time before competitors develop their own improved versions of a new brand is shrinking, due in no small part to companies' understanding and appreciation of technology.

When looking to protect their brand's competitive advantage, some managers attempt to stay ahead of competitors by concentrating, as Porter has shown, on operational effectiveness – performing the same activities better than rivals. A far more effective route to sustain a brand's competitive advantage is by concentrating on performing different activities from its rivals or performing similar activities in different ways.

It is much harder to sustain a brand's competitive advantage with the first route since usually this is technology driven and competitors soon learn to emulate this. For example, one can track over time how some firms have sought to sustain their advantage of a lowest cost sales force through taking advantage of PDAs, Laptops and use of Skype. Yet all of these, plus others such as route planning software, are soon adopted by other competitors. Furthermore, as firms become more receptive to learning about best practices from other industries, there is a trend for the players in a market to adopt operationally effective processes to enhance their brand's performance — but in so doing, to cancel out competitive advantages.

Notable successes have been recorded by brands, such as Southwest Airlines and IKEA, which have sought to sustain their competitive advantages by undertaking different activities from rivals or performing similar activities in different ways. Southwest Airlines is a good example of a cost-driven brand. It concentrates on short distance, low ticket price flights between midsize cities and secondary airports in large cities. While others may fly into primary airports in major cities, Southwest chooses to be different. To drive down its costs, it doesn't offer in-flight meals or pre-assigned seats. It concentrates on rapid turnarounds at gates, typically of 15 minutes, keeping its planes in the air longer and emphasising its on-time performance. In the US airline industry, the company also pride themselves on the lowest rate of complaints per passenger boarded.

By following a strategy of focusing on different activities from its rivals, managers need to make trade-offs. Southwest firstly decided that it would not compete on long haul routes or offer different classes of service or fly into the primary airports of major cities. Taking a firm line in terms of what it would not do, it had in effect decided about its critical brand resources, process configurations, staff recruitment and behaviour.

Being clear about what the brand does and, equally importantly, what the brand does **not** stand for, managers can sustain their brand's competitive advantage through the way its activities fit and reinforce each other. These two factors are important contributors to sustaining competitive advantage. The fit between different resources and processes needs to be assessed in terms of:

- The degree of consistency between each activity. To offer lower prices, IKEA decided upon modular furniture designs enabling choice from a wide variety of potential suppliers who could easily manufacture and supply all year. Furthermore, modular designs can be transported easily and carried relatively quickly from delivery lorries into the backs of stores.

■ The degree of reinforcement between activities. For example, in order to drive costs down, IKEA decided that consumers should transport the furniture kits themselves. Therefore suburban sites in big conurbations were sought, since consumers would be more likely to transport their furniture kits over short distances. However, to achieve good economies of scale, these sites needed to be linked by good road systems to encourage high traffic in visits. Furthermore, the sites needed to be on a flat terrain, to ease the task of consumers pushing their trolleys from the store to the cars. In addition, the company appointed a group of independent home delivery transport companies, available for hire by customers. This offers customers choice and assistance in delivery.

By looking at the whole system of activities, a more integrated investigation of fit can be addressed, enabling a greater likelihood of sustaining competitive advantage.

Focusing on different activities from rivals or performing similar activities in different ways necessitates appreciating the whole of the value chain; and it is through the way that activities are linked together that a barrier is erected, which can prove daunting for competitors. Systems thinking, integration and teamwork are the underpinnings of effective planning for sustainability. Some argue that having a cost advantage is a sustainable basis for a brand's competitive advantage, particularly because it necessitates a challenger achieving major improvements in levels of production or sales, to reach the same point on the experience curve. However, a large relative market share is necessary for the cost driven leader to sustain their cost advantage.

Inherent in any strategy for winning brands are the low number of core competencies that give the brand its competitive advantage. A core competence is a skill or knowledge that a firm has developed to such an extent that it is outstandingly good in its skill/knowledge domain, and uses this to develop and extend its brand portfolio. For example, one of Nike's core competencies is marketing, which it so effectively employs. However, it does not regard production as being a core competence and it outsources its trainer production. For Canon, one of its core competencies is its knowledge of optics, which it leverages, enabling it to extend its portfolio across cameras into photocopiers and other fields. The key to successful branding is not just focusing on developing a low number of core competencies, but then making it difficult for competitors to gain the learning and experience of such competencies. Gehlar et al. (2009) advise firms to identify their unique resource base, which comprises specific expertise and competencies. Firms that recognise their specialised resources are less likely to 'chase the market' when the market changes.

Instead, they tend to specialise in products, which make best use of their competencies. Gehlar et al. (2009) provide the example of Heinz. The company has a portfolio of brands that occupy the Number 1 and Number 2 market positions in 200 countries across the world. Heinz differentiates itself from lower cost competitors by improving the quality of its products, for example, through packaging innovations such as the Heinz Easy Squeeze ketchup bottle, and product innovations such as new soup ranges. The company has also removed complexities from its product portfolio and supply chain; for example, divestiture allows greater focus on core products. Such activities are supported by marketing communications that advise the customer to 'Insist on Heinz'.

SUSTAINING A SERVICE BRAND'S COMPETITIVE ADVANTAGE

The challenge facing the marketer is how to sustain their brand's competitive advantage. Zheng-Zhou et al. (2009) explain that within services, firms adopting a market orientation should distinguish between customer orientation and competitor orientation. Customer orientation requires a focus on target buyers. Competitor orientation emphasises better understanding of the strengths, weaknesses, capabilities and strategies of competitors. From their work in the hotel industry, they advocate customer orientation in particular as a means to achieving a differentiation advantage. As they explain, if a firm does not know what a customer values, they cannot define a suitable value proposition. Financial performance is only enhanced when customers are satisfied and this arises only when differentiation advantages are exploited. In markets where customers are price-sensitive, services may adopt a competitor orientation. Within the services sector, competitive responses can very quickly appear. A study by Easingwood proves particularly illuminating. He analysed 36 successful new services that competitors had difficulties in copying. Interviews with managers associated with these brands indicated 10 factors that impeded rapid competitive responses.

The company's reputation was regarded as presenting the most effective barrier to rapid competitor 'me-too' brands. In consumers' minds, the Automobile Association is strongly associated with services for cars and the launch of its credit card for car-related expenditure was instantly accepted. Were the card to have come from a financial institution, it is thought that it would have had a lot more resistance to overcome.

One of the more widely accepted definitions of reputation is that from Fombrun and Rindova (1996) which is slightly adapted for a brand as 'Brand reputation is a collective representation of a brand's previous actions and results that describe the brand's ability to deliver future valued outcomes to multiple stakeholders'. In other words, while a brand's image is the perception that someone has at a specific point in time of the brand, a reputation is a view someone takes over time of the brand, which they then use to anticipate the brand's future performance. People believe that the history of a brand has a habit of repeating itself. Thus, if over time, a brand has been well managed and has consistently been able to deliver outstanding customer satisfaction, a new entrant striving to steal some of the brand's share will have problems overcoming the reservoir of reputational goodwill. In addition, reputation is not easily changed. Veloutsou and Moutinho (2009) explain that consumers classify brands into categories and hold fixed opinions about those categories. There is also a time lag effect on the influence of any changes made on consumer opinion, which means that a brand' s reputation is durable.

Brand reputation therefore plays a key role in creating consumer-brand relationships and ultimately, helps to create brand loyalty (Veloutsou and Moutinho, 2009). Studying attitudes about soft drinks brands, Veloutsou and Moutinho (2009) explored reputation in terms of perceived trustworthiness and reputation as well as the sustainability of the brand's image and values over time. Their findings highlighted the importance of reputation in creating sustainable relationships with consumers. However, they also advocate full consideration of the stronger influence of the 'Brand Tribe'. Brand tribes are networks of societal micro-groups in which people share emotional attachment to specific brands. Tribes are connected through shared passion or emotion and members are not just consumers, they are brand advocates. With the internet providing consumers with greater opportunities for communicating with others, the role of the brand tribe and consumer advocacy will be critical to the creation of future competitive advantage.

Internally, the culture of an organisation is regarded as a particularly difficult competitive advantage to emulate. Taking the perspective of a brand as a cluster of functional and emotional values, technological advances make it hard to sustain a brand's functional advantage. Shifting attention from *what* customers get to *how* customers get it, i.e. the emotional component, makes it more difficult for competitors to emulate a brand. One way of enrobing a brand with emotional values is to use resources such as advertising and packaging. However, a key contributor to the way stakeholders interpret a brand is the staff working on the brand. They join particular organisations or are committed to delivering specific brands, because some of their personal values align with the brand's values. The greater the coherence between personal and brand values, the greater the likelihood that they will be committed to delivering the brand promise. This can be recognised by customers who see through those employees who are superficially supporting the brand, doing little more than paying lip service and those aligned to the brand promise who genuinely

DYSON

Dyson differentiates itself on its high quality design. In this advertisement, it highlights its design advantage: a Dyson vacuum never loses suction. (Reproduced by kind permission of Dyson)

believe in what they are paid to do and almost "live the brand" in their customer interactions. Through carefully recruiting staff on the basis of the consistency of their values with the desired brand values, then regularly reminding staff of the brand's values and reinforcing brand supporting behaviour through various reward schemes, there will be a more genuine interaction between employees and stakeholders. Building a brand-supporting, organisational culture takes time and is particularly difficult for competitors to copy. Furthermore, different types of cultures result in different types of staff/stakeholder interactions, differentiating brands.

Relating to the previous point about organisational culture, customer service is a further way of sustaining a brand's positioning. Disney, Federal Express and McDonalds are all legendary in the way that they have developed training programmes to ensure that their employees give a unique type of customer service. Provided that the employees are sufficiently briefed about a new brand, their contribution to sustaining its added values can inhibit competitor responses.

A further way of helping the brand remain competitive is to be the first into a market. Being the first to exploit a market opportunity leads to the cost advantages of economies of scale and the learning curve effect. If the firm has a policy of not allowing its employees to present case studies at trade conferences, any learning is proprietary and competitors find it hard to appreciate how to gain from 'sitting on the side' and observing brand developments. Being the pioneer brand offers scope for gaining an advantage through opportunistic marketing, provided the firm monitors progress and rapidly incorporates any learning back into brand enhancements.

ANTICIPATING COMPETITOR RESPONSE

One of the ways in which firms and financial analysts evaluate brand strength is on the basis of their market share. Strong brands tend to be market leaders who have capitalised on the opportunities from economies of scale and the experience effect. Likewise, the brand strategist is concerned with evaluating the brand's performance against key competitors. They will normally have a clear view about who are their prime and secondary competitors. As they formulate brand strategy, they will consider how they can group their competitors, according to their characteristics.

While economists have historically conceived of an industry as a clearly-defined group of competitors, research shows that managers in the same firm do *not* have similar perceptions about who the main competitors are. As part of a major research programme sponsored by the ESRC to evaluate managers' perceptions of strategic groupings, de Chernatony, Johnson and Daniels interviewed senior managers in firms supplying pumps to the North Sea offshore oil industry and repeatedly found differences between members of the same management team concerning who their competitors were and the basis which they were using to form strategic groupings. When undertaking similar interviews amongst senior managers in firms providing mortgages for the first time homeowner market, results again showed different perceptions amongst the management teams in each firm as to the composition and nature of competition. While there was some commonality between managers concerning many competitors, several competitors were unique to particular managers. By debriefing the management team about their views on who the competitors are and what bases they use to formulate their strategic grouping, a better appreciation can be gained as to how each manager is formulating strategy and ensure a more unified approach to competitive brand positioning.

If companies are going to become more effective at formulating competitive brand strategies managers in the same firm will need to discuss more openly amongst themselves who they perceive as their competitors and what basis they are using to group competitors. Without this debate, individual senior managers may be incorrectly directing their departments to follow diverse routes, which may not necessarily best support the intended strategy.

After all those working on the brand have debated who they perceive their competitors to be and what the bases for strategic groupings are, they then need to evaluate which competitors are likely to respond fastest and in the most aggressive manner as a result of the firm's changes to its brand strategy. Attitudes can be an indicator of likely competitor response. For example, Mars, Unilever and Procter & Gamble have always striven to be leaders in their particular markets. It goes against their corporate cultures to accept that a newcomer to their market can go unchallenged in its quest to rob them of their brand shares. History may be another pointer to competitive responses. In some firms, for example, there may be a deep resentment against a particular geographical market or type of product, because the firm tried and failed to succeed with its brand. It does not want to reopen old wounds and may take no action when a new competitor appears to be succeeding. Alternatively, the size of the investment in a plant may indicate what response to anticipate. In the tin packaging market, leading firms have considerable investment in machinery producing high volume runs of standard-sized containers They cannot afford to let their lines run slowly and any new competitor brand launches are quickly evaluated and challenged.

Brand strategists need to consider how important the market is to each of their competitors and what their degree of commitment is. Particularly when competitors have a wide variety of brands across many different markets, their interest and commitment to protecting their brands is likely to vary. For example, commitment to part of the range might be particularly high because these markets are regarded as having significant growth potential and have historically enabled their brands to achieve healthy returns, as well as enabling them to be highly visible players, gaining spin-offs for some of their other brands. In the UK breakfast cereals market, Kellogg's, with its wide range of brands, would never let its flagship brand, Kellogg's Corn Flakes, fall against other competitors. For too many consumers, this particular brand *is* the firm. Any weakening on this brand might be read as a weakening of the Kellogg's firm and, by inference, a deterioration of the rest of its range.

A considerable amount of *data* may be held on competitors about such things as their plant capacities, labour rates, organisational structure, discount structures and suppliers used. However, we question whether sufficient *information* is available about their brand strategy. To position a brand and anticipate competitors' responses, marketers should either know, or be able to gauge for each competitor:

- its main brands, their size, profitability, growth and the importance of each brand to that firm and its commitment to that brand;
- its brand objectives and the strategy being followed — whether, for example, they are trying to enter, improve, maintain, harvest or leave the market;
- its brands' strengths and weaknesses;
- the competitive position of each brand in terms of having either a leadership, strong, favourable, tenable or weak position;
- it's marketing activities, including media share of voice and any trade or consumer promotion activities;
- its consumer profile and customer perceptions of the brand.

Ho and Lee (2008) offer an interesting analogy for the analysis of a business and its competitive environment. They liken the business environment to the ecological system, which means that the firm is a species in changing surroundings. Therefore, they suggest that firms act in response to environmental changes as well as to actions or interactions of other species in the system. They view the firm in terms of its DNA, as it has embedded traits and patterns of behaviour which do not change easily. In order to identify a market's competitive DNA, Ho and Lee (2008, 18) advocate the following value-creating activities:

- determine the scope of the industry and identify major competitors;
- define the value chain and value activities for the target industry;
- collect business intelligence and describe the value DNA of each competitor;
- explain the competitors' behaviour using the value DNA;
- assess the competitors' responses to your firm.

As illustration, Ho and Lee (2008) offer the example of the mobile phone industry.

The key players in the mobile phone market would be identified. The value activities of selected key players would then be explored. The company would source competitor information using available market data and explore the relative strengths of each competitor within each segment. This information would provide the competitive DNA of each firm and would be used to explain competitors' behaviour to date and make predictions about future activities.

When a firm has knowledge of competitors and its competitive environment, it must then consider its strategic response to opportunities or threats. Bowman and Gatignon (1995) used PIMS (Profit Impact of Marketing Strategies) data to explore competitor response time to a new product introduction. They found that competitor reaction was faster in markets characterised by higher growth rates and tends to be more aggressive in these markets. In addition firms with a higher market share react more quickly to competitor activity. Parabhu and Stewart (2001) explored the signalling strategies firms use and examined how competitors interpret signals. They emphasise that signals depend on interpretation within a specific context. Receivers' prior beliefs influence the interpretation of a competitive signal. For example, managers in Japan may respond to American competitor activity differently, based on their cultural experiences and context; and American firms must take this into account when planning marketing activity overseas. However, Parabhu and Stewart (2001) explain that firms may also exploit the contexts in which they send their signals. For example, a firm might time their activity to coincide with external explanations for the signals. Williams (2007) developed a model of competitive dynamics, which are those moves and countermoves that firms pursue to enhance profits. He explains that firms must have certain 'actor characteristics' to facilitate competitive response. These include awareness of opportunities or threats in the competitive environment, as well as the ability and motivation to take strategic action. In addition, rival characteristics will facilitate competitive response. These characteristics include firm size, (as larger firms are more likely to respond to competitive challenge), their level of dependence on the threatened market, their level of success to date and the nature of the strategic action. For example, he suggests that competitors who are threatened by a new product launch are more likely to respond than those who perceive a lower threat. The consequences of competitive action and responses include market share gain or loss, and profit gain or loss. In the next section, we will discuss the meaning of brand share in more detail.

THE MEANING OF BRAND SHARE

When marketers try to outflank competitors' brands with their own brands, they use market share data to track their performance. Unfortunately, all too often market share is used predominantly as a monitoring device rather than as

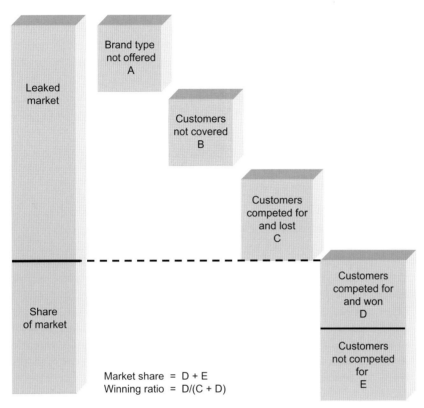

Market share = D + E
Winning ratio = D/(C + D)

FIGURE 9.6
The strategic meaning of market share (based on Ohmae 1982)

a further aid to brand strategy formulation. Kenichi Ohmae's (1982) considerable experience of strategy consulting throws a lot more light on the way that brand share can help strategic thinking. Fig. 9.6 shows the strategic meaning of brand share.

When a firm has achieved a presence in a market, its share of the market reflects the extent to which its brand is meeting consumers' needs better than other brands. Particularly with a high brand share, marketers are prone to complacency and, short-sightedly, do not consider the two components of market share. One element is those consumers actively competed for and won, i.e. area D in Fig. 9.6. This can be regarded as **active brand marketing**. An often overlooked constituent of the brand's market share, however, is represented by area E in Fig. 9.6, i.e. those consumers buying the brand who were not competed for. This component, resulting from **passive brand marketing**, may well be those consumers, for example, intent on buying Comfort fabric softener. When doing their grocery shopping, they discover that this particular brand is out of stock and, rather than go to another store, they choose an alternative brand. The more successful the brand, the closer the ratio D:E approaches 1. Marketing research can help identify this ratio by asking buyers just before the purchase what brand they intended buying and then recording the brand actually bought. By questioning those who bought a different brand from the one intended and why they did this, marketers can develop ways of marketing their brands more effectively.

We believe the model in Fig. 9.6 can be further developed, in particular those customers competed for and won, shown in area D. In our opinion, these customers can be broken down into four further groups:

- D1 — customers who are satisfied with the brand and who will actively select the same brand on the next occasion, regardless of competitive activity.
- D2 — customers who are satisfied with the brand but who can be enticed to a competitor's brand on the next occasion, if there is an attractive incentive, such as a price discount.
- D3 — customers who regard the brand as being adequate, but not fully satisfying their needs. They stay with the brand since it is inconvenient to switch to a different brand. An example of such inconvenience might be changing bank accounts, when all the standing orders have to be changed.
- D4 — customers who have tried the brand but on the next purchase occasion will switch to an alternative brand, even though this may involve more effort on their behalf.

D1 customers, the 'loyalists', need to be nurtured and frequently consulted about their views on the brand. Any dissatisfaction needs to be quickly resolved. D2 customers, the 'swingers', exist in all markets. The trade-offs they are making when choosing a particular brand need to be understood and incorporated as a further factor to consider when formulating brand strategy.

D3 customers, the 'apathetic', are offering the brand a chance for survival. If the marketer can evaluate what aspects of the brand need fine-tuning to satisfy this group, they are then better able to consider implementing corrective action rapidly. Otherwise, a competitor will find a way of taking care of consumers' perceptions of inconvenience and encourage them to switch brands. For example, some financial services organisations offer to take care of notifying the new consumer's employer as well as arranging the smooth transfer of standing orders from their previous financial services organisation.

D4 customers, the 'doubters', gave the brand a chance and are skeptical about whether it could be changed to meet their needs. Their perceptions about what the brand could do for them or say about them were not realised. The reasons for these perceptions need to be assessed and, if possible, changes considered.

Area C in Fig. 9.6, (customers competed for and lost), was an important part of Sir John Egan's strategy when trying to halt the decline of Jaguar cars in the mid-1980s. Market research studies with previous Jaguar owners who had switched to a different marque, indicated problems with Jaguar's quality and reliability. Changes were made to improve quality and reliability and these were communicated in a new promotional campaign.

Some brands do not fully capitalise on their capabilities as they are not available to all of the potential market, as represented by area B in Fig. 9.6. This may require a more intensive distribution push to cover new areas.

Area A represents a brand opportunity, since there is a group of consumers who share needs similar to those of the core market. However, they are looking for something extra which the brand is not yet offering. For example, the novice sailor gains basic sailing skills from the stable Topper dinghy but then switches to the more finely-balanced Laser dinghy, offering greater scope for racing. The challenge here is being able to develop new variants that capitalise on the core brand's heritage, yet do not damage the image of the core brand.

STRIVING FOR PROFITABLE BRANDS

The core of a successful brand is that it offers benefits to consumers in a way that other brands are unable to meet. However, profitability doesn't only result from a brand's unique competitive advantages, as research has shown.

Profitable brands are leaders

From the highly respected PIMS database, information from approximately 3000 business units has been analysed and one of the key findings is that large share brands are more profitable than small share brands. On average, the Number One brands achieve pre-tax returns on investment three times that of brands in the fifth and lower ranked positions. There are many reasons why leading brands are more profitable.

Leading brands have lower costs than followers. Economies of scale are one way that costs are reduced. For example, one source estimates that a 90-million-tonne oil refinery costs only one and a half times as much to build as a 45-million-tonne refinery. Running costs per unit of output are also lower for larger production processes. The larger refinery needs less than double the number of employees of the smaller refinery. Production efficiencies are achieved with some techniques, once a critical volume level is exceeded.

The learning curve presents a further opportunity for leading brands to curtail costs. For example, production managers soon learn the best way of configuring their employees on the production line and engineers soon begin to appreciate how to better harness new technology. Taken together, economies of scale and the experience effect cause a constant reduction in costs, each time cumulative production is doubled.

Leading brands instill more confidence in risk averse consumers. They also attract higher quality employees, who are proud to be associated with a winning brand and willing to stretch their own involvement with the brand, so that it maintains its dominant position.

Profitable brands are committed to high quality The PIMS database has shown that those brands that offer superior perceived quality relative to competitors' brands are far more profitable. Having a brand that consumers perceive to be of superior quality than other brands makes it easier for the marketer to charge a price premium. Part of the extra revenue should be used for R&D investment to sustain the quality positioning for future earnings. Also by committing everyone in the organisation to doing their job in the best possible way, there is a greater conformance to standards, which results in less rejection, less brand recalls, less reworking and, ultimately, greater profitability. Higher quality results in higher brand shares and all the benefits that this brings.

Two points need to be stressed. The first is that it is *perceived* rather than *actual* quality which counts. Engineering may set internal specifications for others to follow but, ultimately, consumers decide whether the quality meets their expectations. Furthermore, consumers are often unable to evaluate the quality of a brand and they use clues to assess its performance. The wattage of stereo speakers, for example, may be taken as an indication of their performance or the price of a brand of wine as an indicator of its taste. In some product fields, they take the core benefit for granted and assess quality by the way the brand has been augmented. Nowadays, consumers automatically accept that any airline will safely transport them to their destination. They assess quality through issues such as the additional costs associated, the politeness of the staff, the in-flight entertainment, choice on the menu, etc.

The second point is that quality is assessed by consumers on a relative basis. They assess brands against other brands. The way they do this might not necessarily be the way that the marketer would expect. For example, even though McDonalds sees Burger King as a competitor in the hamburger market, to the consumer McDonalds may be competing against Pizza Express, since they are interested in places serving fast meals.

For these and other reasons, consumers should be interviewed regularly to enable managers to understand who they see a particular brand competing against and what criteria they use to assess relative quality. Only consumer relevant attributes should guide quality improvement initiatives. Employees should be encouraged to go out and listen to consumers' views about their brands. A corporate culture, such as that at Google, which encourages employees to innovate and continuously make improvements, can raise consumers' perceptions of quality.

Having achieved a reputation amongst consumers for a quality brand, marketers need to work continuously to improve this. Over time, there is a danger of the quality positioning eroding. As markets become more mature, competitors try harder to emulate the leader. Some organisations become complacent and underestimate the potential of new competitors. Others fail to protect their quality image and do not invest in product enhancement. New technologies may make existing product formats redundant, which could have an adverse effect on the brand's perceived modernity and, consequently, on its quality.

Profitable brands capitalise on their environment

The PIMS database shows that different market characteristics offer different levels of profitability for brands. By actively seeking markets that have the right characteristics, the marketer can more successfully utilise resources to nurture a profitable brand. Some of the factors that have an impact on brand profitability are:

- market evolution − brand profitability is highest in fast growing markets and lowest in declining ones;
- markets with a high level of exports are more profitable than those with a significant imported element;
- markets with frequent new product launches are subject to lower returns on investment, although this is less so in the services and distribution sector.

CHARACTERISING SUCCESSFUL BRANDS

To achieve admirable levels of profits, long-term investment is needed in brands. Keller and Lehmann (2009) emphasise the benefits of having a strong brand as a platform for growth. They cite the example of Nike, whose revenue was US $693 million in 1982 and grew to almost $18 billion in 2007. They explain that long term brand value is dependent on a strong brand vision, which is about recognising the brand's potential and brand actualisation or using the brand potential to generate revenue. We highlight in this section some of the findings about factors associated with brand success and sustainability.

Keller (2000) identified 10 characteristics that enhance the likelihood of brand success:

(i) Excelling at delivering customer welcomed benefits. One of the reasons for the enduring success of Rolls Royce is their commitment to excellence.

(ii) Staying relevant. Winning brands track changes in their consumers' needs, refining their brands to continually meet their evolving needs. Relevance is about ensuring appropriate user imagery, usage imagery, brand personality, relationships and evoking suitable feelings amongst target consumers. Unilever's Dove team continually monitors consumers' attitudes, behaviours and expectations of beauty products; investing in R&D, packaging and communication to match the dynamic nature of their market. For example, their Dove Men+Care range is created for men and is a response to changes in male grooming behaviour and competitor activity.

(iii) Pricing on consumers' perceptions of value. There are several ways of devising a price, for example, covering cost then adding a margin, matching competitors' prices, etc. Brands that have been developed with excellent customer welcomed benefits often price on the basis of reflecting the value consumers perceive. For example, when the Toyota Corporation developed its Lexus range of cars, the up-market positioning was reinforced by a view about the value inherent in this range of cars. The pricing policy is more akin to luxury brands such as Mercedes, rather than mass-market brands. Market research can help unearth how much value consumers perceive.

(iv) Effective positioning. Successful brands own a space in consumers' minds as clearly standing for a low number of benefits. When British Airways entered the budget air travel market, they did not extend the BA brand but rather developed the Go brand to deliberately create a different positioning. Over time this brand grew, but the BA management wished to concentrate on their core business and sold it to easyJet in 2002.

(v) Consistency. Marketers face the challenge of staying relevant but not changing the brand to the extent that consumers are confused about what the brand has stood for over time. The American beer brand Michelob hit problems partly due to the way that consumers were told that the brand was for weekend events, then for evenings, then for special days.

(vi) Sensible brand portfolio hierarchy. It is important that the corporate brand creates a unifying umbrella and that there is a logical presentation of the brand portfolio. For example, BMW has taken care to ensure they have a consistent approach to linking their line brands to the BMW umbrella and their hierarchical brand architecture (Series 1, 3, 5 and 7). It logically communicates the increasing benefits across the range.

(vii) Coordinated support. As earlier noted, consumers draw inferences about brands from sources other than advertising. Successful brands have a consistent message across all of the elements supporting the brand. For example, the Disney management is clear about its brand values. Whenever internet, sponsorship, direct response, advertising, promotions and merchandising activities are developed, these must always reinforce the brand's values.

(viii) Understanding what the brand means to consumers. Brand success is more likely when managers understand the core associations consumers make with their brands. Bic is associated with inexpensive products and this allowed it to extend from pens to cigarette lighters and disposable razors. Extending the brand into perfumes went outside the boundary of the core association and was not as successful as was hoped.

(ix) Proper and long term support. Successful brands get backing from all parts of the organisation and the temptation to take a short term, 'on/off' approach to brand investment is ignored.

(x) Monitoring sources of brand equity. Unless there is a monitoring system in place that regularly assesses the different aspects of the brand's equity (addressed in Chapter 12), management find it difficult to know where they need to take corrective rather than reinforcing actions.

In the UK, Burkitt and Zealley (2006) explored research conducted by Accenture and The Marketing Society amongst pre eminent marketing executives across a range of industry sectors. They highlight three characterteristics of high performers, which emphasise the importance of measurement of marketing activities:

1. **They live in a measurement culture**. Successful brands come from companies that are 'hard wired' from the top to measure and evaluate marketing activities and outcomes systematically and performance indicators are shared. For example, Procter & Gamble has systematically developed a database of advertising copy pre-test scores, which allows them to enhance advertising copy and better predict the impact of their advertising.

2. **They invest in the right skills and capabilities**. Successful firms recruit and train individuals with a commercial perspective and an analytical mindset. For example, Diageo offered on-the-job coaching to ensure that their 'Diageo way of building brands' was instilled across the company. In addition successful companies allow managers to distil the data and take a holistic view, accepting that one measure, for example, the single measure of ROI may not capture everything. Instead, a composite assessment using several measures may be more accurate.

3. **They measure intelligently and comprehensively.** The focus of measurement should be on creating insight, not on generating information. Therefore, high performers focus on business outcomes and allow the data to link back to a common key performance indicator. For example, in Toyota, a number of metrics are used to test its direct campaigns but they are ultimately tied back into measurements of prospects' engagement with the brand.

In practice, brands become great over time. Collins (2001) explored the performance of 1435 good companies over 40 years. In his findings, he noticed that companies who go from 'good' to 'great' have no name or programme for their transformation. Instead, improvement happens slowly, with effort from every employee. To better explain the process of achieving great performance, Collins uses the analogy of a bus. Great leaders, he explains, get the right people on the bus, get the wrong people off the bus and get the right people in the right seats. When the right people are on the bus, the leader will achieve success even if their direction has to change as the world changes. He explains that leaders achieve a vision by boiling down complexities into simple but highly effective ideas. The vision is created when companies can distil the answers to three questions: *'What can we be the best in the world at?'*; *'What is the economic denominator that best drives our economic engine?'* and *'What are our core people deeply passionate about?'*. The intersection of the answers to these questions is the 'Hedgehog Concept', which allows leaders to better steer their 'bus' towards success. A challenge for all firms is to identify that which they are best in the world at; and for leaders to 'steer the bus' with the right people in the right seats, to achieve goals within changing market contexts. For example, the Apple brand is currently led by Steve Jobs, who adopts a visionary leadership style. Ranked 20 in the Interbrand 'Best Global Brands' report in 2009, the company is deeply passionate about design and launches products to anticipate and create market needs. While the company was initially established as a computer company, the question *'What can we be the best in the world at?'* was more recently answered by Jobs in a different way. In his keynote address at the launch of the Apple iPad in January 2010, Jobs declared that Apple was in the 'mobile devices' market. This market, therefore, encapsulates the highly successful iPod range of products and the iPhone, as well as Apple laptops. The iPad is a new addition to this 'mobile devices' range and illustrates Apple's innovation and design capabilities. This perspective of Apple as a mobile devices company requires a new look at the market the firm operates in. Apple would now consider brands such as Nokia mobile phones and Amazon's Kindle reader as competitors, in addition to traditional competitors including Microsoft for its software or Dell for its laptops.

Larreche (2008) explains that success requires momentum. To ignite growth, he advocates the 'twin engines' of momentum, which are momentum design, and momentum execution. On one hand, the company needs to design a product that is so compelling that customers will want to adopt it without heavy investment in promotion. In addition, a 'vibrant buzz' must be created to keep the momentum going. Fig. 9.7 illustrates Larreche's (2008) twin engines concept.

Larreche explains that momentum feeds on itself, so success provides energy for the next phase of growth. The connection between momentum design and

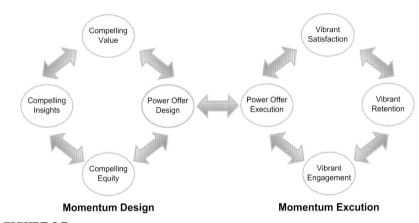

FIGURE 9.7

The twin engines of momentum *(source Larreche, J., 2006, p.33)*

execution is 'power offers', which are offers so carefully crafted that consumers find them irresistible. If we continue with our earlier example of Apple, we can see that the Apple design provides a power offer which creates compelling customer value, has compelling consumer insights and offers compelling equity. Over the longer term, Apple design offers vibrant satisfaction and retention, in addition to vibrant engagement. For example, websites such as MacRumors (www.macrumors.com) are created by Apple fans who are fully engaged with the brand and wish to discuss news and rumours about Apple products. This level of enthusiasm facilitates momentum, new product acceptance and long term brand success.

Ultimately brands succeed because the management identifies an attractive customer value proposition that engages everyone inside an organisation. Through an integrated, coherent approach, a business delivery model is developed that capitalises on the core competencies of the organisation. Regular monitoring enables changes to be instigated, which better meet consumers needs.

NEED FOR PROMOTIONAL SUPPORT

Brands win the minds of consumers because they are distinctive and stand out as having relevant added values. In other words, it is not just because they are heard above the noise of competitors but because they are making the *right* noises. A brand could be scoring well on awareness but if consumers abhor its brashness it will not succeed.

Communicating a brand's added values is an essential component contributing to long-term success. A communication strategy, however, needs to be carefully devised. It is wrong to assume automatically that advertising is the prime route to follow. For example, Singapore Airlines, while an advocate of the benefits of advertising, has not relied solely on advertising since it recognised that there are many other appropriate ways for brand communication. Consumers assess their airline from the interactions they have with the staff, the in-flight service, the décor of their business lounge, the uniforms of staff, the locations they serve, the newness of their aircraft and a host of other factors. Word-of-mouth endorsement is also an important ingredient in their success. The task marketers face is to assess how the myriad of communication clues interact to reinforce their brand's identity. Regular assessments help identify how to better coordinate brand communication devices.

Winning brands have a supporting communication strategy that results from a deep understanding of the myriad clues that consumers use to interpret them. For those brands differentiating themselves primarily through their unique image, advertising is invaluable. It reinforces the essential images amongst consumers and their peer group alike. By establishing the brand on a unique and highly valued pedestal, the marketer is yet again able to charge a premium price.

Deciding how to use advertising to support a brand and the level of this support is a complex issue on which it is difficult to give generalities. The starting point needs to be an assessment of the role expected from the advertising in support of the brand. Is it to create awareness, to give consumers new information, to build the emotional character or to regularly remind buyers? This takes but a few possible roles into account. The problem then deepens, as there needs to be some basis for understanding how the advertising will work. Ehrenberg and his research team (for example Ehrenberg et al. 2004) mainly working on 'steady-state' established markets, argue that the role of advertising as a weak force is that of a brand reminder, keeping the brand in consumers' brand repertoires. By contrast, as White (1999) argues, there are advertisers who regard advertising as a strong force having notable impacts on buying behaviour through building emotional character. While being aware of the ongoing debate, advertising that is creative can generate brand salience, particularly when clearly tied back to specific brand building objectives.

Hallward (2008, 339) explains that marketing is 'evolving with a growing number of media choices, the emergence of digital entertainment, the challenge to TV advertising, increasing consumer generated content, media mobility and many challenges presented by the 'long tail of media'. He explains that in this new world, advertising needs to get attention and achieve a desired response, yet recall may not follow media consumption. For example, he notes that children may have higher recall, while watching less TV than adults. In addition, he challenges the traditional assertion that share of voice (SOV) is positively correlated with recall and impact. Instead, he emphasises the importance of advertising creativity in achieving recall. He asserts that message content typically explains 75% of advertising success or failure. Hallward (2008) advocates 'consumer mix modeling', where marketers evaluate the impact of all brand touch points on brand objectives at the individual customer level, to facilitate target analysis. For example, when Unilever launched the Lynx Click variant, they used multiple touch points to build the brand message. The company had two main target markets: young men who were users or potential users of the product and mothers who would purchase the product for their sons. TV advertising featuring the actor Ben Affleck was supported by billboard, ambient and print advertising, as well as variant point of sale material, online promotions and product sampling. Using Hallward's (2008) recommendation, the company could ask respondents about the brand first and then about each touch point. Measures of success would include brand equity, familiarity, and trial and usage frequency, and these could be used to evaluate the role of each touch point in creating different brand measures among different target markets.

SHOULD A COMPANY BUILD BRANDS OR BUY BRANDS?

Some have argued that the increasing cost of communicating and building new brands caused marketers to shift their attention towards buying other firms' brands rather than building their own. Another reason for the attractiveness of brand buying was the ability to extend a brand's franchise across countries. Where a multinational felt it to be strategically appropriate to have a particular range in each major country, it was not unusual to find a team evaluating different brand purchases and also assessing how easily the new brands could be integrated. A brand acquisition is particularly attractive when it has built a considerable amount of goodwill with a particularly strong reputation.

Brand acquisitions are often seen as a quicker way of gaining entrance to a new market. However, the difficulty of integrating the new brand into the company's structure is often overlooked. Furthermore, the potential for synergy can sometimes be overestimated. The takeover of the American Howard Johnson hotel and restaurant chain by the Imperial Group floundered because Imperial's knowledge of consumer markets was irrelevant to the American hotel and fast food market. This unsuccessful acquisition was sold off five years later.

By contrast, brands acquired in product or services sectors that are closely related to the firm's current expertise stand a greater chance of succeeding under the new owner than those in sectors where the owner has little expertise. There are many stories of firms unsuccessfully diversifying this way. For example, BP bought its way into the coal minerals and information technology sectors, all of which are markedly different from its core competence in oil exploration, transportation, refining and distribution. It soon recognised the problems of this diversification route and sold off these interests.

Where a brand is purchased by a larger organisation, the smaller firm must retain a level of control to ensure that its mission and values are upheld in future activities. For example, when the Body Shop was taken over by L'Oreal in 2006, its strong mission and values were upheld, leaving the brand as a stand-alone company and retaining Anita Roddick as consultant. The ice cream brand Ben & Jerry's was acquired by Unilever in 2000. However, although it became part of a portfolio of 400 brands, Unilever recognises the heritage of Ben & Jerrys, its authentic culture and values and, in particular, its association with Vermont, where the brand originated. Nearly three-quarters of the brand's ice cream is made in Vermont plants and the company employs over 500 people in the state. Its CEO, a Unilever employee, explains that the area of origin is a large part of the brand — 'People are buying a little piece of Vermont all over the world'.

It is critical that firms give consideration to competitor reaction to mergers and acquisitions. For example, in the highly competitive space occupied by Google and Microsoft, where the internet offers a significant platform for growth, acquisitions that 'lock in' partner brands stimulate aggressive competitor response and reaction from competition authorities.

EXTENDING BRANDS

If the firm finds that it is unable to penetrate the market further with its current brands, it may consider moving into a related market. It could argue that the best way to overcome the risk of a new brand failing, of consumer apathy and competitive resistance would be to stretch its existing name. While the inherent goodwill and awareness from the original brand name may help the new brand's development, there is also the danger that the new brand could dilute the strength of the original brand and convey the wrong perceptions, with consequent detrimental effect on the original brand. For example, the economics of establishing new brands are pushing companies more towards stretching their existing names into new markets. Daunted by the heavy R&D costs and more aware of the statistics about failure rates for new brands, marketers are increasingly taking their established names into new product fields. Hennig-Thurau et al. (2009) explain that extensions offer financial benefits in two ways. First, there are 'forward spillover' effects as brand extensions generate higher revenues and reduce project specific risk. Second, there is 'reciprocal spillover' to the parent brand. They examined brand extensions in the movie industry and found that sequels complemented parent films. For example, more than 50% of

DVD sales were explained by the parent's brand awareness and image, and up to 70% were explained when the success of the sequel was included.

Tauber reviewed a sample of 276 brand extensions to evaluate the different ways of extending brands and concluded that there are seven types of brand extensions:

1. *Same product in a different form* – for example, Persil Biological washing powder extending into Persil 'Small and Mighty' concentrated liquid, tablets or liquigel.
2. *Distinctive taste, ingredient or component* – an ingredient or component of the current brand is used to make a new item in a different category. For example, Kraft extended the distinctive taste of their Philadelphia cream cheese into Philadelphia Snacks, co-branded products with Ritz crackers.
3. *Companion product* – where some products are used with others, these lend themselves to brand extensions. For example, Oral B electric toothbrushes will often be used at the same time as Oral B dental floss and the brand has perceived expertise in both areas.
4. *Same customer franchise* – marketers develop different brands to sell to their loyal customers. For example, the AAA is primarily known for its heritage as a roadside assistance service to motorists, yet it markets a variety of AAA products, such as car insurance, driving lessons, travel guides, hotel ratings and the AAA Visacard.
5. *Expertise* – brands are extended into areas where consumers believe the original brand has connotations of special knowledge or experience. For example, Canon's perceived expertise in optics was extended into photocopiers.
6. *Unique benefit, attribute or feature owned by the brand* – some brands stand out for their uniqueness on a particular attribute, which is extended into a related field. For example, aerospace physicist Dr Max Heuber developed a cream to smooth scars received in a lab accident. His product, Miracle Broth is the elixir present in the Crème de la Mer moisturiser and the premium brand has been extended into related products such as Crème de la Mer lip balm and eye treatments.
7. *Designer image or status* – some consumers feel that their Saab cars have a higher status when they know that Saab also work in jet aircraft.

When considering extending the original brand name into a new sector, the benefits that come from stretching the name must be weighed against any negative connotations with the original name and any damage that may be done to the core brand. Martinez, Polo and de Chernatony (2008) investigated the role of brand extension strategy on brand image. They found that a parent brand with strong brand image will be less at risk when they launch an extension with the same brand name. Where the firm stresses the name of the

MARMITE CEREAL BAR

In this advert, Marmite challenge consumers' perceptions about its brand meaning. By suggesting an implausible concept extension such as a fabric softener, the actual brand extension, a cereal bar, appears much more acceptable.

brand over the attributes, they are more likely to succeed with extensions, even if those extensions differ from the parent brand.

A checklist adapted from Aaker (1990) should enable marketers to decide whether it is wise to extend the brand name.

Possible benefits from extending the brand name

1. **Awareness**. Are consumers aware of the brand name? To take advantage of global opportunities the UK brand of household cleaners, Jif, was changed in the early 2000s to Cif, bringing it in line with other countries. Whirlpool, the white goods manufacturer entered into a joint venture with Philips in 1989 as part of its expansion programme into Europe. Philips was not prepared to sell its well known name and Whirlpool employed a dual branding policy, knowing that, as a result of the contract, they did not have many years before they had to drop the Philips name.

2. **Brand associations**. Will the consumer consider the extension to be a credible move? Does the name bring the right sorts of associations to mind? Baileys Irish Cream's use of its name in the ice cream market brings to mind automatically an expectation about a certain type of taste and customer experience.

3. **Quality associations**. Will the name stretch to give the correct perception of quality? Bulgari started out as a fine jeweler in Rome. Over time it gradually extended its name, continually majoring on its high quality philosophy — for example, extending into perfumes, silk scarves, leather bags, eye wear and giftware.

4. **Encourage trial purchase**. Will the name give the needed reassurance to the risk averse consumer? The Apple iPad offers the reassurance of Apple's brand heritage and its success with brands such as the iPod and iPhone to customers wishing to adopt multi-touch technology.

Possible weaknesses from the core brand name

1. **No value-added**. Will the name add value to the new line? Some would argue that the designer label status from the name Pierre Cardin adds little to its line extensions — particularly when it marketed bathroom tiles in Spain.

2. **Negative associations**. Will the wrong associations result? For example, Levi Strauss were known and respected for their jeans. The extension into Levi Tailored Classics suits failed because of the wrong associations.

3. **Name confusion**. Does the name imply the type of product about to be marketed? When McDonalds revitalised its menu to offer a salad range, it created a new sub-brand name 'Changing Tastes' to distinguish the products from the traditional McDonalds fast food product offering.

Possible damage to the core brand

1. **Undesirable associations.** Will the image of the core brand be damaged? Black & Decker's acquisition of the small appliances range from General Electric caused much internal debate. It was questioned if the heavy duty image that Black & Decker had would be weakened by stretching their name across the new range.

2. **Perceived quality deteriorates**. Will the perceived quality of the core brand fall? With its acquisition of Skoda Volkswagen retained the original name, thereby protecting VW if, as was not the case, the acquired brand failed'.

3. **What about disasters?** Will an unforeseen threat damage the core brand? When Persil launched Persil System 3, it did not anticipate dermatological problems and the resulting short-term market resistance.

Besides these three broad areas of questioning, it may be wise for the marketer to address three further questions that focus on the economics of brand extensions. Are there limitations on the size of the marketing budget? If the brand marketer is constrained by company cutbacks, brand extensions may be the only viable tactical route. However, of more importance may be the question of cannibalisation. The presence of the core brand name may result in sales of the new brand coming not just from new consumers, but also from those who used to buy the core brand. While it is inevitable that some cannibalisation will occur, the marketer needs to anticipate the likely extent of this. Further, marketers should consider the influence of culture on the acceptance of brand extensions and use consumer insights to determine whether standardised extensions are the most appropriate strategy. In research across Norway, Spain and the UK, Buil, de Chernatony and Hem (2009) found that culture influenced both consumer reactions to brand extensions and the feedback effects of extensions on brand equity. With an increasingly global marketplace, managers must therefore consider whether modification or adapting the extensions would increase the likelihood of success of new extensions and reduce potential damage to the parent brand.

Researchers have investigated the extent to which brands that stress functional or prestigious connotations can be extended — for example, consumers regard Timex watches as being more associated with functional benefits, while Rolex watches have more prestigious associations. They found that a brand name that is strongly associated with functional benefits can be more easily extended into product areas bought mainly for their functional benefits. So, for example, a branded wristwatch positioned primarily as offering functional excellence could stretch its name with little difficulty into stopwatches. By contrast, they found that if the original brand name majored on the dimension of prestige, it could be more easily extended into a product field known for its prestige rather than its functionality. In other words, the Rolex name could more likely be extended to grandfather clocks than to stopwatches.

It was also reported that when consumers perceived a brand being extended into a product field that many firms would find relatively easy to produce, the brand extension would not be accepted. This may be because consumers felt that in the new product field there was little difference between competing brands and that the new entrant was going to do little more than use its image to charge an unnecessary price premium.

Within the services sector, Pina, Martinez, de Chernatony and Drury (2006) found that corporate image associations are dependent on consumers' initial beliefs about the brand. It is critical that when firms launch a service, the existing service is positively evaluated in the market and that there is a perceived fit between the existing service and the new service. Perceived fit between the corporate brand and the extension may be more important in services, as their intangible nature means that consumers experience difficulties in evaluating quality. A virtuous circle exists as a high corporate image will increase acceptability of the extension when the new service is similar to the parent. In addition, the perceived fit will also prevent image dilution.

CONCLUSIONS

By analysing environmental opportunities and threats as well as thoroughly appreciating the nature of the brand's competitive advantage, marketers are able to develop strategies that position their brands to achieve the best return, while being protected from competitive attacks. The lifespan of brands will depend on the sustainability of their competitive advantage.

The two broad competitive advantages inherent in successful brands are based on either delivering similar benefits more cheaply than the competitors, i.e. cost-driven brands and/or delivering benefits superior to the competition at a price premium, e.g. value-added brands. It is crucial that a decision is made about the strategic path that the brand will follow and that everyone is informed of this. Each of these two routes makes different demands on employees, resources and processes. With everyone in the firm aware of the branding route being followed, they can all contribute by being vigilant in cost-curtailing activities and by being creative and original in devising added value. Strategies appropriate for each type of brand need developing and must avoid the commodity brand domain, which is characterised by little added value and, frequently, no cost advantage.

A useful device for identifying the competitive advantages of brands is Porter's Value Chain. This helps managers to consider the processes and supporting services involved in transforming commodities into highly respected brands. It also acts as a check as to whether the competitive advantages of brands are being reinforced by considering the linkages in each part of the value adding process.

A more sophisticated way of developing brand strategy is to consider both the type of competitive advantage inherent in the brand and the competitive scope of the market it will be targeted at, i.e. a narrow or broad group of consumers. From the resulting matrix, four generic strategies were documented.

Having identified the brand's competitive advantage and the strategic direction to be followed, the marketer then needs to anticipate competitors' responses and develop ways of sustaining the brand's advantage. An important issue is to agree, as a brand team, who the competitors are and to use the concept of market share to evaluate future opportunities. Firms must be vigilant to anticipate likely competitor response time and be accurate in their interpretation of competitive signals.

Profitable brands focus on adopting leadership positions in specific markets, offering superior perceived quality when compared to the competition and taking advantage of environmental opportunities. Brand profitability arises from a long-term brand investment programme and factors that drive brand success, and ultimately brand profitability, have been presented. The potential advantages offered by customer interaction with every touch point of brand communication must be considered, as different customer segments will have different needs that are met by different interactions with the brand message.

More recently, high advertising costs have forced many firms to look more favourably at buying, rather than building brands from scratch. Companies that buy already strong brands must give due consideration to their mission and values to ensure that the brand message is upheld. In addition, managers should take care that acquisition and merger strategies do not initiate competitor response or incur allegations of monopolistic behaviour. Another way that marketers seek to reduce costs and possibly minimise risk when they have a favourable brand reputation is by extending the core brand's name into new markets. The dangers inherent in this strategy and the challenges faced by cultural interpretations of brand meaning need to be carefully evaluated.

MARKETING ACTION CHECKLIST

It is recommended that, after reading this chapter, marketers undertake the following exercises to test and refine their brand strategies:

1. Write down what you believe to be the reasons why consumers buy your brand and your competitors' brands. For each of these reasons, evaluate which of the brands comes closest to satisfying consumers' needs. If this information is already available from an accurate marketing research study, it should be used instead of management judgment. On a separate sheet of paper, summarise these findings by stating for each brand which attributes it best satisfies. This is one way of identifying each brand's

competitive advantage(s). Those brands which do not have any 'success attributes' are unlikely to succeed for long.

2. How easy is it for your competitors to copy any of your brands' competitive advantages? What actions are you taking to sustain your brands' competitive advantages?

3. For each of your brands separately, use the matrix of value-added advantage versus cost-driven advantage in Fig. 9.3 to plot your brand and those of your competitors. Whichever quadrant your brand occupies, ask how appropriate your current strategies are. Those competitors' brands falling in the same quadrant as your brand represent the greatest potential threat. What are you doing to protect your brand against these?

4. Using the example of the value chain in Fig. 9.4, identify all those activities you undertake in transforming low value goods or services into high value, finished goods or services. Which of these activities do you believe you do, or could do, better than competitors? How do these 'doing better or cheaper' activities relate to the competitive advantages identified in Question 1 above? Try to assess what proportion of costs can be allocated to each part of the value chain for a particular brand. How do these cost components compare with competitors'? How well-linked are each of the processes and support activities on the value chain?

5. For each of your brands separately, use the matrix of competitive scope versus competitive advantage in Fig. 9.5 to plot your brand and those of your competitors. Whichever quadrant your brand occupies, check the appropriateness of your current strategy. Which competitive brands fall in the same quadrant as your brand? How appropriate are their current strategies? What plans have you to protect your brand against those brands in the same quadrant?

6. Take one of your brands and evaluate whether you are striving to sustain its competitive advantage by concentrating on operational effectiveness or by concentrating on performing activities different from your competitors or by performing similar activities in different ways. If you are reliant upon operational effectiveness, be aware of the difficulty of staying ahead of competitors and consider how you could follow either of the latter two strategies. To what extent do each of the activities in your brand building (a) show consistency and (b) reinforce each other? Take your value chain and consider how you could restructure any of the activities to give a stronger fit and therefore greater sustainability.

7. As individuals (rather than as a team), select a brand and write down which competitors it competes against. Then, as individuals, group competitors together into discrete clusters so that those competitors grouped together show similarity in terms of the strategy being followed. Then meet as a management team and discuss your rationale for the named competitors and the different bases for strategic groupings.

8. From Question 7, which competitors did the management team identify as following a similar strategy to your brand? How much relevant information do you hold about these competitors (and how able are you to anticipate their future impact on your brand)?

9. Using Fig. 9.6, which shows the components of brand share, what marketing research data do you hold to evaluate the volume of sales that is or could be made to customers in the Blocks A to E? In terms of those customers competed for and won in block D, what proportion of customers falls into each of the categories D1, D2, D3 and D4 (as outlined in the section 'The Meaning of Brand Share')? What actions could be taken to maximise the number of customers in category D1?

10. What marketing research data is held about consumers' perceptions of the quality of your brand and those of competitors? From qualitative research, are you able to establish accurately how consumers evaluate the quality of your brand and competitors' brands? Based on these marketing research reports, what actions are needed to improve consumers' perceptions of your quality?

11. On the last occasion that you followed a policy of using an existing brand name for a new addition to your range, evaluate:
    - the strengths of using the core brand name;
    - the advantages for the new line, of carrying the core brand name;
    - the effect the new line had on the core brand.
      Taking all of these points into consideration, was it wise to have followed a brand extension policy?

STUDENT BASED ENQUIRY

1. Select an example of i) a cost-driven brand and ii) a value-added brand you are familiar with. Explore the differences in their branding strategies that suggest their strategic approach.

2. Fig. 9.3 classifies brands on a strategic basis. Apply this classification to either a) the car industry or b) the computer industry, identifying examples of each brand type within your selected industry.

3. Apply the value chain framework in Fig. 9.4 to a brand of your choice. Discuss how the stages in the value chain can potentially enhance or detract from the brand's competitive advantage.

4. Using the Internet, identify a brand acquisition or merger that has occurred within the past six months. Discuss two possible strategic advantages and two potential challenges that the merger or acquisition may have created for the parent brand.

5. Select a brand you are familiar with. Consider the options for brand extensions presented by Tauber. Select two options which would be applicable to the brand you have selected. Using Aaker's checklist, would

you advocate that they retain their brand name for the two extensions that you propose?

References

Aaker, D. (1990). Brand extensions: the good, the bad and the ugly. *Sloan Management Review, 31* (4), 47–56.

Aaker, D., & Keller, K. (Jan. 1990). Consumer evaluations of brand extensions. *Journal of Marketing, 54*, 27–41.

Bowman, D., & Gatignon, H. (February 1995). Determinants of competitor response time to a new product introduction. *Journal of Marketing Research, XXXII*, 42–53.

Buil, I., de Chernatony, L., & Hem, L. E. (2009). Brand extension strategies: perceived fit, brand type and culture influences. *European Journal of Marketing, 43*(11/12), 1300–1324.

Burkitt, H., & Zealley, J. (2006). *Marketing Excellence*. Chichester: John Wiley.

Collins, J. (2001). *Good to Great*. London: Random House.

Ehrenberg, A. S. C., Uncles, M. D., & Goodhart, G. J. (2004). Understanding brand performance measures: using Dirichlet benchmarks. *Journal of Business Research, 57*(12), 1307–1325.

Ewing, M. T., Jevons, C. P., & Khalil, E. L. (2009). Brand death: a developmental model of senescence. *Journal of Business Research, 62*, 332–338.

Fombrun, C., & Rindova, V. (1996). *Who's Tops and Who Decides? The Social Construction of Corporate Reputations*. Working paper, New York University, Stern School of Business.

Hallward, J. (September 2008). 'Make measurable what is not so', Consumer mix modelling for the evolving media world. *Journal of Advertising Research*, 339–351.

Hennig-Thurau, T., Houston, M. B., & Heitjans, T. (November 2009). Conceptualizing and measuring the monetary value of brand extensions: the case of motion pictures. *Journal of Marketing, 72*, 176–183.

Ho, J. C., & Lee, C. S. (2008). The DNA of industrial competitors. *Research Technology Management, 51*(4), 17–20.

Keller, K. (Jan-Feb 2000). The brand report card. *Harvard Business Review*, 147–157.

Keller, K. L., & Lehmann, L. (2009). Assessing Brand Potential. *Journal of Brand Management, 17*, 6–17.

Larreche, J. C. (2008). *The Momentum Effect*. Upper Saddle River NJ: Wharton School Publishing. See also. www.themomentumeffect.com.

Martinez, E., Polo, Y., & de Chernatony, L. (2008). Effect of brand extension strategies on brand image: a comparative study of the UK and Spanish markets. *International Marketing Review, 25* (1), 107–137.

Ohmae, K. (1982). *The Mind of the Strategist*. Harmondsworth: Penguin.

Parabhu, J., & Stewart, D. W. (February 2001). Signaling strategies in competitive interaction: building reputations and hiding the truth. *Journal of Marketing Research, XXXVIII*, 62–72.

Pina, J. M., Martinez, E., de Chernatony, L., & Drury, S. (2006). The effect of service brand extensions on corporate image. *European Journal of Marketing, 40*(1/2), 174–197.

Porter, M. (1985). *Competitive Advantage*. New York: The Free Press.

Veloutsou, C., & Moutinho, L. (2009). Brand relationships through brand reputation and brand tribalism. *Journal of Business Research, 62*, 314–322.

White, R. (1999). Brands and Advertising. In J. P. Jones (Ed.), *How to Use Advertising to Build Strong Brands*. London: Sage Publications.

Williams, S. D. (2007). Gaining and losing market share and returns: a competitive dynamics model. *Journal of Strategic Marketing, 15,* 139–148.

Zheng Zhou, K., Brown, J. R., & Dev, C. S. (2009). Market orientation, competitive advantage and performance: a demand-based perspective. *Journal of Business Research, 62,* 1063–1070.

Further Reading

Aaker, D. (1991). *Managing Brand Equity.* New York: The Free Press.

Apple. (2010). *ipad Keynote Address. (Online).* Available at. http://www.apple.com/ipad/. Accessed 13 March 2010.

BBC. (2006). *Body Shop Agrees L'Oreal Takeover.* Available at: http://news.bbc.co.uk/2/hi/4815776.stm. Accessed 24 March 2010.

Biel, A. (Nov 1990). Strong brand, high spend. *ADMAP,* 35–40.

Brown, A. (1995). *Organisational Culture.* London: Pitman Publishing.

Buday, T. (1989). Capitalizing on brand extensions. *Journal of Consumer Marketing, 6*(4), 27–30.

Burlington Free Press. (2010). *New Ben & Jerry's CEO Promises to Keep Company Anchored in Vermont.* Available at: http://www.burlingtonfreepress.com/article/20100324/NEWS01/100323019/New-Ben-Jerry-s-CEO-promises-to-keep-company-anchored-in-Vermont. Accessed 23 March 2010.

Buzzell, R., & Gale, B. (1987). *The PIMS Principles.* New York: The Free Press.

Clifford, D., & Cavanagh, R. (1985). *The Winning Performance: How America's High Growth Midsize Companies Succeed.* London: Sidgwick & Jackson.

CNN. (2009). *Microsoft Takes Gloves Off Against Google.* Available at: http://news.cnet.com/8301-30684_3-10460829-265.html. Accessed 24 March 2010.

Daniels, K., Johnson, G., & de Chernatony, L. (2002). Task and institutional influence on managers' mental models of competition. *Organization Studies, 23*(1), 31–62.

Davidson, H. (1997). *Even More Offensive Marketing.* Harmondsworth: Penguin.

de Chernatony, L. (1991). Formulating brand strategy. *European Management Journal, 9*(2), 194–200.

de Chernatony, L., & Daniels, K. (1994). Developing a more effective brand positioning. *Journal of Brand Management, 1*(6), 373–379.

de Chernatony, L., Daniels, K., & Johnson, G. (1993). Competitive positioning strategies of mirroring sellers' and buyers' perceptions? *Journal of Strategic Marketing, 1,* 229–248.

de Chernatony, L., Daniels, K., & Johnson, G. (1995). *Managers' Perceptions of Competitors' Positioning: a Replication Study.* Paper presented at TIMS Marketing Science Conference. Sydney: Australian School of Management.

Doyle, P. (1989). Building successful brands: the strategic options. *Journal of Marketing Management, 5*(1), 77–95.

Doyle, P. (2001). Building value-based branding strategies. *Journal of Strategic Marketing, 4*(1), 255–268.

Easingwood, C. (1990). Hard to copy services. In A. Pendlebury, & T. Watkins (Eds.), *Marketing Educators Group Proceedings* (pp. 325–336). Oxford: MEG.

Easton, G. (1988). Competition and marketing strategy. *European Journal of Marketing, 27*(2), 31–49.

Fortune. (2009). *World's Most Admired Companies.* Available at: http://money.cnn.com/magazines/fortune/mostadmired/2010/full_list. Accessed 8 March 2010.

Gehlhar, M. J., Regmi, A., Stefanou, S. E., & Zoumas, B. E. (2009). Brand leadership and product innovation as firm strategies in global food markets. *Journal of Product and Brand Management, 18*(2), 115–126.

Grossberg, K. (Sept–Oct 1989). How Citibank created a retail niche for itself in Japan. *Planning Review*, 14–17, 48.

Hall, R. (1992). The strategic analysis of intangible resources. *Strategic Management Journal, 13*, 135–142.

Hamel, G., & Prahalad, C. (May–June 1989). Strategic intent. *Harvard Business Review*, 63–76.

Johnson, G., & Scholes, K. (2002). *Exploring Corporate Strategy*. Harlow: Pearson Education.

Jones, B., & Ramsden, R. (Sept 1991). The global brand age. *Management Today*, 78–83.

Jones, J. P. (1999). *How to Use Advertising to Build Strong Brands*. London: Sage Publications.

Karakaya, F., & Stahl, M. (April 1989). Barriers to entry and market entry decisions in consumer and industrial goods markets. *Journal of Marketing, 53*, 80–91.

Karel, J. (May–June 1991). Brand strategy positions products worldwide. *Journal of Business Strategy*, 16–19.

Kelley, B. (May 1991). Making it different. *Sales & Marketing Management*, 52–55, 60.

Lorenz, C. (4 March 1988). Unrelated take-overs spell trouble. *Financial Times*, 22.

Park, C. W., Milberg, S., & Lawson, R. (Sept 1991). Evaluation of brand extension: the role of product feature similarity and brand concept consistency. *Journal of Consumer Research, 18*, 185–193.

Peters, T., & Waterman, R. (1982). *In Search of Excellence*. New York: Harper & Row.

Porac, J., & Thomas, H. (1990). Taxonomic mental models in competitor definition. *Academy of Management Review, 15*(2), 224–240.

Porter, M. (1980). *Competitive Strategy*. New York: The Free Press.

Porter, M. (Nov–Dec 1996). What is strategy? *Harvard Business Review*, 61–78.

Prahalad, C., & Hamel, G. (May–June 1990). The core competences of the corporation. *Harvard Business Review*, 79–91.

Tauber, E. (Aug–Sept 1988). Brand leverage: strategy for growth in a cost control world. *Journal of Advertising Research*, 26–30.

Thornhill, J. (17 September 1991). Ratner gloomy as recession drives group into £17.7m loss. *Financial Times*.

Thornhill, J. (January 11/12 1992). Jewellery innovator and the textile veteran. *Financial Times Weekend, 6*(1), 22.

Tyrepress. (2010). *Snap-on Reports Record UK Sales for 2009*. Available at: http://www.tyrepress.com/News/1/6/18695.html. Accessed 8 March 2010.

Urban, G., & Star, S. (1991). *Advanced Marketing Strategy*. Englewood Cliffs: Prentice Hall.

Wachman, R., & Fairbairn, S. (10 January 1992). Ratner steps down as sales slump. *The Evening Standard*, 5.

Woodward, S. (1991). Competitive marketing. In D. Cowley (Ed.), *Understanding Brands*. London: Kogan Page.

CHAPTER 10

The Challenge of Developing and Sustaining Added Values

OBJECTIVES

After reading this chapter you will be able to:

- Understand what is meant by the term 'value' in the context of branding.
- Describe the nature and role of added values for brand managers.
- Explore how brand value can be created and sustained over time.
- Consider how the four-level model of brands can be applied to well-known brands.
- Identify the role of customisation and relationships in creating brand value.
- Appreciate how brand value can be created in other contexts, for example, in SMEs.
- Apply the concept of the value chain to brands to determine sources of brand value.
- Evaluate the impact of counterfeit brands on brand value and consumer behaviour.

SUMMARY

The aim of this chapter is to consider the challenges that marketers face in developing and sustaining brand added values. The chapter begins by making the point that it is only worth developing added values if they are relevant to the target market and noticeably different from those of competitors. Any marketing activity then needs to integrate these added values and present brands as **holistic** offerings. In other words all, rather than one single aspect, of the brand's assets should be developed, enabling customers to appreciate their points of difference, the way they satisfy both functional and emotional needs, reduce perceived risk and make purchasing easy.

One way of identifying possible added values for brands is to consider a four-level model of a brand as a **generic** product or a basic service with an

371

Creating Powerful Brands. DOI: 10.1016/B978-1-85617-849-5.10010-8

expected, **augmented** and **potential** branding surround. We describe the development of brands using this conceptual model in both product and service sectors. Moreover we identify further ways of adding value, such as consumer participation and customisation. We also consider the problem of sustaining brands' added values over time and against counterfeit imitators. Counterfeiting, however, is but one of the challenges facing brands and we conclude the chapter by considering some of the other challenges.

POSITIONING BRANDS AS ADDED VALUE OFFERINGS

As we explain in Chapter 2: '*A brand is a cluster of functional and emotional values that enables organisations to make a promise about a unique and welcomed experience*'. Brands succeed because customers perceive them as having value over and above that of the 'equivalent' commodity (if there is such a thing!) or value in excess of the sum of the price of the product's or service's constituent parts. This makes them *noticeably differentiated* with *relevant* and *welcomed* attributes.

Added value is integral to brands. Interviews with 20 leading-edge branding experts (de Chernatony, Harris and Dall'Olmo Riley, 2000) revealed added value to be a multi-dimensional concept, which includes functional and emotional benefits as perceived by customers and which exists relative to the competition offerings that are primarily different because their emphasis on name distinctions satisfy consumers' core needs in a tightly defined sense. By contrast, successful brands are differentiated because of their added values which go beyond just satisfying a core need and offer augmented benefits. In addition, those added values which are more sustainable are those intangible psychological values inherent in the brand's essence (de Chernatony, Harris and Dall'Olmo Riley, 2000).

The added value that brands bring to the purchase or consumption experience is not restricted just to consumers. In business-to-business markets, customers will select brands they know and trust. For example, Caterpillar (CAT) focuses on customer loyalty and they have trained over 10,000 employees in their 'one voice' programme in order to demonstrate CAT's trusted personality to all customers. The concept of added value in business-to-business markets goes back a long time, for example, during the 1980s, IBM claimed that 'no one ever got fired for buying IBM', reinforcing reassurance about IBM's performance.

In this advert, IBM illustrates how its ECM compliance solutions limits the impact of compliance on physical records management.

IBM
In this advert IBM emphasis the added value of their ECM Compliance Solutions in terms of a reduced physical records management burden and cost reductions

When managers develop a brand through adding extras to enhance consumers' perceptions of greater value, they need to recognise that, as value 'is in the eyes of the beholder', any extras must enhance customer satisfaction. Continually striving to engender repeat buying behaviour, plus a more favourable consumer disposition to a particular brand, is the goal of all marketers, but from an added value perspective, this presents challenges. Repeat buying behaviour means that, as consumers experiment more, they start to gain more confidence and their expectations rise. With repeated brand buying, what is regarded as an added value by the marketer becomes an expected necessity to consumers, who start to take it for granted. The implication of this is that marketers should continually strive to add further value. The second issue is that any added value is judged by the consumer *relative* to competing brands. It therefore behoves the marketer to track their competitors' strategies. It is interesting to note the way that the airlines seek to add value through enhancements such as in-flight entertainment, sleepers, shower facilities, collect from office service or express check-in, yet as soon as one firm augments their brand, others are tracking their actions and rapidly follow.

The issue of brand sustainability must also be considered. In their interviews with brand experts, de Chernatony, Harris and Dall'Olmo Riley (2000) found that core brand values are sustainable but physical added values may not be. Therefore, while the symbolic and emotional meaning of the brand endures, attributes may change over time. Those familiar with the BBC 'Doctor Who' TV series will know that over time the doctor 'regenerates' as he takes on a new human form, appearing as a new actor. While the doctor's physical appearance changes, his qualities, such as intelligence and ethical stance are unchanged. Others' emotional responses to the doctor, such as feelings of hope and trust, also remain unchanged whenever the doctor changes. Physical products also experience many changes in their physical values. For example packaging or serving formats evolve or the product's range might change. However, the core brand meaning, and the emotional experience people have with the brand will remain the same. For example, Nike has expanded into new sports areas such as golf, with new product designs, yet the emotional meanings encapsulated by the legendary slogan 'Just Do It', applies across all its products. Those who grew up watching Disney movies in the 1950s and 1960s will be familiar with traditional animated characters such as Mickey Mouse and Snow White, which are very different to more recent Disney productions such as High School Musical or Hannah Montana. Yet, Disney's added value remains the same and over the years, Disney has successfully sustained a brand experience 'for the child in all of us'.

There are different interpretations about the meaning of added value. The perspective we have reviewed so far relates to added value in terms of providing extra benefits beyond that of the basic offering. In some cases the consumer may not be fully aware of the extra benefits included when buying the brand. For example, when signing up for an American Express card, members are aware of the associated backup services but not their details. American Express plays on this in its advertisements showing examples of how purchases can result in membership points, which subsequently translate into gifts. Therefore, they explain that a bowl of pasta, paid for with an American Express card, can translate into an iPod through membership points. Ever interested in providing more enhancements, they have moved on to devising almost concierge services, such as present buying for someone emotionally close to the card holder.

Another interpretation takes a slightly different slant, considering added value as being a feeling consumers have that the brand offers more than competitors, regardless of whether the feeling is based on a real or perceived issues. This consumer-focused interpretation draws to our attention the importance of consumers' perceptions. Thus while two financial advisers working for the same consultancy have the same technical knowledge about financial services, a high net worth consumer may prefer to deal with one of these advisers, primarily due to the way they get on with this particular advisor, who goes beyond their expectations to provide outstanding service. A further implication from this interpretation is that it recognises the way that brand personality and reputation are sources of added value. An interesting discussion of types of added values that adhere to this interpretation is provided by Jones (1986). For example, there are added values which:

- come from experience of the brand, due to factors such as familiarity, reputation or personality;
- come from the sorts of people using the brand;
- come from beliefs that the brand is effective, as is the case with cosmetics;
- come from the appearance of the brand or its distinctive packaging.

Another interpretation of added value blends both the consumer and managerial perspective, regarding it as satisfying consumers' needs better than the competitors at an attractive price. Adding extra benefits to a brand results in extra costs and this perspective is concerned with ensuring that consumers are prepared to pay a greater premium than the cost of adding such benefits. One of the problems managers face when following this perspective is that they over-engineer the brand with non-relevant consumer benefits and the price consumers perceive to be a fair representation of its value is lower than the economic costs of developing it. For example, some mobile phones have a notable number of features, yet some consumers only use a proportion of these.

The previous interpretation introduces the idea of moving from price to value. Prices are what marketers set, while value is what consumers perceive. By shifting focus away from a company-centric perspective, i.e. pricing, to a consumer-centric perspective, i.e. value, marketers can further consider how to add value. One way of broadening the possibility for new ideas is to draw on the research that Zeithaml (1988) undertook to understand consumers' views about the concept of value. Her findings are that consumers conceive value in four ways, i.e.:

- value is low price — for example, Lidl offers customers value through low grocery prices;
- value is whatever I want in a product — for example, Jaffa Cakes offer a low calorie treat;
- value is the quality I get for the price I pay — an Aston Martin offers the customer pride though ownership of a prestigious car;
- value is whatever I get for whatever I give — donating to the Red Cross gives the customer a feeling of goodwill and altruism.

These clusters of consumer interpretations provide a basis for adding more value to each customer segment. For example the first segment ('low price') may be particularly responsive to those cost-driven brands which find further ways of driving out unnecessary costs. By contrast the second cluster ('whatever I want') will appreciate brands that are more tightly focused on their need or where possible, enable the consumer to fine tune them to their needs, as we explore in the Chapter on internet-based branding. The third cluster represents trade-off consumers, who are balancing the perceived enhanced quality gains against growing costs. For this group, if the marketer can use their knowledge of consumers to notably leverage the brand's quality at an incremental cost, there is scope for a higher likelihood of satisfaction. The final cluster goes beyond considering just quality but rather a broader array of benefits. In addition, they do not regard the only cost as being financial but rather they take a wider view that looks at any sacrifices, for example, time and concerns about whether their choice will be socially acceptable. Good consumer research can help stimulate new ideas amongst the brand's team about providing added value. Managers are faced with the challenge of defining and quantifying what customers value. Kothari and Lackner (2006) offer a three-step approach to assist this process. They express value creation as a cycle, which incorporates three stages. Fig. 10.1 illustrates the value creation cycle.

As illustrated, managers first need to identify the ways in which they currently create value for customers. They suggest that there are four elements to this value. These are the product or service itself, the extent to which products and services are accessible, the overall experience in using products and services and the extent to which the customer incurs costs in consuming the product or service. Second, managers should understand how value gets delivered through the firm's value chain. As a result of this, instead of managers focusing on internal processes such as the supply chain, they should consider the demand

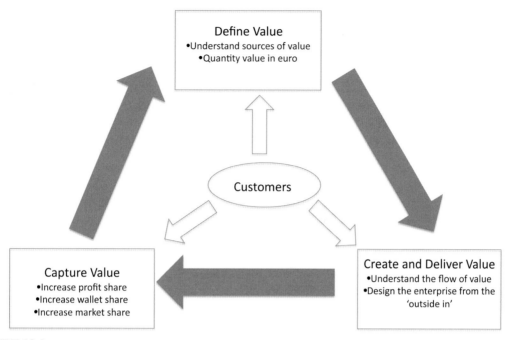

FIGURE 10.1

Value Creation Cycle (*after Kothari and Lackner 2006*)

flow from the customer back through the entire value chain. Third, companies need to capture value for their shareholders. Kothari and Lackner (2006) explain that managers must measure this value based on three dimensions, i.e. in terms of customer profitability, share of wallet and number of customers.

We explore later in this chapter formal models that enable managers to identify ways of adding value. As we have seen in the value cycle, each department in a firm contributes to building a brand's added values and managers from different departments therefore need to be aware and understand, how as a *total system* they all contribute to adding value. A classic problem occurred in one firm that attempted to position itself as adding value through excellent service. When any of its clients reported a product failure, an engineer was on-site within a few hours. Having diagnosed the fault, the engineer then contacted the warehouse parts staff to order a replacement part. But this was where the problem arose — the engineer spoke in terms of engineering parts, while the warehouse staff spoke about part numbers. Until a common communication protocol was agreed, inefficiencies arose from the warehouse staff dispatching incorrect parts.

McPhee and Wheeler (2006) offer a framework that considers added value in the context of the entire firm. Building on Porter's Value Chain (2006), they postulate an 'Added-Value Chain', where the concept of value is extended to incorporate 'brand, reputation and the relationship-based drivers of the firm, like social capital or goodwill' (McPhee and Wheeler, 2006, p. 40). Within this concept, companies such as Google create value through intangible assets such as leadership quality, innovation, brand equity and excellence in strategic alliances. In addition, firms should consider the intertwining of buyer and supplier, their product's use and the end of primary use of the product as source of value. For example, a product's use could incorporate developing networks of users or augmenting products with services. When the customer is finished using the product, the firm can add value through recycling for example, or offering lease-holders reductions for replacement products, thereby ensuring customer loyalty. To create the value chain, McPhee and Wheeler (2006) recommend three steps:

- change the definition of value to include intangible elements, such as reputation and brand value;
- change the firm's perspective outward to a view of the firm as part of the community;
- add activities involving teams of employees across several areas.

For smaller firms that may be curtailed by size and budget, O'Dwyer et al. (2009) highlight the importance of innovation. They explain that marketing in SMEs is often driven by innovation. Within the SME context, the knowledge/experience of the owner/manager and other employees is critical to value creation. In addition, the added-value chain could be applied more easily in SMEs. O'Dwyer et al. (2009) explain that less formal structures in SMEs are conducive to innovation and, in particular, where there is a match between the environment, organisational goals and individual values. Therefore, where values are clear and congruent and structures are more informal, all areas of the SME could work together to innovate, creating value for the customer.

NOTICEABLY DIFFERENT, RELEVANT AND WELCOMED ADDED VALUES

The architect who publicises the fact that his staff have access to an excellent network of builders and tradesmen is more likely to win in the construction market than the undifferentiated firm of architects, that tries to negotiate lower priced design. The winner in the low cost accommodation market is the hotel that offers late night room service for late arrivals. Competing hotels may have better furnishings, which are upgraded every six months, but ultimately it is the smile, civility and genuine personal concern that often prove to be the real discriminators that make some firms more successful than others. These added values contribute to the brand's success because they make these two firms noticeably different on attributes that are customer-relevant and welcomed.

To be successful, it is crucial for firms to have a clear view about precisely what added values their brands offer, as well as understanding the relevance of these added values to consumers. Macmillan (2006, p. 266) provides a helpful example by using the illustration of a building's design. He describes the diversity of values a building can create according to different needs, as well as the different measures of success or metrics, which would apply for each scenario:

- **Exchange Value** – the building is created to be traded, therefore its metrics include its book value or yield.
- **Use Value** – the building supports an organisation and supports a work environment that facilitates productivity and teamwork. Measures of success include satisfaction or retail footfall.
- **Image Value** – the building contributes to corporate prestige, brand image and identity. Its metrics include the PR opportunities or the 'wow' factor it creates.
- **Social Value** – the building facilitates social interaction, encouraging inclusion and contributing to civic pride and increased neighbourliness. Its measures of success include reduced crime and vandalism.
- **Environmental Value** – the building helps limit pollution and is prudent in the use of finite resources. Its metrics include its ecological footprint and its environmental impact.
- **Cultural Value** – the building contributes to the local town or city and its intangibles, such as symbolism and aesthetics, offer cultural value. Its measures of success include press coverage and positive critical opinion.

Within the services sector, managers must also ensure that they understand customers' requirements for added value. For example, a discount offered to a new customer for a broadband product will not have value for the customer if that customer does not trust that the service engineer will turn up when scheduled. Companies should seek to go beyond customer expectations to deliver exceptional value. Within the services sector, customers experience increased risk, particularly when the service is high in credence qualities and the customer does not have the knowledge to evaluate the service. A pharmacist offering a 24-hour emergency service and health screening is adding value, which customers appreciate. In turn, this added value creates positive word of mouth and loyalty. In a world where customers express concern for the environment, SAS airlines offer their customers an online CO_2 calculator. This tool allows them to calculate the CO_2 emissions associated with their travel, as well as an option to buy carbon offsets, through their portfolio, of international wind products.

Once there is a clear internal appreciation about the brand's added values, a holistic strategy then needs to be developed, integrating the added values into every part of the value chain. For example, the added value of reliability in a new brand of testing equipment starts by having good quality components and stringent testing procedures at every stage of production. This is followed through by having everyone who works on the brand committed to satisfying this goal. This means that if the testing equipment does fail in the customer's factory, there is a facility to provide a rapid temporary replacement while the instrument is being repaired. This enables the firm's brands to be differentiated from those of their competitors by positioning them in such a way that their added values are clearly appreciated.

To succeed, a holistic approach is needed when developing a brand's added values, which would need to be recognised as:

- being **differentiated** from competition in such a way that the name is instantly associated with specific added values, as in the case of Nike sports gear, which urges consumers not to be content with being second but rather to aim to win, inspired by the Greek Goddess of Victory after whom the brand has been named.
- having added values that don't just satisfy **functional** needs but also meet **emotional** needs. For example, Virgin Atlantic not only provide the added value of transatlantic flight but also combine it with a sense of fun.
- being perceived as a **low-risk** purchase; for example, Johnson & Johnson emphasise their credo and commitment to quality, which offers reassurance for their baby products and contact lenses.
- making purchasing easy through being presented as an effective **shorthand** device.
- being backed by a registered trademark, **legally guaranteeing** a specific standard of consistency.

In other words, successful brands do not stress just one part of the brand asset. They blend all of these components together. Furthermore, they ensure that a coherent approach is adopted, with each component reinforcing the others.

Brands succeed because they have clearly-defined added values that address consumer challenges at a particular point in time. For example, brands of coffee face challenges from a primarily tea-drinking population in certain countries. Therefore, coffee brands need to position themselves as a symbol of change, to capture a young audience and bring users into their brand.

While these points show the need to use as many elements of a brand's assets as possible when adding value, a consideration of some branding approaches in our sophisticated society shows an excessive reliance on only a few or, worse still, just one component of the brand. The most frequent branding error appears to be an undue emphasis on using the brand name purely as a differentiating device. However, when the brand name is abstract, such as Apple or Samsung, the companies use marketing communications to infuse brand meaning.

We can draw on the work of Interbrand to appreciate how the values added to brands work together to build brands. Value is added to brands through superior technology or superior systems to give **functional** values beyond those of competitors. The problem is that competitors can emulate these. Just consider the race Intel finds itself in, as it continually develops more powerful computer chips.

Expressive values can be added to brands, enabling consumers to make statements about themselves that more clearly express aspects of their individual personality. For example, confident about the Ray-Ban Wayfarer's functional values of style and authenticity, a considerable investment in advertising this brand has added expressive values such as being individualistic, freedom-loving, rebellious and, some groups might argue, more attractive. The Apple iPhone user may well appreciate this brand's expressive values of creativity and being personable. **Central** values can also be added to a brand, showing what the soul of the brand is. These values make it clear what the brand believes in and represent deeply-held philosophical, ethical, political and nationalistic beliefs. In the case of Body Shop, consumers are buying into the central value of environmentalism.

Gluing together and harmonising these three types of values, which have been added to the brand, is the brand's **vision**. One aspect of the brand's vision represents a view about how the brand can make the world a better place. When thinking about Apple's vision of freeing people from being constrained to a static environment, it becomes easier to appreciate how software and hardware engineers have jointly developed easy to use functional benefits, enabling consumers to display something about their own creativity.

IDENTIFYING ADDED VALUES

When faced with the need to find a removal company, the homeowner may initially perceive very little difference between removers. However, when a few firms are asked to give quotations, differences start to become apparent. There are those estimators who call at a time convenient to the homeowner, while others cannot be definite. Some will appear with a brochure describing their firms' capabilities and provide advice on how to minimise the packing effort. Others will wander round the house, making comments about the difficulty of having spiral staircases and the irritation of having to wait for the key to the new home to be released. In other words, while the basic service is akin to an undifferentiated commodity, the way that it is offered is recognised by consumers as having added values superior to competitors.

By recognising that buyers in consumer, service and industrial markets regard products and services as clusters of value satisfactions, marketers can start to differentiate their brands by developing relevant added values. For example, removal firms can differentiate themselves by adding values such as responsiveness to unusual handling requests, politeness, the confidence they give homeowners through the care they take when packing, reliability and the guarantees they offer. All these features present opportunities for differentiation.

As a further example, many would argue that the salt market comes close to being a commodity market, yet Saxa refused to accept this and marketed its brands of salt in order to satisfy consumers' needs for either health or flavour. For example, for health-conscious consumers they launched 'So-Low' Salt (sodium reduced) and Sea Salt Flakes for 'fine diners'.

The model in Fig. 10.2 is a helpful conceptualisation of the way in which further value can be incorporated into brands to satisfy consumers' needs.

The generic level

At its most basic, there is the **generic** product or service functionality that enables firms to be in the market — for example, the cars produced by Ford, General Motors and Toyota; the home loans advanced by Barclays Bank or

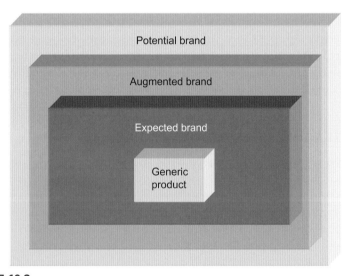

FIGURE 10.2

The four levels of a brand (*after Levitt 1980*)

HSBC; the laptops from Sony VAIO and Dell. At this level, it is relatively easy for competitors to develop 'me-too' versions — just think how many manufacturers produce a small, hatchback, town car. In developed countries, the generic product or service functionality is rarely the basis for sustaining brands, since functional values can be emulated.

The expected level

Buyers and users have a perception about the minimum characteristics that differentiate competing brands in the same product field and value can be added by going just beyond this. At the **expected** level, products and services are made to satisfy purchasers' minimum requirements for attributes such as name, packaging, design, availability, price, quantity and so on. One of the most helpful ways of identifying what these characteristics should be is through depth interviews with buyers and users. The expected level of brand competition is typically seen when buyers do not have much experience of competing brands. In such cases, they look at product attributes to assess how well different brands will satisfy their motivational needs. For example, some of the motivating reasons for having a hot drink are relaxation (Ovaltine), an indulgent yet health conscious treat (Cadbury's Highlights), stimulation (Nescafé) and warmth (Bovril). Buyers consult the brand names, the packaging details, price and promotional details to form an overall assessment of the extent to which competing brands may satisfy different motivational needs. At an early stage in any market's development, it is unlikely that several brands will be perceived as being equally able to satisfy the same motivational need. The added values here again tend to be functional characteristics, reinforcing the positioning of what the brand *does*.

The augmented level

With more experience, buyers become more confident and experiment with other brands, seeking the best value and they begin to pay more attention to price. To maintain customer loyalty and price premiums, marketers **augment** their brands through the addition of further benefits, such as, for example, the inclusion of a self-diagnostic fault chip in washing machines. The chip was something that a less experienced user might perceive as having minimal value but as they become more dependent on the brand, they learn to appreciate the way it reduces delays, by the service engineer arriving with the right spare part. To assess what types of added values would enhance their brands at this stage, marketers should arrange for depth interviews to be undertaken amongst experienced users. They should be asked to talk about the problems they have with different brands and the sorts of ways in which they could be improved.

At the augmented stage, several brands may well come to be perceived as satisfying the same motivational needs. For example, Sprite and 7-Up both

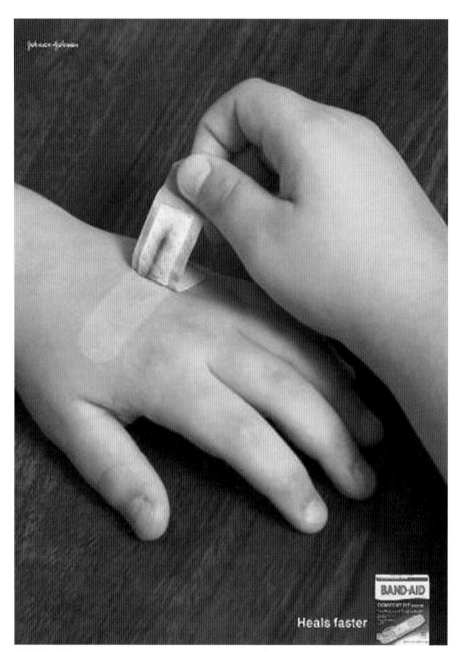

BAND-AID
Band-Aid emphasise the brand's functional value in the simple copy: 'Heals Faster'

offer lemon/lime-flavoured carbonated refreshment. So consumers focus on the **discriminating factors**. These may be functional features such as size, shape, colour and availability or emotional ones reflecting different brand personalities. Several different ways of positioning brands using functional discriminators are:

- with respect to use – 'once a day' dosing for a pharmaceutical brand;
- with respect to the end user – Nurofen for children, Nurofen Plus for adults;
- with respect to the competitor – the number two brand, Avis, trying to take share from Hertz, the brand leader through its campaign 'we try harder' which extends into a 'we try harder' brand blog;
- with respect to a specific attribute – the flavours of Ben & Jerry's Ice Cream or the long life of Duracell Batteries.

Possibly an even more powerful discriminator is endowing the brand with a personality. Both Ford and Renault produce similar hatchbacks, which satisfy the motivational need for cost-effective town transport; but they also have unique brand personalities that differentiate them.

For example, some might perceive the personality of the Honda Civic as a young male, who has many friends and likes to drive fast. By contrast, the Nissan Micra may be a kindly, elderly lady who uses her car as a runaround, has very low mileage and is a cautious driver. In effect, at the *augmented* stage, buyers have narrowed down the list of suitable brands by considering those that match their *motivational* needs; then they differentiate between these brands on the basis of *discriminators* relevant to their particular lifestyles.

The potential level

Eventually, however, buyers and users come to regard such augmentation as a standard requirement for brands. To stop the augmented brand slipping back to the expected level, where buyers would be more interested in prices, the brand marketer needs to become more innovative and develop new added values to push the brand into the **potential** phase.

This is a more challenging task, inhibited predominantly by the creativity of the brand's team and their financial resources. One way of identifying new added values for these highly experienced buyers is to map out the channels through which the brand passes, from manufacturer to end user. At each stage in this chain of events, the brand marketer needs to appreciate exactly how the brand is used and who is using it. A sample of those individuals coming in contact with the brand should be interviewed to assess their likes, dislikes and views about improvements. The opinions of all levels of staff are important in this work.

Kim and Mauborgne (2005) argue that managers should focus their activity at the brand's potential level and they propose the concept of 'Blue Ocean

Strategy'. They explain the concept of value innovation, which means that firms should not focus on beating the competition. Instead, firms should make competition irrelevant by opening new market space through a leap in value. 'Red' oceans, therefore, are industries already in existence in the market where there are battles between competitors. 'Blue' oceans, by contrast, are those other uncontested market spaces that are ripe for growth. Within these new oceans, firms create new demand. For example, Kim and Mauborgne (2005) explain that customers thinking about the circus believe the following to be true:

- A circus has clowns;
- A circus is in a tent;
- A circus has acrobats.

Cirque du Soleil understood these factors and then thought differently about them to identify an uncontested market space. The company shifted clown humour to incorporate more sophistication; they glamourised the traditional circus tent or moved to rented venues; and they built artistic flair into their acrobatics. They also added factors such as themes and storylines, to create more of a theatrical performance. Rather than playing in a traditional three-ring circus with sawdust and benches, Cirque du Soleil plays in venues such as Madison Square Gardens in New York, the Bellagio in Las Vegas and the Albert Hall in London. Furthermore, adding new shows has given consumers a reason to visit the circus more often, which has increased demand for Cirque du Soleil tickets.

Another example of a brand that was managed from the **augmented** to the **potential** level is American Airlines (AA). With deregulation in the American air travel market, the barriers to entry for new airlines were lowered. This resulted in new airlines increasing travelers' choices, along with more price competitive routes. As a brand, AA could have slipped to the expected level, competing against the others on price. Instead, it evaluated how people used its services and identified every point where it came into contact with customers. Based on this, it identified a few areas where it believed customers and consumers would welcome new added values. It undertook a review of its in-flight service and improved its quality. It recognised the need for being on time, both when departing and arriving, and, by assessing all of the events that influence flight operations, developed systems which resulted in it becoming respected for punctuality. The problem of over-booked seats was lessened through better information technology. Through the use of its frequent flyer card, AA was able to understand the needs of its consumers more clearly and develop more tailored services for them. Finally, it communicated all of these added values to its customers and consumers.

Eventually, competitors will follow with similar ideas and buyers will gain more confidence, switching between brands that they perceive as being similar.

Yet again, the brands may slip back to the expected level, unless the brand marketers recognise that they must continually track buyers' views and be prepared to keep on improving their brands.

The problem with continually enhancing the brand is that a point may be reached at which the extra costs may not be recovered through increased sales and competitors may soon find ways of copying the changes. In such cases, an audit needs to be undertaken to evaluate whether the brand has a viable future. The audit needs to look at consumers' views, competitors' activities and the firm's long-term goals. If it is not thought viable to enhance the brand further, a different strategy for its future needs to be identified. This could entail selling the brand off, freezing further investment and reaping profits until a critical sales level has been reached, withdrawing the brand or becoming an own label supplier if there is sufficient trade interest.

The problem of competitors copying a brand's added value can be better analysed if we consider the 'coding' or the building blocks that constitute the brand and its added values. For example, a restaurant owner may find that their restaurant is so successful that they wish to open a second one and let one of the managers run it. To ensure that the new manager is effective, the owner needs to reveal all the codes that constitute the successful formula. On the other hand, they may wish to hold something back, for fear of the manager leaving with the formula or of competitors copying it. Researchers have argued that successful brands are difficult to copy even if the nature of all the component parts are well understood, as long as secrecy is maintained about the manner in which these components are integrated. As a result of this, the owner could explain to the manager all the systems that support the restaurant, such as types of menu, the pricing policy, staff recruitment and so on. They need not insist on always being consulted when decisions are necessary on operational issues. But the development of the restaurant's image or brand personality should be the sole responsibility of the owner. The brand personality is the unifying device that integrates the component parts and it is this that competitors find difficult to copy.

The previous four-level model can be applied to both product and service markets. Grönroos (2000) has developed a more refined version specifically for service markets, as shown in Fig. 10.3.

Grönroos' model shows the different levels on which the service can be offered. Managing the service offering requires the development of:

- a **service concept** — the company's intention and mission for being in the business;
- a **basic service package** to fulfill customer needs — the core, facilitating and supporting services;

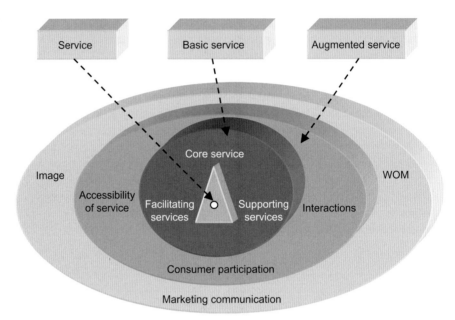

FIGURE 10.3
The service offering (*after Grönroos 2000*)

■ an **augmented service offering** — the processes and interactions between the company and its customers for the production and delivery of the service;

■ a system to manage **image** and **communication** so as to enhance the perception of the augmented service offering.

In the **basic service package,** the core service clarifies the main benefit being offered. For example, a hotel's core service could be lodging as opposed to catering for conferences and business meetings. Facilitating services make it possible for customers to use the core service. For instance, customers need an efficient and friendly clerk at the hotel reception or at the airport check-in desk. Supporting services can be used to increase the value of the core service. Examples are the provision of shampoos and hair-dryers in hotel rooms and meals served at the airport lounge instead of on the aircraft, which allows business travelers to rest during intercontinental flights.

In the **augmented service offering,** the accessibility of service depends on the number of employees and their skills, the office hours, the locations and the ease of use for consumers — all of which provide ways to add further value. Designing new, valuable systems to increase the accessibility of service is not sufficient if employees are not trained to implement or to use them. It is useless for intercontinental airline passengers to wear eye-masks with the label 'do not disturb' if the cabin staff still wake them up to serve the meal. Interactions with the service firm take several forms, all of which offer value adding opportunities:

- interactive communication between employees and customers – how they behave towards each other;
- consumers' interactions with physical resources – for example vending machines and self-service check-ins;
- consumers' interactions with systems, such as waiting, delivery and billing systems;
- consumers' interactions with other customers involved in the process.

The correct type of customer participation is fundamental when striving to enhance the service brand. For example, at a hairdresser's, the client's ability to describe their ideal haircut correctly influences the amount of cutting the stylist will do!

Using vibrant advertising creates an identity for the service brand and the resulting **image** perceived by consumers will either enhance or diminish their perception about the likely services. As they experience the brand, they will talk to others and will start to affect their views about the service. Effective brand management should capitalise on these opportunities.

ADDING VALUE THROUGH CONSUMER PARTICIPATION

In the traditional view about value creation, a firm adds value to the inputs received from its suppliers and passes them on to its customers. This approach is based on the notion that a firm 'does things' to its customers, (either to another firm or to the final consumers) who are passive receptors. Within this context, the firm's strategy is based on finding the right positioning within the value chain.

Value is also created when consumers are regarded as being actively involved in creating value and added values can be jointly tailored more closely to their needs. A forerunner to the concept of the Service Dominant Logic discussed in Chapter 6 (Vargo and Lusch, 2004), Normann and Ramirez (1993) are advocates of this perspective by which firms do not just add value, but rather reinvent it. This strategy focuses on a value-creating system in which economic actors (suppliers, firms, consumers) work together to co-produce value. Managers' key strategic task is to reconfigure the roles and relationships among all the actors in the value-creating system.

A good example of value creation according to this view is provided by IKEA, the home furnishing retailer. The key elements of IKEA's formula are: simple, high-quality Scandinavian designs; global sourcing of components; customer kits that are easy to assemble and huge suburban stores. The result is products offered at prices 25% lower than the competition. However, IKEA's real innovation is the redefinition of the actors' roles and of organisational practices. This company has invented value by matching the capabilities of participants more efficiently and effectively.

IKEA involves customers in the selection, transportation and assembly of its products and designs its business system to support them. For example, clear catalogues explaining each actor's role, stores developed as family-outing destinations, measuring aids at the front door (catalogues, pens, paper, tape measures), product groupings to offer designs for living rather than the traditional product groupings, detailed labels on each item and car roof-racks for hire or sale.

IKEA helps customers to create and then consume value. It does not aim to relieve customers of their tasks but rather to mobilise them to do tasks never done before. With the same purpose of creating value, IKEA mobilises numerous suppliers around the world, chosen for their ability to offer good quality at low cost. A chair, for example, is assembled with a Polish seat, French legs and Spanish screws! All parts are ordered electronically with EPOS systems and the warehouse functions as a logistics control point, consolidation centre and transit hub.

In this new logic of value creation, roles and relationships are redefined and the distinction between products and services disappears. A visit to an IKEA store is not just shopping but also entertainment. The new value is more 'dense' insofar as more information and knowledge is provided with each product to allow the appropriate actors to create their own value. There are three strategic implications from this. First, the firm must be able to mobilise customers to take advantage of the 'denser' value and help them create value for themselves. Secondly, the firm can rarely provide everything by itself and should therefore aim to develop relationships with all actors. Thirdly, it can achieve competitive advantage from reconceiving the value chain as a value-creating system.

Some brands already build on the opportunity of adding value through active participation. Build-a-Bear workshops encourage customers to create their own customised teddy bears or stuffed animals. Consumers can also actively add value in service brands. They are not necessarily passive receptors of the service but can actively perform what is required to satisfy their needs – preparing their own salad in a restaurant as opposed to being served by a waiter, or booking their cinema tickets online in advance and using their credit card to print them from a self-service vendor. In this case the producer of the service is the consumer, termed the **prosumer**. In the traditional view, companies focus on *relieving* the customers of certain functions and activities, such as providing a house cleaning service. The new view shows companies targeting prosumers and focusing on *enabling* them to perform particular activities. In the case of house cleaning, the enablers would be vacuum cleaners. The following equation (Michel 1996) shows how firms can add value to customers by either concentrating on supplying them with a 'reliever' or encouraging them to buy an 'enabler' and perform the activity themselves at a time and in a way that best suits them.

$$\text{Reliever} = \text{Enabler} + \text{Prosumer}$$

There are several reasons for prosuming. It is cost effective and can save consumers time; for example, using an online banking site to transfer money, rather than queuing in a bank. More importantly, it allows consumers to control the timing, duration and pace of their own activities and by being more involved, they are more likely to show higher satisfaction levels.

ADDING VALUE THROUGH CUSTOMISATION

Companies can adapt each of their products and services to suit each individual customer's unique needs but while this adds value, it may be uneconomic. Instead, a smaller variety of customised brands can be offered to different groups of customers. Using traditional segmentation variables may still result in customers in the same segment showing differences, since each one of them is in a different market, at different times, in different places. Just think of passengers flying long-haul: are they travelling for business or leisure? If they are flying on business, will they have the same refreshment needs going out to their meeting (e.g. Coca-Cola, Schweppes Tonic) as when they return (for example, Bacardi and Coca-Cola, Bombay Sapphire Gin and Schweppes Tonic)? The challenge facing managers is getting the right balance between adding value through a small or large degree of customisation and the payback they can expect from consumers trading off extra benefits against higher costs.

To help managers appreciate alternative ways of customising their brand, Gilmore and Pine (1997) devised a useful two-dimensional matrix (shown in Fig. 10.4), whereby standardised or adapted forms of the product can either be presented or portrayed in a standardised or adapted format, according to four customer types. We are all aware of examples of ways of changing product forms and additionally this matrix brings out the need to consider changing the product's representation through aspects such as its packaging, name, terms and conditions, promotional material, distribution and merchandising. From this matrix, they identify four approaches to customisation – collaborative, adaptive, cosmetic and transparent.

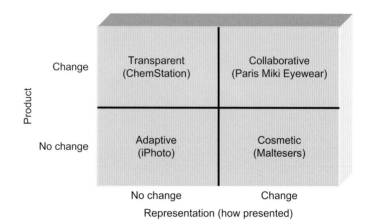

FIGURE 10.4
Approaches to customisation (*after Gilmore and Pine 1997*)

Collaborative customisers conduct a dialogue with individual customers at the design stage to help them describe their needs and identify the best solution. The Japanese eye-wear retailer Paris Miki spends time understanding how consumers would like to look, then produces a digital picture of the consumer's face and within minutes, the consumer can see how their face appears with lenses of different shapes and sizes. Consumer and optician jointly adjust the shape and size of lenses and the type of bridges and hinges, until both are pleased with the final result. A technician then grinds the lenses and builds the glasses within an hour. There is a high degree of collaboration to tailor the product and its presentation.

Transparent customisers provide products and services unique to each customer without letting them know explicitly that they have been customised. This approach is the most appropriate when customers' needs are predictable or can be easily forecast and saves repeatedly asking customers about their reorder quantities. ChemStation supplies industrial soaps and after analysing the usage patterns of each of its customers, provides them with the most appropriate formulation in the quantity and frequency of delivery, best suited to their needs. Customers do not have to reorder or specify the desired characteristics. This type of customisation depends on having a good system for collecting data on individual customers and a flexible design and delivery system.

Adaptive customisers offer one standard product designed so that users can customise it themselves. The iPhoto software package provided on Apple's iLife allows users to organise, edit and share their photographs, according to their needs. For example, with the same photo album, users can choose to create a slide show, design custom photo albums or upload photographs to Facebook to share with friends. This type of customisation transfers to consumers the ability to design and use the product or service in a way that best suits them.

Cosmetic customisers present standard products differently to different customers by changing only the package and not the product itself. Maltesers provides different retailers with different pack sizes according to their customers' likely preferences. For example a larger share pack retails in cinemas, whereas fun-size multipacks are distributed through grocery multiples, where mothers purchase weekly treats for their children. This type of customisation is appropriate when the standard product satisfies almost all customers and only the presentation needs changing.

The previous four examples show how companies gained competitive advantage by selecting the most suitable stage during which customers value an individual approach: Paris Miki customises during the design stage, Maltesers during packaging, ChemStation during both production and delivery and iPhoto lets customers do the customising.

ADDING VALUE THROUGH BUILDING RELATIONSHIPS

As the previous sections indicate, adding value can be achieved through a variety of ways. In today's world where firms strive to build stronger relationships with customers, adding more value to the generic offering should increase customer satisfaction and more strongly bond the customer link. By delving into the pricing literature, another view on the concept of value emerges as the ratio of perceived benefits relative to perceived costs (financial as well as non-financial). This echoes one of Zeithaml's (1988) findings about the way some consumers conceive value, as discussed earlier. From this ratio it is apparent that brands can be perceived to have even more value when either more has been added (as has been addressed so far in this chapter) or when customers experience less cost.

In addition, a consumer's perceptions of value are time and place specific. For example, a teenager perceives value from their Swatch because of its accuracy but when with a group of friends, perceives even more value because of its noticeably attractive design. Furthermore, as the teenager wears the watch over a number of years, they begin to form a closer relationship with Swatch. Thus, as Ravald and Grönroos (1996), argue, we need to have an expanded concept of value, i.e.:

$$\text{Total episode value} = \frac{\text{Episode benefits} + \text{relationship benefits.}}{\text{Episode costs} + \text{relationship costs}}$$

If the teenager values the relationship with their iPod and is committed to the brand, they are more likely to tolerate the occasional episode problem, such as a battery losing power. Positive episodes build consumers' perceptions of the brand's credibility, increase the total value of the relationship and enhance the total episode value.

When consumers judge a single episode, they consider factors such as superior product quality and supporting services. On the other hand, when they reflect on the value of the entire relationship, they normally consider whether the supplier has offered them safety, credibility and continuity. Fig. 10.5 shows

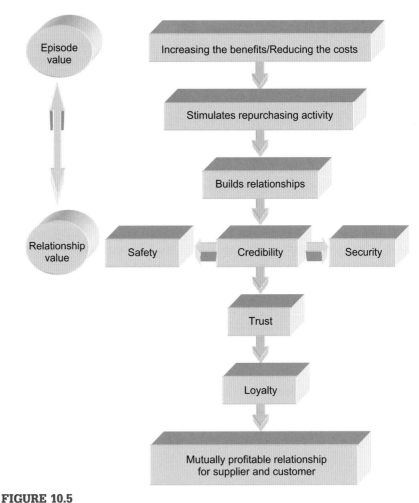

FIGURE 10.5
The effect of value-adding strategies in a long-term relationship (*after Ravald and Grönroos 1996*)

how the perceived value can be increased both on an episode level and on a relationship level.

Often, added value is considered in terms of the extra benefits that augment the offering but less thought is given to devising ways of reducing consumers' costs. When a customer selects a new supplier offering a better product at a lower price than competitors, the supplier needs to work on each episode in order to build the relationship by addressing ways of reducing the customer's costs. For example, providing online order tracking and on-time delivery accurately reduces the user's costs, which, along with other benefits, contributes to building a cost-effective relationship.

PROTECTING BRANDS AGAINST COUNTERFEIT BRANDS

By blending all of the assets constituting brands, marketers are able to develop brands which build goodwill between the brand producer and the consumer. As one advertising executive commented, 'Powerful brands are just like families. They persist through thick and thin'. Brands are intangible assets that contribute to the equity of a firm. The goodwill that Kellogg's has built up over the years is such a valuable asset that if all its production facilities were destroyed, it could get adequate funds to rebuild them using the goodwill from the brand name as security. Likewise, even though Mars Confectionery does not own property, hires its distribution vehicles and leases its machinery, it would cost any potential acquirer hundreds of millions of Euro to buy this firm, since what is being bought are not the tangible assets but the goodwill and reputation from the Mars name.

Unfortunately, the success of some brands has driven certain competitors to respond by developing counterfeits. The global market for counterfeit goods is estimated to exceed $600 billion or 7% of the world's trade and is linked to threats such as narcotics, terrorism and human trafficking (Wilcox et al., 2009). Staake et al. (2009, pp. 321-322) define counterfeiting as 'the unauthorised manufacturing of articles which mimic certain characteristics of genuine goods and which may pass themselves off as registered products of licit companies', and counterfeit trade as 'trade in goods that, be it due to their design, trademark, logo or company name, bear without authorisation a reference to a brand, a manufacturer, or any organisation that warrants for the quality or standard conformity of the goods in such a way that the counterfeit merchandise could, potentially be confused with goods that rightfully uses this reference'. Their definition of counterfeit trade includes, for example, the unauthorised use of brand names or characteristic shapes but excludes, for example, private users copying digital content or bootlegging. They explain that counterfeiting is becoming more common due to the growth of Asian markets, a general dismantling of border controls and the integration and interaction of organisations in disparate locations. On one hand, they explain, counterfeiting can benefit a brand as it may increase brand awareness and create demand. However, the negative consequences outweigh the benefits. Counterfeiting results in a loss of revenue for the original brands. Where the counterfeit products are difficult to distinguish from the genuine brand or are available at a cheap price, substandard products can diminish a brand's perceived quality and reduce its perceived exclusiveness.

Klara (2009) warns that the concept of counterfeiting has gone from 'nuisance' to 'epidemic'. In 2008, he explains, the US government seized over $270 million of counterfeit merchandise, which was a 38% increase on 2007. He presents a number of reasons for the rise in counterfeiting. First, weaker economies encourage consumers to make their money go further and buying counterfeit goods can appear to be a prudent means to attaining luxury. Second, the 'wild west' quality of the internet means that brand owners have little control over fake products, which are often sold from obscure locations. Third, genuine brands sometimes choose to produce mass market versions of their brands and these products can be easier to copy. In addition, manufacturing has shifted to less developed countries. The last two factors help to furnish factory owners with the manufacturing know-how to create a counterfeit product.

Brands must be vigilant to prevent counterfeiting at best or minimise its impact at worst. Many leading brands are victims to copying. For example, in 2006, iLounge reported that Apple warned its service partners about illegally manufactured digital music players which had 'strikingly similar' features to the company's successful iPod product, including product packaging, logos and serial numbers. However, there were telltale signs that the products were not genuine. Product details such as the dock connector and the Apple operating system were missing. In South Africa in 2008, Levi Strauss set up its own Brand Protection Department to 'stem the tide of losses to the business' caused by counterfeit products. The department collaborates with product design to include features that make Levis more difficult to copy. These features include washing instruction labels and packaging. Klara (2009) notes that Louis Vuitton developed their world-famous 'LV' monogram partly to discourage copying. Yet the brand conducted more than 7600 anti-counterfeiting raids globally in 2007. On the brand's website Louis Vuitton claimed 'In 2004, the brand's firm stance led to over 13,000 legal actions, more than 6000 raids, over 947 arrests and the seizure of fake printing cylinders'. The company advised customers that their brands are only available through exclusive Louis Vuitton stores and on their website.

There have also been collective responses to counterfeiting. For example in 2008, New York City ran an advertising campaign to educate consumers about the costs of buying counterfeit brands. Their slogan was *'when you buy counterfeit goods, you support child labour, drug trafficking, organised crime and even worse'*. Another advert stated *'The sale of counterfeit goods costs New Yorkers $1 billion in lost tax dollars each year — less money to improve schools, staff hospitals and make our streets safer'*.

A critical problem caused by counterfeit trade is its impact on consumer relationships with genuine brands. Commuri (2009) explains that counterfeit brands exist most where signalling is important to consumers. Signalling may include exclusivity, prestige and privilege. Consumers who are able to afford genuine brands at premium prices are seen as models of aspiration. Commuri (2009) conducted in-depth interviews with consumers of genuine-item brands across Thailand and India. Findings suggested that consumers adopted one of three strategies when counterfeits were prevalent — 'flight', 'reclamation' or 'abranding'. 'Flight' or abandoning a brand happened among younger consumers who were well educated and did not tend to have a brand relationship longer than two years. When their chosen brand was counterfeited, these consumers abandoned the brand for new ones. 'Reclamation' was practiced by older consumers, unsettled by the existence of counterfeit goods, yet reluctant to terminate their brand consumption. Instead, they showcased their brand relationship and emphasised their correct choice in purchasing the genuine item. 'Abranding' involved consumers disguising brand cues such that others could not understand its identity or meaning. These consumers hid brand affiliations and expressed distain for imitators. Instead, they viewed a tendency to flaunt brand names or imagery as an expressive lack of confidence. Commuri (2009) notes that customers of genuine items did not express sympathy for a brand when it was counterfeited. Brand managers must consider their customers' likely response to counterfeit products in order to better respond to the challenge imitators. For example, those consumers who would be likely to flee a brand when it is copied, might accept an alternative, competitive, genuine brand at a higher price.

Brand managers must also consider *why* consumers might purchase counterfeit brands. Wilcox et al (2009) examined the motivations guiding consumers' propensity to buy counterfeit brands. They distinguish between deceptive and non-deceptive counterfeiting. In the latter case, consumers are aware that they are purchasing counterfeit goods and such counterfeiting typically occurs in the luxury goods market. Across a series of studies, they found that consumers' favourite luxury brands were Louis Vuitton, Gucci and Rolex. Consumers' desire for counterfeit goods was related to the extent to which the brand fulfilled social goals. In addition, the meaning created by advertising for a genuine luxury brand made visible through product design

UNICEF

Unicef warn about the dangers of buying counterfeit brands, using the imagery of an adapted version of the well-known Lacoste logo influenced consumers' desire for counterfeits. Where logos were more prominently displayed, consumers could use counterfeits to display the brand's aspirational characteristics to others. Wilcox et al. (2009) highlighted that managers would need to consider the implications of making their brands less conspicuous. In addition, their research illustrated that consumers were more likely to desire a counterfeit brand where advertising connoted status and supported social-adjustive goals. A challenge for marketers therefore is to balance the extent to which their brand's conspicuous attributes and status connotations encourage counterfeit purchases, yet at the same time, retain the motivators of luxury brand consumption, which create demand for their genuine item.

Whilst counterfeiting is likely to pose a continuing threat to manufacturers, it is only one of the many challenges that face marketers, who must continually devise added value strategies.

In the UK, if consumers are misled by representations, for example oral statements or implied statements such as similarly designed packaging, into thinking that one product is made by or approved by another, then action may be taken under the law of 'passing off'. An example of this is look-alike packaging, where products are so similar in appearance to a competitor's brand that consumers may mistake them for the genuine article. Research indicates that approximately a fifth of British shoppers have erroneously selected similarly looking brands (Mitchell and Kearney, 2002). There have been several highly publicised cases involving look-alikes. For example, in 2007, Karen Millen Ltd successfully took the Irish retailer Dunnes Stores to court because a jumper and two skirts sold by Dunnes were almost identical to the Karen Millen designs. While it can take time and resources for the brand owner to fight their case, they need to instigate action fast to halt the erosion of their brand equity.

CONCLUSIONS

Brands succeed because they have real added values that are relevant to the target market and that the market welcomes. The longer it takes competitors to develop an equivalent, if not better, added value, the longer the brand has to capitalise on the goodwill it builds with customers and consumers. Added values are not about superficial, one-off issues such as a smile or the greeting, 'Have a nice day'. Instead, they are about integrating relevant ideas into every experience through every point of contact that the brand has with consumers. So, for the car hire company differentiating itself through customer concern, it means such things as checking each car before the hirer is given the keys, recruiting staff with a genuine interest in consumers, training staff, manning a 24-hour breakdown recovery service and making car reservations as simple as possible. All aspects of the brand's assets are employed to communicate rapidly the point of difference from competition, so that the consumer instantly associates the brand with functional reliability and 'no hassle' administration, reinforcing the car hire service as a low-risk event that only the firm with this specific logo can provide.

One of the problems marketers face is finding added values that are relevant to the brand's stage of development. Even if the marketer erroneously thinks they have a commodity, it can still be developed as a brand through the way it is offered to consumers. The sorts of added values appropriate for both young and mature brands can be identified by considering a brand as growing from being solely a generic product to being a product surrounded by an expected, an augmented and a potential layer. The generic product is the commodity form which enables firms to compete. The same considerations apply to services, that can be developed from a basic concept into an augmented service offering.

The expected brand represents the minimum criteria needed for purchasers to perceive sufficient added value to warrant a price premium in excess of the commodity costs of producing the brand. Here, the added values tend to be functional, reinforcing what the brand can do. However, functional values can be easy for competitors to imitate. In addition, as consumers gain more experience, they expect more from competing brands and these are challenges of the augmented level.

A strong brand personality can be an effective added value at this stage. Over time though, competitive activity and consumers' variety seeking behaviour necessitates new added values to push the brand to the potential level. Once at the potential level, a time will be reached when consumers become disenchanted with the brand. At this stage, an audit is needed to evaluate whether the brand has come to the end of its useful life or whether there is scope for the future through introducing a new added value. Value can also be added to a brand in the following ways: by stimulating customers to participate in the value-creating activity, by offering the most appropriate degree of customisation and by building long-term relationships with customers.

If the brand augmentation is successful, competitors may try to develop similar versions; so marketers need to protect their brands against blatant imitators. Counterfeiting negatively impacts on brand equity and market share and can be detrimental to consumer's relationships with the genuine brand. An insight into the reasons for the rise of counterfeiting, the steps taken by leading brands to combat this activity and consumers' responses to counterfeit brands are explored. Marketers need to be continually vigilant to new challenges, some of which have been highlighted in this chapter.

MARKETING ACTION CHECKLIST

To help clarify the direction of future marketing activity, it is recommended that the following exercises are undertaken:

1. List the added values you believe your brands have. Evaluate the strength of these added values by undertaking interviews with your customers and consumers to assess how relevant and unique they are and how much people appreciate these values, compared with those of your competitors.

2. Scan your company brochures and check whether you use very broad terms to describe your brands' added values — for example, 'quality', 'dependable', 'caring'. If you have some of these all-encompassing added value terms, get your team together and clarify amongst yourselves what you mean. Once you have reached a consensus, evaluate the strength of these added values and incorporate the most appropriate new added values in your brochures.

3. Map out the main groups of people who are involved with each brand as it evolves from raw material entering the factory, right through to the point of consumption. Check the extent to which each group knows about each brand's added values. Have any of these groups ever been asked for their comments about how their task could be changed to better contribute to the brand's added values? Do the tasks of all of these groups contribute to the brand's added values?

4 For each of your brands, assess the extent to which your marketing programme incorporates these added values in order to:
   - differentiate you from competition;
   - satisfy customers' functional and emotional needs;
   - reduce customers' perceptions of risk;
   - aid rapid selection;
   - be backed by a registered trade mark.

   If your assessment shows an excessive reliance on only one of these points, consider the relevance of developing a more balanced programme.

5. If there is a view in your firm that you are competing in a commodity market, evaluate the differing needs of the channels that your brand passes through and consider ways of tailoring your offerings to better satisfy the needs of each group.

6. For each of your brands, identify from the model shown in Fig. 10.2, the level on which each brand is competing. Have you a clear view about the motivational needs that your brands satisfy and the discriminators people use to differentiate between you and your competitors? How have each brand's added values developed? What plans do you have

to enhance your brands when customers become more demanding and competition becomes more intense?

7. Do customers purchase your product or service to be relieved of undertaking particular tasks or because they want to use the product or service themselves, enjoying the task? How could the brand be modified to reinforce its relieving or enabling function more strongly?

8. For each of your brands, compare the extent to which they are customised against your customers' requirements. With your colleagues, assess whether all of the different variations are really necessary and whether fewer variations would satisfy customers' needs as efficiently. Alternatively, is there a need for a greater degree of customisation that consumers would welcome? What are the cost implications of any changes and how do these compare against the different prices you could achieve?

9. Consider the list you have already prepared, from the first question, of value adding factors associated with your brand. Assess whether these solely focus on 'adding' benefits to customers or whether they look at reducing customers' sacrifices. With your team, analyse each of your major customers' value chains and identify opportunities to provide benefits and reduce customers' sacrifices that affect the long-term relationship with your customers.

10. What systems do you have to identify when another firm is illegally copying your brand? How might you use marketing communication to educate customers about your brands' features, to ensure that they do not buy an imitation product?

STUDENT BASED ENQUIRY

1. Select i) a clothing brand or ii) a food brand that you are familiar with. Identify its added values as you perceive them. Distinguish between its functional and emotional added values and discuss whether these values have meaning and relevance for you as a consumer.

2. Examine how the four levels of a brand presented in Fig. 10.2 might be applicable to ice cream brands. How would you describe the 'generic' level for this sector? Select a brand in this sector that you believe has been 'augmented' and identify three new added values that its brand managers could apply to push the brand into its 'potential' space.

3. Discuss what 'value' means for you as a consumer. How can firms communicate this value? Select a brand that you use and admire. Explore how it communicates value in a meaningful way to you. To what extent does the brand use participation, customisation and/or relationships to enhance the value added for you?

4. Identify a luxury brand that you are familiar with and that may be subject to counterfeiting. Explore how counterfeiting could affect the brand's image and brand value. Examine the company's website. Have they taken any steps to inform consumers about counterfeiting or to mitigate against counterfeit products in other ways?

References

Commuri, S. (May 2009). The impact of counterfeiting on genuine-item consumers' brand relationships. *Journal of Marketing, 73*, 86–98.

de Chernatony, L., Harris, F., & Dall'Olmo Riley, F. (2000). Added value: its nature, roles and sustainability. *European Journal of Marketing, 34*(1/2), 39–56.

Gilmore, J. H., & Pine, B. J. (Jan–Feb 1997). The four faces of mass customization. *Harvard Business Review,* 91–101.

Jones, J. (1986). *What's in a Name? Advertising and the Concept of Brands.* Lexington: Lexington Books.

Kim, W. C., & Mauborgne, R. (2005). *Blue Ocean Strategy: How to Contest Uncontested Market Space and Make the Competition Irrelevant.* Boston: Harvard Business Press.

Klara, R. (2009). The fight against fakes. Available at: http://www.brandweek.com/bw/content_display/news-and-features/direct/e3i344418db676344f061e2b8a71119963e (Accessed 4 August 2010).

Kothari, A., & Lackner, J. (2006). A value based approach to management. *Journal of Business & Industrial Marketing, 21*(4), 243–249.

Levitt, T. (Jan–Feb 1980). Marketing success through differentiation of anything. *Harvard Business Review,* 83–91.

Macmillan, S. (2006). Added value of good design. *Building Research and Information, 34*(3), 257–271.

McPhee, W., & Wheeler, D. (2006). Making the case for the added-value chain. *Strategy and Leadership, 34*(4), 39–46.

Michel, S. (1996). *Prosuming Behavior and its Strategic Implications for Marketing. The 9th UK Services Marketing Workshop.* Stirling: Stirling University.

Mitchell, V. W., & Kearney, I. (2002). A critique of legal measures of brand confusion. *Journal of Product & Brand Management, 11*(6), 357–379.

Normann, R., & Ramirez, R. (July–Aug 1993). From value chain to value constellation: designing interactive strategy. *Harvard Business Review,* 65–77.

O'Dwyer, M., Gilmore, A., & Carson, D. (2009). Innovative Marketing in SMEs. *European Journal of Marketing, 43*(1/2), 46–61.

Ravald, A., & Grönroos, C. (1996). The value concept and relationship marketing. *European Journal of Marketing, 30*(2), 19–30.

Staake, T., Thiesse, F., & Fleisch, E. (2009). The emergence of counterfeit trade: a literature review. *European Journal of Marketing, 43*(3/4), 320–349.

Vargo, S. L., & Lusch, R. F. (2004). Evolving to a new dominant logic for marketing. *Journal of Marketing, 68*, 1–17.

Wilcox, K., Min Kim, H., & Sen, S. (2009). Why do consumers buy counterfeit luxury brands? *Journal of Marketing Research, 46*(2), 247–259.

Zeithaml, V. (July 1988). Consumer perceptions of price, quality and value: a means-end model of synthesis of evidence. *Journal of Marketing, 52*, 2–22.

Further Reading

Angell, L. C. (2006). Apple warns of counterfeit iPods. Available at: http://www.ilounge.com/index. php/news/comments/apple-warns-of-counterfeit-ipods/P20. Accessed 8 April 2010.

Anholt, S. (1996). Making a brand travel. *Journal of Brand Management, 3*(6), 357–364.

Anon. (15 September 1990). Coke's Kudos. *The Economist*, 120.

Benady, D. (21 November 2002). Deadly ringers. *Marketing Week*, 24–27.

Bidlike, S. (7 Feb 1991). Coca-Cola changes ad strategy to bolster its spin-off brands. *Marketing*, 4.

Blois, K. (1990). Product augmentation and competitive advantage. In H. Muhlbacher, & C. Jochum (Eds.), *Proceedings of the 19th annual conference of the European Marketing Academy*. Innsbruck: EMAC.

Brownlie, D. (1988). Protecting marketing intelligence: the role of trademarks. *Marketing Intelligence and Planning, 6*(4), 21–26.

Cohen, D. (Jan 1986). Trademark strategy. *Journal of Marketing, 50*, 61–74.

Davies, I. (1995). Review of the Trademarks Act. *Journal of Brand Management, 2*(2), 125–132, and 2(3), 187–90.

Davies, I. (1997). The Internet: some legal pitfalls. *Journal of Brand Management, 4*(4), 273–277.

de Chernatony, L., & McWilliam, G. (1989). The varying nature of brands as assets: theory and practice compared. *International Journal of Advertising, 8*(4), 339–349.

de Chernatony, L. (2001). *From Brand Vision to Brand Evaluation*. Oxford: Butterworth-Heinemann.

Doyle, P. (1989). Building successful brands: the strategic options. *Journal of Marketing Management, 5*(1), 77–95.

Drucker, P. (1990). *The New Realities*. London: Mandarin Paperbacks.

Grönroos, C. (1990). *Service Management and Marketing*. Chichester: John Wiley & Sons.

Hemnes, T. (1987). Perspectives of a trademark attorney on the branding of innovative products. *Journal of Product Innovation, 4*, 217–224.

King, S. (1991). Brand-building in the 1990s. *Journal of Marketing Management, 7*(1), 3–13.

Kochan, N. (Ed.). (1996). *The World's Greatest Brands*. London: Macmillan Business and Interbrand.

Leadbeater, C. (14 March 1991). Moles unearth spare part scam. *Financial Times*, 1.

Lee, J. (2008). What's 'Even Worse' about buying fake handbags? The New York Times, May 16th. Available at: http://cityroom.blogs.nytimes.com/2008/05/16/whats-even-worse-about-buying-fake-handbags. Accessed 8 April 2010.

Levis'. (2008). Top Brand creates its own 'protection' unit. Available at: http://www.levis.co.za/Press/PressDetail.aspx?id=612. Accessed 8th April 2010.

McKenna, R. (Jan–Feb1991). Marketing is everything. *Harvard Business Review*, 65–79.

Moore, K., & Andradi, B. (1997). Who will be the winners on the Internet? *Journal of Brand Management, 4*(1), 47–54.

Olins, W. (1995). *The New Guide to Identity*. Aldershot: Gower.

Porter, M. (1985). *Competitive Advantage*. New York: The Free Press.

RTE. (2007). Dunnes Stores found to have copied designs. Available at: http://www.rte.ie/news/2007/1221/dunnes.html. Accessed 9 April 2010.

SAS. Emission Calculator. Available at: http://sasems.port.se/Emissioncalc.cfm?sid=simple&utbryt=0&res=Result&lang=1. Accessed 7 April 2010.

Thomas, T. (19 Oct 1990). A new golden gate for great pretenders. *The European*, 24.

Wilkie, R., & Zaichkowsky, J. (1999). Brand imitation and its effects on innovation, competition and brand equity. *Business Horizons, 42*(6), 9—18.

Zeithaml, V. A. (July 1988). Consumer Perceptions of Price, Quality and Value: A Means-End Model and Synthesis of Evidence. *Journal of Marketing, 52*, 2—22.

Brand Planning

OBJECTIVES

After reading this chapter, you will be able to:

- Explain the concept of core values and appreciate the importance of sustaining values.
- Use the concept of bridging to discuss how a brand's component values can be synthesised to form a composite, powerful brand.
- Describe the concept of brand experience and discuss how customers evaluate core values.
- Identify opportunities presented by brand communities as sources of innovation.
- Discuss the options for managing brands through their growth, maturity and decline phases and suggest methods to rejuvenate brands in decline.

SUMMARY

This chapter considers some of the issues in brand planning. It opens by explaining the meaning of core values, stressing that consumers welcome consistency and that, as such, a brand's core values should not be tampered with. Understanding the core values of a brand is essential but managers should also consider how these synergistically blend together to form a more holistic brand. The use of bridging is described as one way of enabling managers to take a more holistic perspective on brand building. We show that consumers evaluate brands primarily by the extent to which they satisfy functional and representational needs. Through an appreciation of brands' functional and representational characteristics, we consider how best to invest in brands. We review some of the issues in developing and launching new brands and consider the role of customers in creating brand innovation. We also address the problems of managing brands during their growth, maturity and decline phases. The financial implications of these different phases are considered. Finally, we look

407

Creating Powerful Brands. DOI: 10.1016/B978-1-85617-849-5.10011-X

at ways of rejuvenating 'has been' brands and offer suggestions to sustain brand performance.

UNDERSTANDING THE MEANING OF CORE VALUES

In Chapter 2, we defined a brand in terms of 'a cluster of values'. Urde (2009) explains that values can be explored from three perspectives:

- **Values related to the organisation** – these are those 'more or less expressed common values, supporting ideas, positions, habits and norms' (Urde, 2009, 620) which converge to provide a corporate culture.
- **Values that summarise the brand** – which are also described as 'brand essence', which reflect the spirit of the brand.
- **Values as perceived by customers** – these values are more explicit and add value for the customer. This value reflects what customers will exchange for the brand.

de Chernatony, Drury and Segal-Horn (2004) offer a further distinction between those core values that are enduring and those that are peripheral and less central, but nevertheless are important at a point in time and may reflect societal change. For example, a coffee brand may have a core value of offering taste and sophistication. Changing social norms may also require that the brand owner purchases only fair trade beans; however, ethical product sourcing may become a peripheral value as consumer demands change and new issues need to be addressed.

Enduring core values define the corporate brand identity and support the brand promise being made to the customer. Therefore, they guide both internal and external brand building. As Urde (2009, 622) explains, 'they are like the melody of the brand that follows through all of the product and service design, supporting behaviour from the organisation, and communications'. He cites the example of IKEA, whose mission is 'to build a better life for the many people'. The brand's core values of 'common sense and simplicity', 'dare to be different' and 'working together' support IKEA's mission. For example, to bring the value 'common sense and simplicity' to life, the company searches for 'the right solution at the right costs' and retains 'a healthy aversion to status symbols'. In Chapter 10 we discussed the role of the value chain in adding value for the customer. In this IKEA example, Urde (2009) explains that the value 'common sense and simplicity' also reflects a strategic approach. It runs through the organisation, as it guides operations, product design, customer relations and internal and external relations.

Furthermore, brand values need to 'speak' to the consumer. In services, de Chernatony, Drury and Segal-Horn (2004) explain that people are drawn to service brands that have an image expressing values which are congruent to their own actual or aspired values. They explain that this image is projected by employees and therefore service organisations must also take care to ensure that all front line employees bring the brand values to life through their performance. In interviews with leading service branding consultants, they found that core values need to be developed in a consultative and democratic process or there is little chance of sustaining them. Peripheral values can evolve over time and should, therefore, be assessed regularly and amended to allow the brand essence to adapt to external changes. In addition, human resource management is critical. They contribute to sustaining brand values through supporting the dissemination of values and rewards for value-appropriate behaviour, as well as attracting employees who are most likely to embody the brand values. However, de Chernatony, Drury and Segal-Horn (2004) also explain that employees self-select roles through reacting to values expressed by the organisation. To attract the right staff, it therefore behoves managers to ensure that core values are correctly communicated and maintained over time.

MAINTAINING THE BRAND'S CORE VALUES

In previous chapters, we stressed the point that brands succeed because marketers have a good appreciation of the assets constituting their brands. By recognising which aspects of their brands are particularly valued by consumers, marketers have invested and protected these, sustaining satisfaction and maintaining consumer loyalty. Any pressures to cut support for these core values have been strenuously resisted.

Over time, consumers learn to appreciate the core values of brands and remain loyal to their favourite brands, since they represent bastions of stability, enabling consumers to confidently anticipate their performance. It therefore behoves companies to have a statement of their brand's values, which is given to anyone working on the brand, whether internal or external to the firm. It's not just having documents prominently displayed stating the brand's values, but it is important to get everyone to enact these values in all their activities. There needs to be a mechanism in place whereby any marketing plans for the brand are carefully considered against the statement of core values, to ensure that none of the core values are adversely affected by any planned activity. It is common for companies to regard brand management as a good training ground for junior managers. They would typically work on a brand for around two to three years and then seek promotion. As a consequence of their junior positions and their focus on achieving improvements in short periods, there is a danger that their concern with short-term rather than long-term horizons results in their changing some of the brand's values. In effect, they are driven more to build short-term market share rather than to build brands. When there

Don't
irritate
your new
boss.

Gentle and thorough cleaning for sensitive skin.

HUGGIES

The brand extension Huggies Sensitive offers the functionality of wipes, with the reassurance that the product is both 'gentle and thorough'

is a system in place whereby any brand plans have to be checked against the statement of core values, there is a greater likelihood of the brand thriving over a longer horizon.

Having a statement of core brand values ensures that changes in any of the firm's agencies, aimed at breathing new creative life into the brand, will be in a direction consistent with the brand's heritage. It can also act as a 'go/no-go' decision gate when managers are faced with the need to respond to an increasingly hostile commercial environment.

When Red Bull was launched as an energy drink, GlaxoSmithKline (then SmithKline Beecham) had to consider the impact this would have on their Lucozade brand. Traditionally, Lucozade was perceived as an energy drink for those who were ill. However, the brand was repositioned in the 1980s through the use of advertising featuring the athlete Daley Thompson. Lucozade built on the brand's energy heritage with the launch of the Lucozade Sport range in 1990. As both Lucozade and Red Bull were sold through the grocery channels, the threat of market share losses to a rival energy drink were considered. However, it became apparent that Red Bull had different brand values, through its positioning as an edgy drink and its association with clubbing and alcohol. If Lucozade changed its positioning as a result of the launch of Red Bull, its core values would have been lost. Instead, Lucozade communicates its sports nutrition values through increased sponsorship of sporting events, celebrity endorsement by leading sports figures and by providing sports nutrition details and running training plans on its website (www.lucozade.com/sport).

From Red Bull's perspective also, core values are critical Logman (2007, 268) explains that Red Bull's 'life is somewhat "newer" than other brands'. However, it must be vigilant about protecting its brand and ensuring that its values are relevant to all of its brand publics. Its association as the drink of choice for the 'hardcore party crowd' may alienate it as a drink for long distance lorry drivers, for example. Red Bull has broadened its association with extreme sports, such as Formula 1, X Fighters and the King of the Hill snowboard competition, to widen its appeal while sustaining its brand positioning: 'Red Bull revitalises body and mind!' These two different brand strategies allow Lucozade and Red Bull to successfully coexist, with very different brand images borne from their diverse brand values.

A further advantage of having a statement of brand values is that it enables managers to check their interpretation of the brand against the agreed view. By so doing they can then evaluate the appropriateness of their planned actions. Managers wisely concentrate on matching their target market's needs and are rightly concerned about not letting their brand fall against competitive actions or changing consumer needs. However, what is often overlooked is the question: 'Do all of the brand's team have the same views about the brand's values?' Brands don't just die because of the external environment — their life can be shortened by a lack of consistent views amongst the brand's team. In a study amongst leading-edge service brand consultants, de Chernatony, Drury and Segal-Horn (2004) identified a number of factors that contribute to employee adoption and therefore sustainability of brand values:

- **Define clear values that are appropriate and that people can identify with** — the best way to achieve this is to involve employees in the identification of brand values and the ways that they should best be enacted.
- **Use internal communication to help employees internalise brand values** — when people are trained to understand the brand history, they are proud to be part of the brand.
- **Use artifacts to illustrate what a brand stands for** — allowing the brand values to become part of the ritual and language of the organisation helps to keep them alive.
- **Advertise** — external marketing communications also reinforce brand values internally.
- **Create offshoot brands** — creating new brands as offshoots helps to retain the values of the parent brand over time.
- **Behave in-line with values** — managers must 'walk the talk' to reinforce brand values for employees.
- **Create Champions** — owners or managers must be passionately committed to the brand and share succession with others who share those beliefs.
- **Use the Human Resources function** — recruitment of people with the same values helps to reinforce employee buy-in to brand values.

In his research in Accenture, Amla (2008) found that a commitment to employee diversity helps to attract talented employees, ensure employee satisfaction and retain experienced staff. In addition, committing to diversity helps to mirror the diversity of local communities which, in turn, increases local identification with the organisation. Furthermore, when managers' commitment to diversity is experienced by employees, they offer higher levels of service and customers benefit from increased innovation.

Further outcomes arising from employee adoption of values are evident in the tourist industry. Freire (2009) found that the evaluation of places as brands ('geo brands') and the development of places as brands are highly dependent on local people. Their study sought British consumers' views about the Algarve region of Portugal. They found that the friendliness of local people, defined as the extent to which they were ready to help, were talkative and also family oriented, was used for important decisions such as buying a property in the area. Furthermore, friendliness could be managed, as much of the customers' experience of friendliness arose through encounters with resort staff, as well as with staff in external services, such as restaurants. The study found that factors such as low staff turnover and cultural traditions such as shaking hands helped to positively reinforce the Algarve's core values and enhance its image among visitors and potential property buyers.

Any plans to cut back on investments affecting the core values of the brand should therefore be strenuously opposed by strong-willed marketers. By ensuring that everyone working on a particular brand is regularly reminded of the brand's values, an integrated, committed approach can be adopted, so that the correct balance of resources is consistently applied. Checks need to be undertaken to ensure that any frills that do not support the brand's values are eliminated and that regular consumer value analysis exercises, rather than naive cost-cutting programmes, are undertaken to ensure that the brand's values are being correctly delivered to consumers.

BRIDGING THE BRAND'S VALUES

There is a danger that when guarding the consistency of their brand's core values over time, managers become too focused on considering their brand in terms of its individual values. Whilst this is an important part of brand analysis, it should be recognised that brands are holistic entities where individual values are integrated into a whole, whose strength comes from its interlinking parts. Managers therefore need to consider how their brand's component values are synergistically integrated to form a more powerful whole. A branding consultancy, Brand Positioning Services, developed a technique which enables managers to appreciate how bridging between these parts makes the brand stronger and enables it to attain optimal positioning.

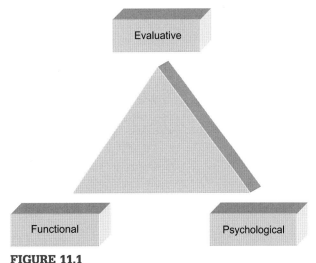

Brand Positioning Services conceptualise the brand as being composed of three components. The **functional component** characterises what the product or service does. The **psychological component** describes which of the user's motivational, situational or role needs the product or service meets and the **evaluative component** considers how the brand can be judged. The brand, as Fig. 11.1 shows, can then be considered as the integration of these three components.

FIGURE 11.1

The integrated brand (*after Brand Positioning Services 1987*)

Consumers do not consider any of these three components in isolation. When a brand of yoghurt is described as being functionally low fat, consumers' perceptions of the brand may evoke thoughts of a healthy lifestyle. Thus, the functional component of the brand is assessed within the perspective of its psychological associations. For an integrated brand, both the functional and psychological components need to work together. When this is the case, they are regarded as bridged and a single word should describe the benefit that both these components satisfy.

When developing a brand of toothpaste, there are many functional needs it could satisfy, such as the desire for white teeth or to fight bacteria. Colgate Palmolive decided that their brand, Colgate toothpaste, would focus on fighting bacteria, thereby reducing the likelihood of problems such as cavities, plaque, tartar and bad breath. Psychologically, some consumers are worried by the prospect of regular visits to dentists and the social embarrassment of bad breath. Analysis indicated that Colgate toothpaste could be positioned in terms of protection, since this word bridged the functional and psychological needs, leading to an integrated brand. Likewise Comfort fabric conditioner is about softness, bridging the functional and psychological components and its advertising often depicts users enjoying the benefits of soft clothes, as well as the nostalgia and associations of home which are attached to Comfort's familiar smell.

Several competing brands may be able to meet consumers' needs in a particular category. To give the firm's brand a lead over competition, managers need to suggest to consumers how to judge competing brands and encourage evaluation along a dimension that their brand excels on. This is the third component of the brand, the evaluator. It was decided that Colgate toothpaste should be about *trusted* protection and that comfort fabric conditioner should be about *loving* softness.

A unique two-word statement for each brand — the evaluator plus the bridged need — not only defines the brand's positioning but also enables managers to consider their brand as a holistic entity. While it is laudable to understand the core values constituting the essence of the brand so that they can be protected over time, these need to be integrated to produce a holistic brand. The procedure that Brand Positioning Services have developed is a helpful way of getting managers to think beyond the component parts to arrive at the integrated whole.

Since a brand is the totality of thoughts, feelings and sensations evoked in consumers' minds and hearts, resources can only be effectively employed once an audit has been undertaken of the dimensions that define it in the consumer's mind. To appreciate this planned use of resources, it is therefore necessary to consider the dimensions that consumers use to assess brands.

DEFINING BRAND DIMENSIONS

When people choose brands, they are not solely concerned with one single characteristic nor do they have the mental agility to evaluate a multitude of brand attributes. Instead, only a few key issues guide choice.

In some of the early classic brands' papers, our attention is drawn to people buying brands to satisfy functional and emotional needs. One has only to consider everyday purchasing to appreciate this.

For example, there is little difference between the physical characteristics of bottled mineral water. Yet, due to the way that its advertising has reinforced a particular positioning, Perrier is bought more for its 'designer label' appeal, which enables consumers to express something about their upwardly mobile lifestyles. By contrast, some may buy Evian more from a consideration of its healthy connotations. If consumers solely evaluated brands on their functional capabilities, then the Halifax and Santander, with interest rates remarkably similar to other competitors, would not have such notable market shares in the mortgages sector. Yet the different personalities represented by these financial institutions influence brand evaluation.

This idea of brands being characterised by two dimensions, the rational function and the emotionally symbolic function, is encapsulated in the model of brand choice shown in Fig. 11.2. When consumers choose between brands, they rationally consider practical issues about brands' functional capabilities. At the same time, they evaluate different brands' personalities, forming a view about them that fits the image they wish to be associated with. As many writers have noted, consumers are not just functionally oriented; their

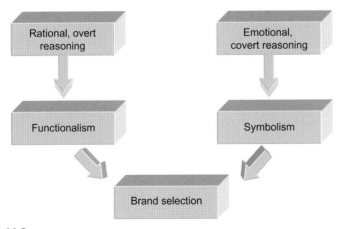

FIGURE 11.2
Components of brand choice (*ater Lannon and Cooper 1983*)

Surprisingly crispy

PRINGLES

In this advertisement, Pringles emphasise the functional crispiness of their product in an image which also captures its flavour

behaviour is affected by their interpretation of brand symbolism, as was shown in earlier chapters. When two competing brands are perceived as being equally similar in terms of their physical capabilities, the brand that comes closest to matching and enhancing the consumer's self concept will be chosen.

In terms of the functional aspects of brand evaluation and choice, consumers assess the rational benefits they perceive from particular brands, along with preconceptions about efficacy, value for money and availability.

One of the components of functionalism is quality. For brands that are predominantly product-based, Garvin's work (1987) has shown that when consumers, rather than managers, assess quality they consider issues such as:

- **performance** – for example, the safety rating of a car.
- **features** – does the car come with an MP3 docking station as a standard fitting?
- **reliability** – will the car start first time every day it's used?
- **conformance to specification** – if the car is quoted to have a particular petrol consumption, when driving around town consumers expect this to be easily achieved.
- **durability** – which is an issue Volvo majored on, showing the long lifetime of its cars.
- **serviceability** – whether the car can go 12 months between services.
- **aesthetics** – which BMW capitalised on with its Mini and DKNY uses to their advantage with the glass bottles of perfume resembling apples, suggesting a fresh smell and the 'big apple' associations of New York.
- **reputation** – consumers' impressions of a particular car manufacturer.

For predominantly service based brands, Parasuraman, Zeithaml and Berry (1988) developed the SERVQUAL instrument to assess consumers' perceptions of service quality. They found that the five core dimensions of service quality are:

- **tangibles**: the physical facilities, equipment and appearance of staff;
- **reliability**: the ability of staff to perform the promised service dependably and accurately;
- **responsiveness**: the willingness of staff to help consumers and provide prompt service;
- **assurance**: the knowledge and courtesy of staff and their ability to inspire trust and confidence;
- **empathy**: the caring, individualised attention provided to consumers.

At a more emotional level, the symbolic value of the brand is considered. Here, consumers are concerned with the brand's ability to help make a statement about themselves, to help them interpret the people they meet, to reinforce membership of a particular social group, to communicate how they feel and to say something privately to themselves. They evaluate brands in terms of intuitive likes and dislikes and continually seek reassurances from the advertising and design that the chosen brand is the 'right' one for them.

In assessing the brand's quality, managers must also consider how consumers experience the brand. Brakus, Schmitt and Zaranatello (2009) conceptualise brand experience as 'subjective, internal customer responses and behavioural responses evoked by brand-related stimuli that are part of a brand's design and identity, packaging, communications and environments'. For example, when they investigated conceptions of experiential brands, consumers' descriptions included the following:

- Abercrombie & Fitch — 'its like a membership in an exclusive, country-clubbish community';
- BMW — 'its just great to drive';
- HBO — 'puts me in a good mood';
- Nike — 'makes me feel powerful';
- Starbucks — 'smells nice and is visually warm'.

Brand managers should consider how brand experience arises in settings where consumers search for and consume brands. Brakus, Schmitt and Zaranatello (2009) propose a measure of brand experience dimensions, which incorporates measures of sensory, affective, behavioural and intellectual components. They advocate that managers should evaluate customers' brand experiences, as they influence customer satisfaction and loyalty both directly and indirectly through brand personality.

THE DE CHERNATONY–MCWILLIAM BRAND PLANNING MATRIX

Building on the previous section, we see that when consumers choose between competing brands, they do so in terms of a low number of dimensions. Further support for this is provided by Sheth and his team's work on values (1991) which we reviewed in Chapter 4, where they found evidence of brand choice being influenced by functional, social, emotional, conditional and epistemic values. In fact, even this could be conceived in terms of fewer dimensions, since by focusing on a given situation for regularly purchased brands (i.e. holding the conditional value constant and where there is no novelty or epistemic value), brand choice behaviour is influenced by functional value and social and emotional values. If we regard social and emotional values as describing a consumer's need for personal expression or representation, we can describe brands in terms of their functional and representational qualities.

There is notable evidence supporting the view that consumers choose between brands using two key dimensions. The first dimension is the rational evaluation of brands' abilities to satisfy utilitarian needs. We refer to this as 'functionality'. Marketing support is employed to associate specific functional attributes with particular brand names, facilitating consumers' decision-making about primarily utilitarian needs such as performance, reliability and taste. Brands satisfying primarily functional needs include WD-40, Band-Aid and Dunlop.

The second dimension is the emotional evaluation of brands' abilities to help consumers express something about themselves; for example, their personality, their mood, their membership of a particular social group or their status. We call this dimension 'representationality'. Brands are chosen on this dimension because they have values that exist over and above their physical values. For example, Hermés scarves and Prada sunglasses are very effective brands for expressing particular personality types and roles, with a secondary benefit of inherent functional qualities.

These two dimensions of brand characteristics are independent of each other but can interact together to create an overall impression. Consumers rarely select brands using just one of these two dimensions. Instead, they choose between competing brands according to the *degree* of 'functionality' and 'representationality' expressed by particular brands. It is possible to use these two dimensions to gain a good appreciation of the way consumers perceive competing brands. For example, Mercedes cars are perceived as being very effective brands for communicating status and personality issues and at the same time also reassure consumers about design and engineering excellence.

Through qualitative market research techniques, it is possible to identify consumer attributes reflecting the dimensions of 'functionality' and 'representationality'. Then, by incorporating these into a questionnaire, a large sample of consumers can be interviewed to gauge their views about the competing brands in a particular product field. Statistical analysis of these questionnaires enables the marketer to characterise each of the competing brands in terms of their 'functionality' and 'representationality'. By knowing the scores of each brand on these two dimensions, a spatial display of the brands can be produced by plotting the brand scores on a matrix, whose two axes are functionality and representationality, as shown in Fig. 11.3. From this matrix, the marketer is able to consider how resources could best be used to support their brand. The reader interested in the details of this market research is referred to the paper by de Chernatony (1993) listed in the 'References' section at the end of this Chapter.

COLGATE

Colgate represents trusted protection; here the advertisement promotes Colgate junior 'For Brighter Smiles'

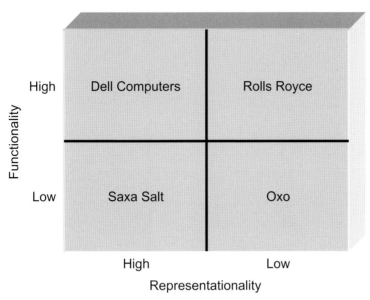

FIGURE 11.3

An example of the de Chernatony–McWilliam matrix

Strategies from the de Chernatony–McWilliam Matrix

If the marketer is satisfied with the quadrant within which the brand is located, as would be Rolls Royce in the high representationality–high functionality quadrant, then the brand strategy needs reinforcing. If, however, the brand is not perceived by consumers in the quadrant desired by the marketer, a strategy appropriate to the desired quadrant should be enacted. To sustain the brand in a particular quadrant, the following strategies are felt to be particularly appropriate.

High representationality–high functionality

Brands in the top right quadrant of Fig. 11.3 are perceived as providing functional excellence and, in the consumer's mind, to be very good vehicles for non-verbal communication about themselves. Qualitative market research needs to be undertaken to appreciate the lifestyle that users wish to project through the brand and a positioning objective subsequently defined and satisfied through the appropriate marketing mix. A creative strategy that reinforces consumers' lifestyle requirements should be developed (for example, using reference group endorsement, appropriate mood and tone of advertising, etc.) and communicated through selective media channels. A continuous promotional presence is essential to reinforce users' brand choice and to communicate symbolic meaning to those in contact with brand users. The quality of the brand needs to be maintained through high standards of quality control and continuous

DANONE

DANONE represents the high functionality, low representationality quadrant of the deChernatony-McWilliam matrix. (Reproduced by kind permission of Danone)

product development. Regular consumer surveys need to assess users' views on product performance and any negative comments need to be investigated. Availability of the brand should be restricted to quality distributors. A strict audit of the way distributors interact with the brand is required and they would need to be dissuaded from any activities that may undermine the brand's value.

Low representationality–high functionality

These brands are in the top left quadrant of Fig. 11.3. They are sought by consumers because of high utilitarian need and a less pressing drive to communicate something about themselves. Product superiority needs to be maintained. Continuous R&D investment is particularly necessary here, since it will be vitally important for the company to maintain functional advantage over the competition. 'Me-tooism' will always be a potential threat to these brands. Promotional support is important in communicating the functional benefits of the brand. The creative strategy would probably focus upon 'product as hero' in the advertising.

High representationality–low functionality

Consumers are primarily concerned about using brands in the bottom right quadrant of Fig. 11.3 as symbolic devices and are less concerned about satisfying functional needs. They would probably recognise that there are small differences between brands in product performance but they would believe that the representational issues are of more importance. The role of advertising for these brands is either to gain their acceptability as 'part of the culture' (e.g. 'the Lynx effect' promoted by this deodorant brand), or to reinforce a lifestyle (e.g. San Pellegrino water). A continuous advertising presence would be needed here. Product development issues would be less crucial compared with those brands which satisfy high functional needs. The product's strategy, however, must ensure a coherent approach to satisfy the positioning objective. More reliance needs to be placed upon the results of branded, rather than blind, product-testing against competition.

Low representationality–low functionality

Brands in the bottom left quadrant of Fig. 11.3 are bought by customers when they are not particularly concerned about functional needs. The development of Spar as a convenience store epitomises this type of brand – a limited range of groceries that satisfy consumers who realise that they have run out of a grocery product and whose primary concern is replacing the product regardless of brand availability. In general, brands in this quadrant must have wide distribution and be very price competitive. To be able to fight on price, the producer needs to strive for cost leadership in the industry. This entails being an efficient producer, avoiding marginal customer accounts, having long production runs and continually monitoring overhead costs. Brands in this quadrant are vulnerable to delisting and to succeed, the supplier must be able to justify an attractive price proposition to the distributor and consumer. It is quite possible for manufacturers to deliberately launch 'fighting brands' into this sector, the purpose of which, for example, would be to complete a range of products, to launch an offensive attack against a particular competitor or as part of a segmented approach to their market.

DETERMINING BRAND PERFORMANCE

In the next section, we discuss the management of brands over their life cycles. While Chapter 12 deals more specifically with brand metrics, this section offers suggestions to characterise brand performance. Lehmann, Keller and Farley (2008) examined studies of soft drinks, toothpaste and fast food in the US and China. They propose that brand performance can be characterised with six factors:

- **Comprehension** – is the brand seen and thought of?
- **Comparative Advantage** – to what extent is the brand favourably differentiated?
- **Interpersonal Relatedness** – how does the brand treat customers? How do others see the brand?
- **History** – what have been consumers' past experiences and emotions about the brand?
- **Preference** – what are consumer attitudes towards the brand?
- **Attachment** – how strong is the consumer connection to the brand?

They advise that these factors help managers tap into different aspects of the brand and help to capture intangible components, such as history and interpersonal relations. In the next section, we demonstrate how brands can be managed at various stages of their performance.

MANAGING BRANDS OVER THEIR LIFE CYCLES

So far, this chapter has focused on clarifying what combination of marketing resources best supports a particular type of brand, *at a given point in time*. It needs to be appreciated, however, that the returns from brands depend on where they are in their life cycle. Different types of marketing activities are needed according to whether the brand is new to the market or is a mature player in the market. In this section, we go through the main stages in brands' life cycles and consider some of the implications for marketing activity.

Developing and launching new brands

Traditional marketing theory argues for a well-researched new product development process. When new brands are launched, they arrive in a naked form, without a clear personality to act as a point of differentiation. Some brands are born being able to capitalise on the firm's umbrella name but even then, they have to fight to establish their own unique personality. As such, in their early days, brands are more likely to succeed if they have a genuine functional advantage, as there is no inherent goodwill or strong brand personality to act as a point of differentiation.

Wireless Application Protocol (WAP)-enabled telephones were launched into a market that had unrealistically high expectations. They proved to be slow and costly to use. Instead of seeing the Internet Explorer displays that they had become accustomed to on their PCs, consumers had small screens with a few lines of text that they had to scroll through. Unable to deliver the anticipated functional benefits, WAP-enabled telephones did not achieve the forecast level of sales.

New brand launches are very risky commercial propositions. To reduce the chances of a new brand not meeting its goals, many firms rightly undertake marketing research studies to evaluate each stage of the brand's development amongst the target market. Sometimes, however, very sophisticated techniques are employed, lengthening the time before the new brand is launched. While such procedures reduce the chances of failures, they introduce delays that may not be financially justified. It is particularly important that delays in the development programmes for technological brands are kept to a minimum. For example, it was calculated that if a new generation of laser printers has a life cycle of five years, assuming a market growth rate of 20% per annum, with prices falling 12% per annum, delaying the launch of the new brand by only six months would reduce the new brand's cumulative profits by a third. Virtual reality techniques provide a means to cut market research time by using 3D computer simulations to immerse consumers in new environments and experience new products. Marketers launching new technological brands need to adopt a more practical approach, balancing the risk from only doing pragmatic, essential marketing research against the financial penalties of delaying a launch. The Japanese are masters at reducing risks with new technology launches, with their so-called 'second fast strategy'. They are only too aware of the cost of delays and once a competitor has a new brand on the market, if it is thought to have potential, they will rapidly develop a comparable brand. A classic example of this was when Sony launched the very successful 'passport sized' CCD TR55 camcorder. This weighed 800 grams and had 2200 components shrunk into a space that was a quarter of that of the conventional camcorder. Within six months, JVC had an even lighter version, soon followed by Sanyo, Canon, Ricoh and Hitachi. Likewise in the 35 mm single lens reflex camera market, Minolta launched the first autofocus model. Realising the potential in this new development, Canon underwent a crash research programme and shortly afterwards launched an improved model, which halved the autofocusing time. However, challenges always remain. For example, consumers often use the video facility of their mobile phone, which offers the advantage of ease of use, convenience, and quick uploading of spontaneous video clips to sites such as YouTube. Managers must be vigilant to always offer new advantages, to avoid consumers replacing the use of camcorders entirely.

It is less common for Japanese companies to test new brand ideas through marketing research to the extent seen in Western Europe. One report, for example, estimated that approximately 1000 new soft drinks brands were launched one year in Japan, with 99% failing. In both low and medium cost goods, more emphasis is placed on testing through selling. In part, this reflects the Japanese philosophy of not just passively listening to the customer but of more actively leading them. The greater emphasis on testing new products through selling reflects the Japanese view that this enables their managers to get more experience of new markets faster. The importance of understanding new markets resulted in one Japanese car manufacturer sending its design team to live in the country where their new brand was being targeted, to appreciate the environment within which their car was to be driven. Time was taken to drive around towns, observing how consumers used their cars. They took photographs of people driving and queuing, to really get to know their new consumers.

Brand management often requires risk taking and innovation. Logman (2007) presents an 'innovation assessment framework', which starts with the concept of customer value and then 'spins into other perspectives'. These perspectives are:

- **Customer value perspective** – focusing on customer utility, the jobs that customers do themselves and the functional and emotional benefits received by the customer. Value innovation may result in 'Blue Ocean' spaces (recall the discussion of Blue Ocean Strategy in Chapter 10).
- **The segment perspective** – focusing on the segment(s) to be served. This may require focusing on the 'bread and butter' of less demanding segments and always requires an ability to listen to the market. Logman (2007) provides the example of the video game launched for children yet loved by adults and explains that the market can dictate its own segments.
- **The internal and external process perspective** – focusing on 'the effects of the outside on the inside'. How will new innovations impact on current internal systems and processes? Both internal capabilities and external change should be aligned.
- **The financial perspective** – focusing on the financial costs of innovation and of attracting customers.

To generate the ideas and innovations, recent literature has suggested the use of brand community members as a source of creativity and passion. Chapter 8 contains a discussion about brand communities on the Internet. In the context of brand building, Füller, Matzler and Hoppe (2008) explain that community members are a positive source of innovation because they are passionate about the brand and its members are interested in new products. They provide the example of Niketalk, where Nike fans can create their own basketball shoe designs or design shoe features. In their study on Volkswagen Golf GTI fans, they discovered that consumer creativity — for example demonstration of task motivation or skills — determined their willingness to participate in open innovation. However, they also noted that identification with the brand would not automatically translate into willingness to participate in innovation projects, as community members may not be willing to share their knowledge with the brand. They explain that instead, community membership may be a 'knock-out' for engaging in an innovation project by a rival brand.

New product failures should not be seen as a hunt for a scapegoat. Instead, analysis is needed to learn from the failures and these results rapidly fed back to improve the next generation of products. Just as the archer's arrow rarely hits the centre of the target the first time but does so on the second trial, so the analogy of using learning to further refine new brand concepts needs adopting.

There are several benefits from being first to launch a new brand in a new sector. Brands which are pioneers have the opportunity to gain greater understanding of the technology by moving up the learning curve faster than competitors. When competitors launch 'me-too' versions, the innovative leader should be thinking about launching next generation technology. Being first with a new brand that proves successful also presents opportunities to reduce costs due to economies of scale and the experience effect.

Brands that were first to market and are strongly supported, offer the opportunity for consumer loyalty. Almost without thinking about it, customers ask for the brand which has become generic for the product field. Xerox copiers have been so well-supported with their innovative developments that they've become part of everyday terminology.

Being the first with a new brand sets habits which are difficult to change. For example, in the drugs industry, GPs prefer to prescribe medication with which they've become confident. To listen to a sales representative talking about a 'me-too' version takes time and introduces uncertainty about efficacy in the GP's mind.

When launching new brands in technological markets, marketers are only too aware of the high costs they are likely to incur. As such, the successful firms look at launching a new technology into several product sectors. One of the nagging doubts marketers have when launching a new brand is that of the sustainability of the competitive advantage inherent in the new brand. The 'fast-follower' may quickly emulate the new brand and reduce its profitability by launching a lower-priced brand.

In the very early days of the new brand, the ways in which competitors might copy it are through:

- design issues, such as, colour, shape, size and unusual functionality;
- physical performance issues, such as quality, reliability, durability and serviceability;
- product service issues, such as guarantees, installation, after sales service, online support;
- pricing;
- availability through different channels, including online websites and auction houses;
- promotions;
- image of the producer.

If the new brand is the result of the firm's commitment to functional superiority, the design and performance characteristics probably give the brand a clear differential advantage but this will soon be surpassed. In areas like consumer electronics, a competitive lead of a few months is not unusual. Product technology issues can sometimes be a more effective barrier. For example, Apple's iPad has created a new platform for media, which has changed the way newspapers are delivered. Consumers in Europe can read the New York Times on an app and interact with its news stories, which creates a competitive barrier to other newspapers that cannot provide the same level of support. However, changing technologies also changes the competitive environment for the New York Times, as it must consider new sources of competition, such as the Internet or even games, which consumers could access via their iPad or iPhone instead of reading the newspaper. Price can be easy to copy, particularly if the follower is a large company with a range of brands that they can use to support a short-term loss from pricing low. Unless the manufacturer has particularly good relationships with distributors that only stock their brand, which is not that common, distribution does not present a barrier to imitators.

Further insights about issues to address when developing and launching new products is provided by research undertaken jointly by IMD and PIMS (Kashani, Miller and Clayton, 2000). They undertook research into 60 consumer product innovations launched by 34 companies across 26 product categories within the European Union.

Those innovations that notably enhance consumers' perceptions of value are more likely to succeed. New, clear plastic packaging (Barex) was developed for Pampryl, the fruit juice division of Pernod Ricard. This prevented oxidation of pasteurised juice and extended shelf life by up to 12 months. In addition, they also improved the juice quality, offering consumers the taste of "real" fruit juice rather than a diluted concentrate.

At one extreme, innovation is seen through incremental changes to the product or service, as shown in the Pampryl example. At the other extreme, there is radical innovation that creates new categories and markets and through the introduction of new products or services, and these have the potential to significantly change consumers' behaviour. An example of radical innovation is the Nespresso System, when Nestlé created a new coffee category, that of single portion coffee. A range of roast and ground coffee packaged in individually-portioned capsules was devised solely for the Nespreso machine. This combination offers consumers individual cups of espresso coffee with speed and convenience. Brand leadership through this innovation has been helped by its technical complexity and patents.

Radical innovation is perceived by managers as riskier than incremental innovation. As such, most of the companies in the study were focusing on incremental innovation.

Inspiration for new ideas is always a challenge, partly because of the problem of there being only a minority who genuinely think differently. As Sir George Bernard Shaw once remarked, 'Some people look at the world and ask why. I dream of things that have never been and ask why not?' The most frequently cited source for innovative ideas came from market research and marketing (44%), then retailers and suppliers (26%), followed by R&D (25%) and other internal sources (5%). The researchers found that successful innovations arise from a combination of market research and R&D. For example, guided by the insight that consumers with heavy colds suffer from soreness caused through using facial tissues, Kimberley Clark scientists coated their tissues with a lotion which softens the skin. Tracking studies showed that once users started to use the new brand, Kleenex Coldcare, they continued using it after their colds had gone. As such, the R&D team developed Kleenex Balsam, which benefited from a new coating process that lowered production costs, yet still offered superior product performance.

The case studies indicated that gaining insight about how much consumers value a new concept and then feeding this back into R&D is more likely to result in successful new brands. Through understanding the new brand's problems, managers can make changes more effectively and better satisfy consumer needs.

This study shows that innovative brands succeed as a result of three characteristics, as shown in Fig. 11.4.

The starting point is a commitment of resources to innovation, through a good understanding of consumer behaviour. Only if the proposed innovation leads to greater customer value should the idea be taken forward — for example, ICI's Dulux Once (offering the convenience of a single coat of paint), Mars Celebrations (creating the category of "miniatures" in the confectionery market)

FIGURE 11.4

A process for winning new brands (*after Kashani, Miller and Clayton 2000*)

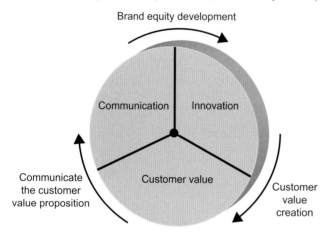

Brand equity development

Communication

Innovation

Communicate the customer value proposition

Customer value

Customer value creation

and Olay Total Effects skin care (which protects skin and claims to 'fight the seven signs of ageing'). Having progressed the innovation to the state that adds greater value to consumers, it then becomes appropriate to communicate this. Provided the promised value is delivered, a trusted relationship should then start to grow between the consumer and the brand.

Managing brands during the growth phase

Once a firm has developed a new brand, it needs to ensure that it has a view about how the brand's image will be managed over time. The brand image is the consumers' perceptions of who the brand is and what it stands for, i.e. it reflects the extent to which it satisfies consumers' functional and representational needs.

As sales rise, the brand's image needs to be protected against inferior, competitive, look-alikes. The functional component of the brand can now be reinforced, either through a problem-solving, specialisation strategy or a problem-solving, generalisation strategy. When Lego building bricks were launched in 1960, this challenge was faced. If the specialisation strategy were to have been followed, Lego would have been positioned solely for educational purposes. It would have been targeted at infant and primary school teachers. The problem with this strategy is that in the long term competitors may develop a brand that meets a much broader variety of needs. By following a problem-solving generalisation strategy, a brand is positioned to be effective across a variety of usage situations. This was the route Lego actually followed. Today for example, Lego offer Lego games, Lego products licenced with strong brands such as Indiana Jones and Toy Story, as well as a Lego iPhone app which allows users to apply the Lego brick style to their photographs.

The original approach to supporting the representational component of the brand needs to be maintained as sales rise. For example, for those brands that are bought predominantly to enable consumers to say something about themselves, it is important to maintain the self-concept and group membership associations. By communicating the brand's positioning to both the target and non-target segments but selectively working with distributors to make it difficult for the non-targeted segment to buy the brand, its positioning will be strengthened.

Managing brands during the maturity phase

In the maturity part of the life cycle, the brand will be under considerable pressure. Numerous competitors will all be trying to win greater consumer loyalty and more trade interest. One option is to extend the brand's meaning to new products. A single image is then used to unite all the individual brand images. This strategy was successfully employed by Polycell in the DIY (Do It Yourself) home improvements market. It established an image of DIY simplicity and reliability with its original range of wallpaper paste and then extended this image across such products as fillers, double glazing and home security, as well as online advice for DIY enthusiasts. When following this strategy in the maturity stage, the firm must continually question whether a new addition to the product range will enhance the total brand image. If the new line has brand values inconsistent with the parent brand, serious consideration should be given to launching the new line without any associations of the parent organisation.

Where the brand primarily satisfies consumers' functional needs, these functional requirements should be identified and any further brand extensions evaluated against this list to see if there is any similarity between the needs that the new brands will meet and those being satisfied by an existing brand. Where there is a link between the needs being satisfied by the existing brand and the new needs being fulfilled by a new brand, this represents an appropriate brand extension. For example, Black & Decker's proposition is that of making DIY jobs easier with the use of high performance, electrically-powered tools. Their extension from hand-held electrical drills to lawn mowers and car vacuum cleaners was entirely consistent with their original brand image. By contrast, when the brand primarily satisfies representational needs, these should be assessed and taken as an essential criteria for future brand extensions. For example, the Chanel range of fashion wear, beauty products and fragrances says a lot about the chic and discerning tastes of a sophisticated person. Their range is successfully expanding by building on this representational dimension, showing how the consumer's lifestyle is more complete with further Chanel brands.

Managing brands during the decline phase

Managers often draw upon the product life cycle when charting the life stage of their brands. However, Thomas and Kohil (2009) explain that this can be a self-fulfilling prophecy; when sales decline, managers withdraw investment and decide instead to cut prices and milk the brand. They investigate the reasons why brands decline. Competitor actions can certainly be a mediating factor. A brand can quickly become obsolete if its technologies or formats are replaced by new technologies or formats offered by a competitor.

Managerial actions can also be responsible for brand death. When a brand is successful, managers may become complacent or neglect the brand and the brand's success creates its failure. Pricing can also affect brands. Raising prices without adding additional value or lowering prices without considering brand image can both damage the brand.

Further decline in brand health comes from neglect of the target market. Thomas and Kohil (2009) provide the example of GAP. The brand developed more edgy advertising in a bid to capture the teenage and young adult market, as they were a growth segment in the 1990s. However, this alienated customers who were older and they felt neglected by a brand which, they perceived, had become more youthful and trendy.

Finally, Thomas and Kohil (2009) cite environmental factors as a transformational force. For example, cigarette brands face challenges from changes in the legal environment that have an impact on sales.

As brand sales begin to decline, firms need to evaluate carefully the two main strategic options of recycling their brand or coping with decline. A recycling strategy was used to revive Head and Shoulders shampoo. When sales for the brand began to decline, the brand was repositioned to meet the needs of customers who wanted more of a cosmetic shampoo than an anti-dandruff shampoo. With repackaging and the introduction of new formulations, Head and Shoulders now appeals to those seeking a shampoo for bounce and shine or silky hair, as well being an anti-dandruff shampoo.

Should the firm feel there is little scope for functional or representational brand changes, it still needs to manage its brands in the decline stage. If the firm is committed to frequent new brand launches, it does not want distributors rejecting new brands because part of the firm's portfolio is selling too slowly. A decision needs to be taken about whether the brand should be quickly withdrawn, for example, by cutting prices or whether it should be allowed to die, gradually enabling the firm to reap higher profits through cutting marketing support.

Before the brand is finally withdrawn, however, managers should consider the implications for other brands in its portfolio. Removing a brand may offer shelf space to a competitor, for example. Brand managers wishing to 'kill' a declining brand may also meet resistance from retailers and customers. A leading make-up brand was considering removing products from its branded ranges as its colours were no longer fashionable. However, sales representatives met resistance in the pharmacy trade, as pharmacists and pharmacy assistants had noticed its popularity with older ladies. Removing the products from the brand would have alienated these customers and negatively affected the valuable relationship between the company and its pharmacy business, which was vital to the success of the company's other brands. Therefore, the complete brand was retained, but investment focused on newer ranges.

Financial implications of brands during their life cycle

According to its stage in the life cycle, a brand needs to be managed for long-term profitability. In the early stages of its life, the brand will need financial support, while in the maturity stage it should generate cash. The matrix in Fig. 11.5 shows these financial implications.

In its introduction stage, the new brand will be fighting for awareness amongst consumers and will depend heavily on the skills of the sales force to win the trade's interest and stock the new brand. Company executives must resist the temptation to try to recoup the large R&D brand investment by saving on promotional support; all this will do is to prolong the period of slow sales. On the matrix in Fig. 11.5, the very early days of the brand are represented by Quadrant A, where there is a need for large cash resources with only a small market share resulting. At this stage, the new brand is a drain on company resources.

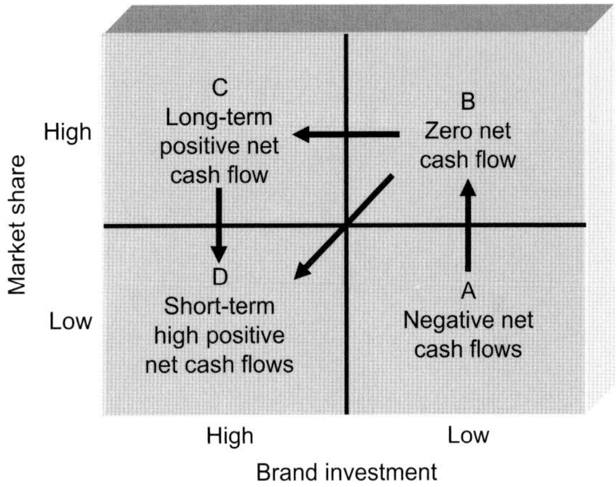

FIGURE 11.5
Financial management of brands (*after Ward et al., 1989*)

As consumer and retailer acceptance grows, sales rise and the brand moves into Quadrant B. Satisfied consumers talk to their friends and more retailers stock the brand as they begin to appreciate its potential. Higher levels of sales begin to cover the continuing brand investment and eventually the brand revenue begins to balance the costs being incurred. This stage represents the transition from marketing investment being required to develop the brand, to that of maintaining the brand.

At this crossroads in its career, the brand can then fall into Quadrant D, or be nurtured into Quadrant C. Where marketers are obsessed with short-term gains, brand investment will be cut very quickly after the brand begins to show a net positive cash flow, the argument being that there is sufficient awareness and goodwill amongst consumers. However, the brand will glide along for a short time without any driving power but like most non-powered aircraft, it will then go into rapid freefall.

Long-term benefits can be accrued by maintaining marketing support and keeping the brand in Quadrant C. As the rate of market growth slows down and competitors become more aggressive, a maintenance strategy should be directed at sustaining, if not increasing, the brand's leadership position. With economies of scale and the benefits of the experience effect, the brand should generate healthy profits. Eventually though, the firm will feel that the market for the brand is less attractive, for reasons such as more aggressive competitors or falling consumer interest. Brand support will be cut, market share will fall and for a short time, the brand will be in Quadrant D prior to being withdrawn.

From time to time, brands are not sufficiently nurtured and they lose their leading market position. This is not an inevitable feature but, should the brand experience a worsening position, there may well be a need to put new life into it. This is discussed in the next section.

REJUVENATING 'HAS BEEN' BRANDS

The cynic within the overly-cautious marketer would argue that from the moment a new brand is launched, it becomes a wasting asset. The technology to produce the brand soon becomes outdated, demography changes and competition becomes intense. The market research for the new brand concept may have been conducted in the most thorough manner, but it investigated attitudes, beliefs and social norms *at a specific point in time*. If the brand is to succeed, it usually has a short time to prove itself. Yet, even with hostile forces fighting against brands, many not only survive but thrive. The reason for their success is a strong, internal commitment to brands, based on a solid belief that they have unique characteristics which consumers value sufficiently to buy. Thomas and Kohil (2009, 383) suggest that there exists 'a significant amount of equity in declining brands, and with proper diagnosis, strategy and execution, a brand can be revived'. They advise that managers should first consider whether the brand is worth reviving and advocate a brand audit as a tool to pinpoint weak areas. Branding is a long-term initiative; therefore they advocate patience as revitalisation can take one or two years. Further, they suggest finding out what is unique about the brand, investing in it and telling the market.

Well thought through brand plans have been prepared, documenting realistic objectives and viable strategies to counter competitive threats. Brand successes abound. For example, brands such as Gillette, Hoover, Schweppes, Colgate, American Express and Hovis have pedigrees going back over 70 years.

It would be naive to assume that the great brand successes have not had problems. What is clear about brands which have a long history is that they have been subtly adjusted to keep them *relevant* to changing market conditions. For some brands, this has resulted from continually tuning their pack designs to keep them contemporary. In other instances, it has been a case of putting the brand's core values in a different context. For example, Polaroid faced brand death with the arrival of the digital camera, as the value added from instant photographs was no longer unique to the brand. However, Polaroid has subsequently revitalised its iconic status; for example, it announced a multiyear strategic partnership with Lady Gaga as creative director of speciality products which, they claim, will blend film with digital. The shared associations between Polaroid imaging and the Haus of Gaga, known for its fashion and design aesthetic, will revitalise the brand for a new generation.

Brands that have successfully stood the test of time have built up a considerable amount of goodwill with their consumers, so much so that when sales start to fall, it should not be automatically assumed that they are in a terminal state and that investment must be cut. It is often less expensive to revitalise an established brand than it is to develop and launch a new brand. Furthermore, consumers are less likely to try a new, unknown brand than a name that they are very familiar with. When years of activity have engendered such strong consumer trust in brands like Kellogg's Corn Flakes and Nescafé, it is almost an abdication of marketing responsibility to ignore the potential for revitalising old brands.

The task of revitalising old brands is less difficult when the core values of the brand have been protected and consistently presented to consumers. It is likely that the main task will be to update the way that the brand is presented. For example, After Eight mints were created in 1962, and are positioned as a luxury post-dinner chocolate. The brand has always been associated with thin mints and the evening time. The packaging has been changed to keep the brand up to date and its single-serving sleeves reinforce its use as a product for sharing. The distinctive dominance of the green packaging and the serving sizes have been subtly changed to reinforce the brand's modernity.

Brands sometimes need rejuvenating when they have not adapted to different social situations. One has only to think of the trainers market, where the leading players are continually monitoring and anticipating changes in social and cultural trends to fine-tune their brands to keep them fresh and relevant.

A SYSTEMATIC APPROACH FOR REVITALISING BRANDS

The case of the Harlem Globetrotters provides insights about how one organisation succeeded through adopting a systematic approach to revitalising its brand. Jackson (2001) explains that when he took over the Globetrotters in 1993, it had an annual attendance of less than 0.3 million, yet with his rejuvenation strategy it increased to 2 million by 2001. His revised strategy had three elements: reinventing the product to become more relevant, showing stakeholders that the brand cares about them and making the organisation more accountable.

To address the first part of the strategy, i.e. making the product more relevant, market research was undertaken to appreciate people's views about the brand and to understand why it was not seen as being relevant. This stimulated two main approaches to making the brand more relevant in a changed society. Firstly, to show that the brand represents top quality, the best players were entered 30 to 40 times a year against first rate teams all over the world. The second route was to play their touring teams against their exhibition teams. These regular events set out to show just how well they play basketball, to provide an exhibition of basketball feats never seen live before and to make the audience laugh and feel good. Everything was carefully rehearsed and choreographed against appropriately paced music.

The second element of the strategy focused on understanding different stakeholders in order to show that the brand cared about them. Arena owners and their marketing staff had not developed strong relationships with the Harlem Globetrotters, so relationships were built with them, showing the brand's plans and how their revival would increase revenue for the arena owners. A considerable emphasis was placed on courting interviews with the media to explain the turnaround strategy. Every event was used to connect directly with the fans. Each year, 25–30 summer camps were run for young fans. To get close to other members of the community, the players regularly visited hospitals, schools and youth clubs. Notable effort was also directed at gaining more sponsors.

The third element was making the organisation more accountable. Staff were told about the new vision for the brand and the changes needed. Early in this process some found the ideas unacceptable and many left as they could not adapt. Training programmes were instigated to give staff the necessary knowledge and skills to run the business more professionally. Financial controls were instigated to monitor and better use resources. Each game had to have a forecast for costs, revenue and profit; then each night Jackson insisted that he received a document showing how the actual figures compared against the forecast.

This case study shows how a rejuvenation strategy can help a stagnant brand overcome the challenge of thriving in a hostile environment. It also indicates the importance of a committed, energetic and enthusiastic leader who has an exciting vision that challenges staff and aligns their interests with the goals for the future.

While the Harlem Globetrotters case exemplifies some of the elements that are important when rejuvenating brands, Berry (1988) shows the key stages that need addressing. The model in Fig. 11.6 provides guidance about how to progress a revitalisation strategy.

A considerable amount of data has been collected by the Strategic Planning Institute (Buzzell and Gale, 1987) looking at those factors that are strongly related to profitability. One of the key findings was that superior quality goes hand in hand with high profitability. But, it is not quality as defined from an internal perspective but from the consumers' perspective, relative to the other brands that they use. The slamming of a car door and the resultant 'thud', says more to many consumers about a car's quality than does a brochure full of technical data. The first stage in any revitalisation programme should, therefore, investigate what consumers think and feel about the brand. This can be done using a minimum of 10, depth interviews, where consumers are presented with the firm's and competitors' brands and their perceptions of relative strengths and weaknesses are explored. It is particularly important that this be done using qualitative research techniques, since this identifies the attributes that are particularly salient to consumers. The findings can then be assessed with more confidence by interviewing a larger sample of consumers with a questionnaire incorporating the attributes found in the depth interviews.

The qualitative market research findings will broadly indicate how the physical characteristics of the brand are perceived, such as product formulation, packaging, pricing, availability, etc. They are also likely to provide guidance about emotional aspects of the brand, such as the type of personality that the brand represents. The issues here may lead to questions such as, 'Is it old fashioned?', 'Is it "fuzzy"?', 'Is it relevant?', 'Is it too closely linked to an infrequently undertaken activity?' Also, by investigating changes in demography, social activities, competitive activity and distribution channels, the marketer should then be able to identify what changes might be needed for the brand's positioning.

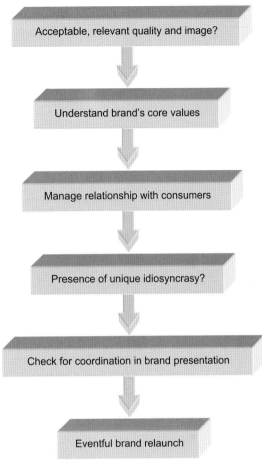

FIGURE 11.6
Stages in rejuvenating brands (*adapted from Berry, 1988*)

With ideas beginning to evolve about positioning and personality changes, the marketer needs to consider how these would affect what the brand has always stood for. Managers need to be clear about their brand's core values. Any changes from the first part of the process need to be considered against the brand's core values. Marlboro stands for dominance, self-esteem, status, self-reliance and freedom. As it faces an increasingly challenging environment, any changes to make it more acceptable should not go against these core values.

The marketer needs to consider what type of relationship their brand has had with consumers and whether this is still relevant. For example, it may have adopted an 'authority figure' relationship, treating the consumer almost as an inexperienced child. Without sufficient knowledge of the product field, the consumer may be content to abdicate responsibility to the brand since it offers an almost paternalistic reassurance. However, as buyers become more experienced they will be looking for a 'husband–wife' type relationship, where they are treated as an equal and with some respect. If the marketer has not allowed the brand to recognise the more mature nature of the consumer, the wrong type of relationship may impede brand sales.

Brands succeed because people recognise and value their personalities. But, just as people are not perfect, so brands need to have some idiosyncratic element to make them more human. If the brand needs revitalising, one way to bring it more alive is to add an idiosyncratic element.

Once a view has been adopted about possible changes, these must be carefully coordinated to ensure that each element of the marketing mix supports the new proposition. What is then required is a promotional launch that rapidly gets the message across about the rejuvenated brand.

CONCLUSIONS

Once a brand has been launched with core values that consumers appreciate, it is important that these core values are clearly communicated to all the people working on the brand and any differing perceptions amongst the brand team about the brand's values should be identified and addressed. It is vital that employees understand and believe in brand values, as their behaviour reinforces the brand message. Further, external marketing communications about brand values help to reinforce these values as cultural norms, as well as attracting new brand-supporting employees to the firm. Any attempt to change core values over time should be strongly resisted. There may be a need to update the brand's presentation but its new design should still communicate that the brand represents those values which consumers have grown to respect and welcome.

Managers need to recognise that while they analyse brands in terms of details about their core values, consumers do not undertake such detailed assessments and when they choose between brands only a few issues are considered. It is therefore important for managers to consider how the component parts of a brand are synergistically integrated into the brand as a holistic entity. By considering the bridged need satisfied by the functional/psychological components of the brand and then thinking about the evaluator consumers can use to judge their brand, a more holistic brand is likely to succeed with a clear positioning.

When consumers choose brands, there are two broad dimensions that are primarily assessed in each situation. They consider the extent to which the brand satisfies their **functional** needs. For example, when wind surfing, they gauge whether certain brands of watches accurately tell the time regardless of how often they are immersed in water. They also evaluate the effectiveness of different brands in communicating something about themselves. In other words, they take into account how well different brands satisfy **representational** needs. The wind surfer may consider the extent to which different brands of watches imply that he or she is sporty and experienced.

In any product field, it is possible to evaluate the extent to which different brands satisfy functional and representational needs. By plotting competing brands on a functional–representational matrix, it is possible to assess from consumers' perspectives which brands are close substitutes for each other by identifying groups of brands that cluster together. Investment strategies to sustain brands can also be developed from this matrix.

An understanding of brands' functional–representational characteristics enables marketers to plan resources at a particular time. In addition, understanding consumers' perceptions about brand experience will provide insights into perceived brand quality. A longer-term planning process, however, needs to anticipate resource requirements over brands' life cycles. During the development stage of the brand, it is crucial that the functional characteristics of the brand are able to meet the expectations raised by any awareness campaign. To ensure that it is functionally superior, and perceived to be of notable value by consumers, market research is needed. However, the marketer needs to be aware of the high costs incurred by delaying the launch due to over-sophisticated testing procedures. Being first to market with a new brand that catches consumers' interest offers several attractive benefits.

As sales of the brand start to grow, its image should be reinforced, enabling consumers to recognise precisely what the brand represents and in which functional situation it is particularly effective. In extending their customer base, managers should take care not to alienate previously loyal customers. In early maturity, when it starts to become a cash generator, its high brand share needs to be guarded to ensure that long-term profitability accrues. Short-sighted marketers may be inadvisably tempted to cut brand support and enjoy only a limited period of brand profitability. Successful brands remain vigilant to changes such as consumer trends and competitor actions to avoid complacency leading to obsolescence.

After a healthy period in its maturity phase, the brand may show a sluggish fall in sales. Rather than incorrectly assuming that the brand is in terminal decline, analysis should be undertaken to evaluate why its popularity is waning and whether it is viable to inject new life. New brands are very expensive to launch and their future is uncertain. By contrast, revitalising brands may well be a less expensive route to follow, with a greater likelihood of success. A brand audit can be a helpful tool to pinpoint any brand weaknesses. The revitalisation process takes time and managers must take care to identify what is relevant and distinctive about their brand. Good analysis should show how the brand's well-established core values can be presented in a way more in tune with developing consumer expectations.

MARKETING ACTION CHECKLIST

To help clarify the direction of future marketing activity, it is recommended that the following exercises are undertaken:

1. Is there a document within your firm that, at the birth of each new brand, states what each brand's core values are? What systems are in place to stop any of your management team changing the brand's core values? Do you have an annual check of each individual manager's views about the core values of the brands they are working on and do you provide feedback about instances where their views are dissimilar from the central statement of core values?

2. Take one of your major brands and write down the most important functional consumer needs it satisfies. What are the most important psychological needs being satisfied? Are these needs interlinked? What one word bridges the key functional and psychological benefit of this brand? What evaluator encourages consumers to judge the brand? From this analysis how could your brand be better positioned?

3. For a particular product field, reach a consensus view with your colleagues about the functional attributes consumers consider when choosing brands. You should not have more than five attributes. Give a score to your brand and to your competitors' brands, according to the consensus view about how well each brand satisfies consumers' functional needs. Do this on a scale where 10 represents 'excellent satisfier of consumers' functional needs' and 1 represents 'extremely poor satisfier of consumers' functional needs'. What could be done to improve your brand's positioning on this dimension? What is the brand leader doing particularly well on the functional dimension? With those colleagues who have been with your firm for some time, repeat the exercise, but this time for a period 12 months ago. Compare the historical and current scores for all the brands on the functional dimension and evaluate why these movements took place.

4. Amongst your colleagues, form a consensus view about what consumers are trying to say about themselves when they buy brands in your product field. Identify no more than five attributes describing this representational dimension. Score your brand and competitors' brands according to the consensus view about how well each brand satisfies consumers'

representational needs. Do this on a scale where 10 stands for 'excellent satisfier of consumers' representational needs' and 1 represents 'extremely poor satisfier of consumers' representational needs'. Are there any improvements that will help your brand's positioning on this dimension? What is the brand leader doing particularly well on the representational dimension? With long-serving colleagues, repeat the exercise for a previous period 12 months ago. Compare the historical and current scores for the brands on the representational dimension and evaluate why these movements took place.

5. From the results you now have from Questions 3 and 4, plot your brand on the functional and representational dimensions of the de Chernatony–McWilliam brand planning matrix. Knowing within which quadrant your brand is positioned, how appropriate is your strategy to support this brand?

6. Using the approach from Questions 3 and 4, with your colleagues, plot on the functional and representational dimensions where your competitors' brands are positioned. Focusing on those competitors who are in the same quadrant as your brand, evaluate the strengths and weaknesses of their strategies. What could you do to better defend your brand?

7. Undertake an audit of all the new brands which your firm has launched in the past five years. Focusing on those brands that have since been withdrawn and those which are regarded as doing badly, identify the reason for their poor performance. Does any one reason continually appear?

8. For any one of your more recently launched new brands, ask your Consumer Insight Manager to produce a timetable detailing the amount of time taken for each piece of marketing research. Does any one piece of marketing research appear to have consumed a lot of time? Are there any reasons why this took so long? How much time elapsed between the completion of fieldwork for each project and the presentation of results? How long could it have taken to a brief 'top line' finding, where only the results on each question are presented without any written report? Are there any implications about how time can be saved on future marketing research reports?

9. How effective is your firm at bringing together groups of people from different functional backgrounds to form a new brand project team? Are there any barriers within your firm that impede the formation of new project teams?

10. Do your brand plans consider how each brand's image will be protected as they pass through their life cycles?

11. For any recent brand extensions, evaluate whether there is a natural link between the functional or representational needs satisfied by the

brand extension and the original brand. Where there is only a weak link, it is worth considering severing any association between the two brands.

12. Using the matrix shown in Fig. 11.5, plot where each of your brands currently resides. If there are no brands in Quadrant C, you should question what has to be done to secure your firm's future over the next three years.

13. Undertake an analysis of the brands in your firm's portfolio and identify any brands that are showing sluggish sales performance. Are there any changes in the external environment that have gradually been impeding these brands' performance? Using the flow chart in Fig. 11.6, how could new life be put back into these rather staid brands?

STUDENT BASED ENQUIRY

1. Select i) a sports brand or ii) a clothing brand and discuss what its brand's core values might be. Use the brand's website and other marketing communication to determine whether these values are stated explicitly. How meaningful are the core values to you?

2. Using the model presented in Fig. 11.1, evaluate how the concept of bridging might apply to customers' evaluation of i) an Apple iPod or ii) McDonalds or iii) Facebook.

3. Select a brand you use regularly that you believe offers a positive brand experience. How do you evaluate the brand experience? Use the model in Fig. 2.2 to discuss why you select this brand instead of its closest competitors.

4. Identify three brands, one of which you believe to be in the launch phase, one in the growth phase and one in the maturity phase of their lives. For each brand discuss why you believe it is in this phase and identify *two* strategies that brand managers could adopt to enhance the brand's performance.

5. Select a brand you are familiar with, which is in decline or which has been withdrawn from the market. Would you revitalise the brand? Outline why, offering *three* strategies that brand managers might utilise to breathe new life into the brand.

References

Amla, I. (2008). Managing and sustaining a world of workplace diversity: the Accenture experience. *Strategic HR Review, 75,* 11–16.

Berry, N. (1988). Revitalizing brands. *Journal of Consumer Marketing, 5*(3), 15–20.

Brakus, J. J., Schmitt, B. H., & Zarantonello, L. (May 2009). Brand Experience: what is it? How is it measured? Does it affect loyalty? *Journal of Marketing, 73,* 52–68.

Brand Positioning Services. (1987). *Positioning Brands Profitably*. London.

Buzzell, R., & Gale, B. (1987). *The PIMS Principles — Linking Strategy to Performance*. New York: The Free Press.

de Chernatony, L., Drury, S., & Segal-Horn, S. (2004). Identifying and sustaining services brands' values. *Journal of Marketing Communications, 10*(2), 73—93.

de Chernatony, L. (1993). Categorizing brands: evolutionary processes underpinned by two key dimensions. *Journal of Marketing Management, 9*(2), 173—188.

Freire, J. R. (2009). 'Local people' a critical dimension for place brands. *Brand Management, 16*(7), 420—438.

Garvin, D. (Nov—Dec 1987). Competing on the eight dimensions of quality. *Harvard Business Review, 65*, 101—109.

Jackson, M. (May 2001). Bringing a dying brand back to life. *Harvard Business Review*, 5—11.

Kashani, K., Miller, J., & Clayton, T. (2000). *A Virtuous Cycle: Innovation, Consumer Value and Communication–. 'Research Evidence from Today's Brand-Builders'. A study by IMD and PIMS Associates for AIM.*

Lannon, J., & Cooper, P. (1983). Humanistic advertising: a holistic cultural perspective. *International Journal of Advertising, 2*, 195—213.

Lehmann, D. R., Keller, K. L., & Farley, J. U. (2008). The structure of survey-based brand metrics. *Journal of International Marketing, 16*(4), 29—56.

Logman, M. (2007). Logical brand management in a dynamic context of growth and innovation. *Journal of Product and Brand Management, 16*(4), 257—268.

Parasuraman, A., Zeithaml, V., & Berry, L. (1988). SERVQUAL: A multiple-item scale for measuring consumer perceptions of service quality. *Journal of Retailing, 64*(1), 12—40.

Sheth, J., Newman, B., & Gross, B. (1991). Why we buy what we buy: a theory of consumption values. *Journal of Business Research, 22*(2), 159—170.

Thomas, S., & Kohil, C. (2009). A brand is forever! A framework for revitalizing declining and dead brands. *Business Horizons, 52*, 377—386.

Urde, M. (2009). Uncovering the corporate brand's core values. *Management Decision, 47*(4), 616—638.

Ward, K., Srikanthan, S., & Neal, R. (Autumn 1989). Life-cycle costing in the financial evaluation and control of products and brands. *Quarterly Review of Marketing*, 1—7.

Further Reading

ABC. (2005). *Do Smoking Bans Really Get People to Quit?* Available at: http://abcnews.go.com/WNT/QuitToLive/story?id=1292456. Accessed 14 April 2010.

Alcock, G., & Batten, C. (25 April 1986). Judging the worth of brand values. *Marketing Week*, 58—61.

Anon. (12 January 1991). What makes Yoshio invent. *The Economist*, 75.

Brierley, S. (2 December 1994). ASA document slams P&G ads. *Marketing Week*, 7.

Brierley, S. (20 January 1995). Lever drops 'accelerator' formulation. *Marketing Week*, 5.

Brierley, S. (24 February 1995). Making way for New Generation. *Marketing Week*, 23—24.

Burke, R. (Aug 1995). Virtual shopping. *OR/MS Today*, 28—34.

Clifford, D., & Cavanagh, R. (1985). *The Winning Performance: How America's High Growth Midsize Companies Succeed*. London: Sidgwick & Jackson.

Connor, B. (20 Feb 1986). How oldies go for black. *Marketing*, 39—43.

Cooper, R. (1987). *Winning at New Products*. Agincourt, Ontario: Gage.

de Chernatony, L., & McWilliam, G. (1989). Clarifying how marketers interpret 'brands'. *Journal of Marketing Management, 5*(2), 153–171.

de Chernatony, L., & McWilliam, G. (1990). Appreciating brands as assets through using a two dimensional model. *International Journal of Advertising, 9*(2), 111–119.

de Chernatony, L., & Took, R. (1994). Team based brand building: questioning the current marketing research role. In *Proceedings of Building Successful Brands* (pp. 265–278). Amsterdam: European Society for Opinion and Marketing Research.

Füller, J., Matzler, K., & Hoppe, M. (2008). Brand community members as a source of innovation. *Journal of Product and Innovation Management, 25*, 608–619.

Gardner, B., & Levy, S. (March–April 1955). The product and the brand. *Harvard Business Review, 33*, 35–41.

Hamel, G., & Prahalad, C. (July–Aug 1991). Corporate imagination and expeditionary marketing. *Harvard Business Review, 69*, 81–92.

Hoggan, K. (3 March 1988). Back to life. *Marketing* 20–22.

Interbrand. (1990). *Brands – An International Review*. London: Mercury Books.

Ismail, A. (2008). Managing and sustaining a world of workplace diversity: the Accenture experience. *Strategic HR Review, 7*(5), 11–16.

Jones, J. P. (1986). *What's in a Name?* Lexington: Lexington Books.

Katz, D. (Summer 1960). The functional approach to the study of attitudes. *Public Opinion Quarterly, 24*, 163–204.

Kim, P. (1990). A perspective on brands. *Journal of Consumer Marketing, 7*(4), 63–67.

Landon, E. (Sept 1974). Self concept, ideal self concept and consumer purchase intentions. *Journal of Consumer Research, 1*, 44–51.

Lawless, M., & Fisher, R. (1990). Sources of durable competitive advantage in new products. *Journal of Product Innovation Management, 7*(1), 35–44.

Lego. (2010). *Lego Games*. Available at: http://games.lego.com/en-US/default.aspx. Accessed 12 April 2010.

Munson, J., & Spivey, W. (1981). In K. Monroe (Ed.), *Advances in Consumer Research. Product and Brand User Stereotypes Among Social Classes, Vol. 8* (pp. 696–701). Ann Arbor: Association for Consumer Research.

Nevens, T., Summe, G., & Uttal, B. (May–June 1990). Commercializing technology: what the best companies do. *Harvard Business Review, 68*, 154–163.

Olshavsky, R., & Granbois, D. (Sept 1979). Consumer decision making – fact or fiction? *Journal of Consumer Research, 6*, 93–100.

Park, C., Jaworski, B., & MacInnis, D. (Oct 1986). Strategic brand concept-image management. *Journal of Marketing, 50*, 135–145.

Polaroid. (2009). *Press Release: Lady Gaga Named Creative Director for Speciality Line of Polaroid Imaging Products*. Available at: http://www.polaroid.com/About/News/Press+Release:+Lady+Gaga+Named+Creative+Director+for+Specialty+Line+of+Polaroid+Imaging+Products/4339. Accessed 14 April 2010.

Rosenberger, P., III, & de Chernatony, L. (1995). Virtual reality techniques in NPD research. *Journal of the Market Research Society, 37*(4), 345–355.

Saporito, B. (28 April 1986). Has-been brands go back to work. *Fortune*, 97–98.

Solomon, M. (Dec 1983). The role of products as social stimuli: a symbolic interactionism perspective. *Journal of Consumer Research, 10*, 319–329.

Superbrands. (2005). *CoolBrands: an Insight into Some of Britain's Coolest Brands.* London: Superbrands International Publications.

Urban, G., & Star, S. (1991). *Advanced Marketing Strategy.* Englewood Cliffs: Prentice Hall.

Urban, G., Weinberg, B., & Hauser, J. (1996). Premarket forecasting of really new products. *Journal of Marketing, 60,* 47–60.

Weitz, B., & Wensley, R. (1988). *Readings in Strategic Marketing.* Chicago: Dryden Press.

Wicks, A. (1989). Advertising research — an eclectic view from the UK. *Journal of the Market Research Society, 31*(4), 527–535.

CHAPTER 12

Brand Evaluation

OBJECTIVES

After reading this chapter you will:
- Understand the concepts of brand equity and brand valuation.
- Understand the multi dimensions that characterise brand equity.
- Be able to define brand loyalty.
- Be able to measure brand awareness.
- Understand the different models for valuing brands.
- Understand brand scorecards.

CONTENTS

SUMMARY

The purpose of this chapter is to examine the issue of evaluating the health of brands. This is done using the concepts of brand equity and brand valuation. We open the chapter by exploring the dynamics of a brand's equity and reviewing some of the interpretations of this concept. Using several models we consider how to grow brand equity.

Brand equity is multi-dimensional and we review some of the dimensions that characterise this concept. Finally, we address some of the challenges when evaluating the financial value of brands and review possible valuation methods.

GROWING BRAND EQUITY

'Equity' is a financial term that has been adopted by marketing people to reflect the fact that the brands they manage are financial assets that create significant shareholder value.

Whilst 'brand equity' is often talked about, it is seldom clearly defined.

The previous chapters in this book have considered how resources can best be employed to develop and sustain powerful brands. Once managers have

Creating Powerful Brands. DOI: 10.1016/B978-1-85617-849-5.10012-1

been successful in using these resources for branding purposes, they will need to monitor the health of their brands. In order to be able to sustain their brands' strengths, they require a method of regularly monitoring performance. Managers are particularly interested in measuring the equity that has been built up by their brands. Delving deeper into this issue of measuring brand equity reveals that it is a multi-dimensional concept, which we discuss in this chapter.

In order to better understand the process in which 'brand equity' increases the financial value of a branded business, Brand Finance defines 'brand equity' as a measure of 'the propensity of specific audiences to express preferences which are financially favourable to the brand'. This can be illustrated below in Fig. 12.1, which highlights the effect 'brand equity' has on each stakeholder group's behaviour and ultimately leads to increased financial value.

A classic example of how brand equity translates into better financial performance can be seen in the cola market. In blind tests Pepsi Cola consistently outperforms Coca-Cola in terms of consumer taste preference. But when Coke branded packaging is revealed, initial preference completely reverses. When still tap water is branded Highland Spring, the price shoots up. Branding persuades consumers to behave by appreciating the added values beyond that of the functionality of identical products and services. Thus, strong brands with high 'brand equity' possess the ability to persuade people to make economic decisions based on emotional rather than rational criteria.

One of the challenges managers face when attempting to measure brand equity is that there are numerous interpretations of this concept, each leading to

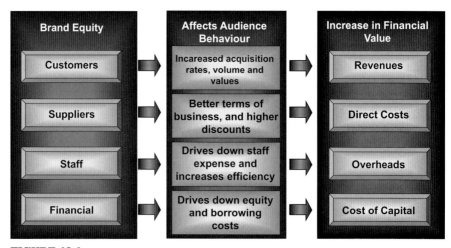

FIGURE 12.1
The financial effect of brand equity on each stakeholder group. *Source: Brand Finance plc 2008.*

a different set of measures. For example, two independent groups of academics in the USA, Farquhar (1989) and the work of Simon and Sullivan (1993), adopt a financial perspective regarding brand equity as the incremental cash flow resulting from associating a brand name with a product. By contrast, Aaker and Biel (1993) take a value-added perspective, conceiving brand equity as the value added to the core product or service by associating it with a brand name. Keller (1993) takes more account of consumer behaviour, regarding brand equity as the result of consumers' responses to the marketing of a particular brand, which depends on their knowledge of that brand. A widely used definition of brand equity is that which was originally devised by Srivastava and Shocker (1991) and was endorsed by the Marketing Science Institute. Brand equity is defined as 'a set of associations and behaviours on the part of a brand's consumers, channel members and parent corporation that enables a brand to earn greater volume or greater margins than it could without the brand name and, in addition, provides a strong, sustainable and differential advantage'. In view of the inclusiveness of this definition and its managerial perspective, we favour this interpretation.

It is worth stressing the point that brand equity describes the perceptions that consumers have about a brand and this, in turn, leads to the value of a brand.

A CASE STUDY – HOVIS: LEVERAGING OFF HISTORICAL 'BRAND EQUITY'

In 1886, Richard 'Stoney' Smith invented a way of retaining wheat germ in flour. In 1887, the brand was created in a newspaper competition in which a student called Herbert Grimes won a £25 prize for the best name – Hovis, which he derived from the Latin phrase, Hominis Vis (Strength of Man). In 1987, Hovis became the first major brand to be valued in the Rank Hovis McDougall takeover defence against GFW of Australia. This started an international debate about accounting for brands.

In 2000, Hovis was owned by Doughty Hanson, a private equity company. Hovis had operated for 115 years as the quintessential wholemeal loaf but now operated in the £1bn+, wrapped bread market. But the problem was that the market segment was declining and the Hovis brand was becoming a 'loss leader' for retailers.

However, everybody still loved the Hovis brand but it was too strongly associated with the traditional wholemeal (brown) product. Consumers were being driven to buy other bread brands such as Kingsmill or Warburtons, as they supplied both white and brown bread.

To escape declining growth, Hovis decided to relaunch by going back to the brand's foundations and leveraging off its historical brand equity as the experts in delicious, high quality, everyday bakery products that are all good for you.

With 53% of its success attributable to econometric modelling work, Hovis relaunched with new packaging. In addition, they developed a new innovative product, Hovis 'Best of Both,' – 'Soft white bread, with all the natural goodness of brown.' Hovis 'Best of Both' re-engineered category value by adding genuine value for consumers and providing them with a real reason to buy Hovis.

As a result of leveraging off their historical 'brand equity', Hovis' wrapped bread category increased by over 32%. Every £1 spent, generated £1.67 extra profit. Sales rose from £150 million in 2002 to £285 million in 2005. The financial value of the Hovis brand increased by over 31% in the first year and by 60% after two years.

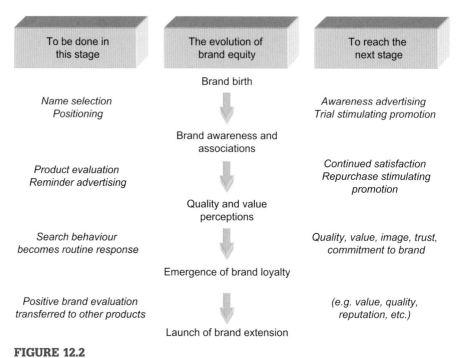

To be done in this stage	The evolution of brand equity	To reach the next stage

Brand birth

Name selection
Positioning

Awareness advertising
Trial stimulating promotion

Brand awareness and associations

Product evaluation
Reminder advertising

Continued satisfaction
Repurchase stimulating promotion

Quality and value perceptions

Search behaviour becomes routine response

Quality, value, image, trust, commitment to brand

Emergence of brand loyalty

Positive brand evaluation transferred to other products

(e.g. value, quality, reputation, etc.)

Launch of brand extension

FIGURE 12.2
The evolution of brand equity (*after Gordon et al. 1994*)

It is essential that managers track their brand equity on a regular basis since, as outlined in Chapter 11, managers are continually striving to fine-tune their strategies over the brand's life cycle.

Readers can begin to appreciate the dynamic nature of brand equity from a particularly helpful evolutionary model developed by Gordon and his colleagues (1994), shown in Fig. 12.2.

When a new brand is developed, it initially exists only through its physical characteristics. For the brand to be **born**, two decisions must be taken: a brand name and a positioning strategy must be agreed on. The former is critical because it contributes to the brand identity and can also communicate information about the product performance or ease of use. For example, the brand name 'Internet Explorer' clearly states the function of the product as a guide to visitors to the Internet and underlines its characteristics of user friendliness and reliability. Likewise, Post-it notes were given a very explicit brand name in order to make clear what the function of the product was, in its early days. Positioning is critical because it determines the brand's desired competitive set and it is against this, if consumers' perceptions also concur with managers', that relative brand strength will ultimately be measured. At this stage, bipolar maps of the functional attributes of competing brands are very useful to appreciate similarity/dissimilarity.

When a brand is launched, managers have three objectives: the attainment of brand awareness, the development of favourable associations and the involvement of consumers so that they will want to try the new brand and purchase it. To achieve these goals managers need to skillfully blend the elements of the marketing mix.

As consumers are repeatedly exposed to the brand, they become more familiar with it. Their degree of **brand awareness** depends on their ability to recall any promotional messages and the brand's availability. At this point, marketers need to focus on efficient communication and gaining distribution. Once consumers become familiar with the brand, their perceptions of it become more detailed. Managers aim at this stage to ensure that consumers include the brand in their consideration set.

The long-term success of a brand, however, is influenced by **consumers' perceptions of its value**, which are often based on functional and psychological attributes. There is a greater likelihood of success when marketers create some unique performance characteristics that appeal to a sufficiently large number of consumers. Consumers' judgements of the brand's quality are based on both objective measures, such as performance and fitness for use, and subjective criteria, such as past experiences and associated cues, for example the packaging colour. Past experiences can have a substantial impact on consumers' perceptions of quality.

Price differentials are an indicator of the relative risk that consumers perceive when they choose one brand over another, particularly when they have no previous experience of a brand. While the competing brands may contain identical ingredients, consumers' perceptions of quality and value for money may differ. Their choice is influenced by their evaluation of whether the price difference justifies the risk incurred in switching from their regular brand to a new one. Eventually the new brand becomes part of the consumer's brand repertoire. At this stage, they rarely compare competing brands, automatically choosing their preferred brand for particular situations. Their behaviour shifts from searching for information in order to choose between brands to a routine response, which in many cases leads to loyalty towards particular brands.

Brand loyalty is a measure of a consumer's attachment to a specific brand and is a function of several factors, such as the perceived quality of the brand, its perceived value, its image, the trust placed in the brand and the commitment the consumer feels towards it. Committed consumers guarantee future income streams as well as facilitating brand extensions by transferring any positive associations to new brands.

The last stage in the evolution of brand equity enables firms to strategically exploit any equity the parent brand has built up. **Brand extension** allows companies to further grow brand equity by gaining loyalty for related brands from existing consumers and existing channels.

COMMERCIAL MODELS OF BRAND EQUITY GROWTH

Models, such as the one just described, have also been developed by commercial organisations. Young & Rubicam have their own interpretation of the brand equity growth process resulting in their Brand Asset Valuater™. According to their model, brand equity growth is achieved by building on four brand elements: differentiation, relevance, esteem and familiarity.

Differentiation represents the starting point of the growth process, as the brand cannot exist in the long run unless consumers can distinguish it from others. To attract and retain consumers, the brand needs to convince them that it is **relevant** to their individual needs. As competition increases, marketers wisely protect their brand and show consumers that it delivers what has been promised. The next challenge is to ensure that consumers have regard and **esteem** for the brand's capabilities. If the brand has established itself as distinctive, appropriate and highly regarded, its ultimate success will depend on **familiarity**; that is, whether the brand is truly well known and is part of consumers' everyday lives. Familiarity does not solely depend on advertising, albeit this is a notable contributor, but also results from consumers recognising that the brand provides more value than other brands.

Young & Rubicam's empirical analysis indicates that scores on relevance and differentiation provide an assessment of the brand's potential for growth and they refer to this as 'brand vitality'. Furthermore, scores on esteem and familiarity measure the brand's current strength, its 'brand stature'. By plotting these values on the matrix, shown in Fig. 12.3, it is possible to consider the equity the brand has achieved and to identify appropriate strategies for its future growth.

Initially a brand begins its life in Quadrant A with low scores on all attributes. For the brand to move upwards into Quadrant B and gain more vitality, managers need to invest in attaining higher levels of differentiation and relevance. Once brands have

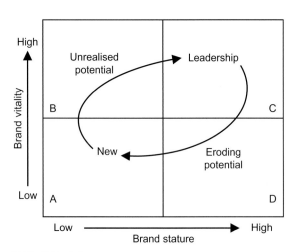

FIGURE 12.3

The strategic direction of brand strength (*after Young and Rubicam 1994*)

reached a higher level of 'vitality', brand owners have two options — maintenance by establishing them as niche brands or investing in building the brand's esteem and encouraging growth into Quadrant C. The top right hand quadrant is home to strong brands that have achieved remarkable brand equity growth, though they still may have potential for further growth. By maintaining the brand's stature and creatively managing its vitality, managers can look forward to the brand having a long lifetime. However, without sufficient maintenance of the brand's vitality, its differentiation and relevance fall, resulting in the brand increasingly selling on price promotions and declining to Quadrant D. In such a situation, the brand becomes vulnerable to price wars. As firms lose confidence in a brand's future, they cut marketing support, resulting in familiarity and esteem falling. Consequently, brand equity falls as the brand slips back to Quadrant A.

The framework by Young and Rubicam helps managers to understand the concept of brand equity and highlights aspects of the brand (differentiation, relevance, familiarity and esteem) that need attention over the short and long terms. Moreover, by comparing the position of the company's brands with that of competitors' brands, the model suggests appropriate strategies to increase brand equity and protect it against competition. Millward Brown International have devised a helpful diagnostic tool that, like the Young and Rubicam approach, enables managers to appreciate the basis for their brand's equity compared with competing brands. Their Brand Dynamics™ pyramid model is shown in Fig. 12.4, portraying the way consumers' value of a brand grows from a distant to a closely-bonded relationship.

To be considered for purchase, a brand must have a presence, both physically in terms of availability and psychologically in terms of awareness. Should people find the promise inherent in the brand to be relevant to their particular needs, they are more likely to progress to trying the brand, forming a view about its performance.

Evaluation of the brand's functional and emotional performance capabilities relative to other brands, leads consumers to a view about its relative advantages. If these advantages are particularly strong, they are likely to continue buying the brand and over time a bonded relationship results.

The benefit of this diagnostic is that by interviewing consumers about competing brands in a market, their profiles can be assessed on these pyramids. Through comparing these profiles, strengths and weaknesses can be identified, enabling appropriate strategies to be devised.

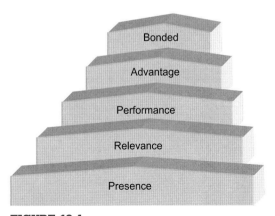

FIGURE 12.4

The criteria to assess the strength of a brand (*after Millward Brown International 1996*)

MEASURING BRAND EQUITY

As this book has sought to clarify, brands are complex concepts which can be characterised by several dimensions. Therefore, evaluating the health of a brand by measuring its brand equity necessitates taking several different measures along several different dimensions. Just as there are several different definitions of brand equity, so there are a variety of measurement methods. We next overview some of these.

Managers need to decide which dimensions are more appropriate to their environment since, for example, according to the extent to which a brand draws on functional or representational characteristics (covered in Chapter 11), perceived performance or image measures are particularly relevant. To have confidence in the results of the brand equity monitor over time, the same dimensions must be used on each occasion. The measures we review are comprehensive, although the practitioner may prefer to use just a few of them. Wisely selected, a few dimensions should enable managers to appreciate the state of health of their brands.

One way to appreciate some of the dimensions is to build on a causal model advanced by Feldwick (2002), as shown in Fig. 12.5.

In essence, the relative attributes of a brand will affect its strength and this, in turn, will be reflected in the financial value of the brand. Measuring brand equity could therefore involve an investigation of the first two components. Drawing on the published literature about the different types of measures, managers are able to choose from the following ways of evaluating each of the linkages.

Brand attributes

The response by consumers to a brand will, as Keller (1993) so cogently argued, depend on their favourable or unfavourable knowledge about it. Their brand

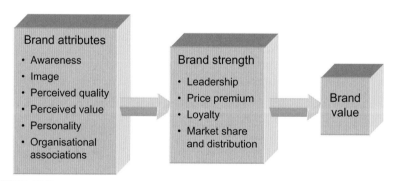

FIGURE 12.5

The causal nature of brand equity (after Feldwick 1996)

knowledge results from their level of awareness and the images they have about the brand. Thus, these two core components are at the heart of any brand attributes.

Brand awareness reflects the salience of a brand and facilitates consumers' abilities to identify the brand with a specific product category. Aspects of brand awareness can be measured, as Aaker (1996) showed, through some of the following ways:

- *Brand recognition* — This refers to the consumer's ability to recall previous exposure or experience with the brand. For example, 'Have you seen this brand before or not?'
- *Brand recall* — This refers to the consumer's ability to retrieve the brand from memory when given the product category as a cue. For example, 'What brands of personal computers are you aware of?'
- *Brand dominance* — This identifies the most important brand in a specific product category. For example, 'Which brand of tea do you drink most often?'
- *Brand knowledge* — This evaluates the consumer's interpretation of the values linked to a brand. For example, 'To what extent do you agree or disagree that the following cars have high accelerations?'

Brand image reflects consumers' perceptions of a brand's characteristics and can be gauged by the associations they hold in their memory. The different types of brand associations can be grouped according to: their level of abstraction, the amount of information held, whether they are product related or non-product related and whether they refer to attributes considered essential by consumers. There are different tools to measure the brand image:

- *Projective techniques* are helpful when consumers are unable or unwilling to express their feelings. These techniques include:
  - sentence completion — 'When I buy a personal computer, I look for…';
  - brand personality descriptors — 'The typical driver of a Jaguar is …';
  - picture interpretation — There may be a picture of a man driving his new BMW into a golf club's car park and two golf players are looking on — what would they say to each other about the driver?
- *Qualitative techniques*, such as free association, are used to explore possible associations along with further investigation during group discussions or depth interviews.
- *Ratings of evaluations and beliefs* are suitable to capture consumers' views on key attributes and the strength of their associations with particular brands.
- *Comparison of brand associations* with those of competitors identifies the relative strengths and weaknesses of the brand. For example, 'Of the fruit

juice brands that you identified earlier, which one do you believe to be the best? Why is it better than the other brands? What don't you like about it compared with the other brands?'

Aaker's work (1996) enables us to delve deeper into measuring a brand's attributes and its strengths as we explore.

Perceived quality is an important brand attribute and can be measured by comparing the brand with its competitors, using scales such as:

- above average, average and below average;
- consistent quality and inconsistent quality.

To measure perceived quality, consumers should first be asked which category they perceive the brand to be in and against whom it competes. Without asking this, there is a danger that inappropriate competitors might be specified by managers who are not relevant to consumers.

Perceived value indicates the extent to which the brand meets performance expectations, given its price. It can be measured by considering value for money and the reasons for purchase, for example:

- 'Do you think that the Toyota Avensis represents good value for money?'
- 'Why did you buy the Toyota Avensis rather than another car brand?'

The brand's **personality** is a useful metaphor to appreciate the brand's values and this shows the brand's emotional and self-expressive capabilities. This is particularly useful for brands that have only minor physical variations and are conspicuously consumed, for example, brandy. In these instances, very few consumers can distinguish between the taste of different brands and the brand is used to make a statement about the user. The brand's personality can be identified through questions such as:

- 'If brand X came to life, what sort of person would it be?'
- 'If brand X were to die, what would be written on its tombstone?'
- 'What type of person do you think would use this brand?'
- 'If brand X were a famous person, who do you think it would be?'

Organisational associations refer to the perceptions of a brand that consumers derive from its parent organisation. This dimension is appropriate when the organisation is particularly visible (as in a service business) or a corporate branding strategy is being used, such as Ford. Positive associations provide a valuable basis for differentiation. Measurements focus on how consumers consider the organisation, for example:

- 'Do you trust this brand, knowing it comes from…?'
- 'How do you feel about this organisation?'
- 'How would you describe the people that work for this organisation?'

Brand strength

As a consequence of its attributes, the strength of the brand can be gauged. Another set of measures needs to be used to assess brand strength.

Leadership not only identifies the most successful brand but also whether it is technologically or socially innovative within its category. This dimension can be measured, besides using market audit data, with questions such as:

- 'Do you regard brand X as being a leading rather than following brand?'
- 'Is brand X the first to break with tradition?'
- 'Does brand X offer you the latest technological development?'
- 'Is brand X a fashion leader in its category?'

The **price premium** reflects the brand's ability to command a higher price or to be less price sensitive than its competitors. This measure needs to be defined relative to those brands that consumers consider as substitutes. A brand's price premium can be identified by informing consumers of the price of competitors' brands, then asking consumers how much more or less they would pay for the brand. A more involved, albeit some argue a more reliable, method is to use trade-off analysis.

Price premium is not a suitable measure in markets with legal restrictions that prevent companies from charging a premium price. Also, it is not appropriate for strong brands, such as Swatch, that intentionally charge lower prices to keep competitors out of the market or for brands such as Mars Bars that have no direct substitutes for their products.

There are numerous measures of **loyalty**, for example, measuring actual purchasing behaviour over time, which reflects the degree of satisfaction that existing customers have with the brand. Loyalty can also be gauged asking questions such as:

- 'Next time you buy this product category, would you buy this brand again?'
- 'Thinking about the few brands of this product category that you often buy, is this brand one of your more frequently bought brands?'
- 'If someone were thinking of buying this product, which brand would you recommend?'

Managers should be aware that the responses to these questions may reflect past behaviour rather than intended future behaviour and that the favourable nature of replies may be more a reflection of brand size than loyalty.

Another method of measuring loyalty is provided by the concept of 'Share of Category Requirement' (SCR). The SCR for Ski yoghurt is all Ski yoghurt volume expressed as a share of all yoghurt bought by consumers who purchase Ski yoghurt during a defined period, such as a year. An alternative is to define loyalty by considering consumers' purchasing patterns over time and estimating the probability of their buying the brand on the next purchasing occasion. However, this analysis should also include data on price variations, as most patterns are strongly influenced by promotions.

Market share and **distribution** data are further indicators of brand strength. To obtain realistic results, however, marketers need to define the market and the competitor set from consumers' perspectives and recognise that market share indicators are often distorted by short-term price and promotional activities.

Before leaving this section on dimensions of brand equity, it is worth noting that the emphasis of these dimensions has been overtly focused upon consumers. Yet, staff are a key stakeholder group affecting the performance of brands, and their 'buy-in' to a firm's philosophy about its brands can be critically important. As such, writers such as Ambler (2000) argue that a matching set of dimensions should be considered to capture the equity arising from the staff's view of the 'employer brand'. For example, some of the employer brand dimensions could assess employees' awareness of brand goals, perceptions of the calibre of their organisation relative to comparable firms, their commitment to delivering the brand promise, etc.

The financial value of brands

In 1988, when RHM moved to valuing its brands for its financial balance sheet, debate started about whether brands should be included in the balance sheet. In December 1997, the UK Accounting Standards Body published a ruling on this matter through Financial Reporting Standard 10 (FRS 10), which was amended in July and December 1998. The objective of FRS 10 is to ensure that purchased goodwill and intangible assets are charged to the profit and loss account. Acquired brands must be included in these accounts at their purchase price. In the past, acquired brands would have been written off as goodwill. Internally created brands cannot be included in the financial accounts. The January 2003 view of the Accounting Standards Board is that due to the difficulties of arriving at a reliable valuation, companies should discuss their internally created brands in the operating and financial review that accompanies their financial statements. Many firms now do this, disclosing, for example, the amount they spent building their brands in the year and/or statistics demonstrating the strength of these brands.

To be acceptable in financial accounting terms, a valid and reliable brand valuation method should apply to both acquired and internally-created brands. One of the problems is that there are different perspectives on the value of a brand at any one time. For example, prior to market bids, Rowntree was worth around £1 billion to its shareholders, yet a few months later it was worth £2.5 billion to Nestlé. Although the value of a brand becomes much more apparent at the time it is acquired by another company, there remains uncertainty about a firm's annual valuation of its brands. In the absence of a generally accepted standard for brand valuation, the internally calculated value is subject to various interpretations.

Nonetheless, the 'authorities' have now addressed the issue of the several methods by which brands are valued by external organisations such as Brand Finance and Interbrand. After three years of discussion, IS 10668—Monetary Brand Valuation—was released in early 2010. This sets out the principles which should be adopted when valuing any brand. The brand valuer must declare the purpose of the valuation as this affects the premise or basis of value, the valuation assumption used and the ultimate valuation opinion, all of which must be transparent to a user of the final brand valuation. Thus, IS 10668 gives brand valuation the institutional credibility which it previously lacked.

At the acquisition stage, the brand's value depends very much on the purchaser, who will probably value it more if the acquisition is expected to bring synergy to the company, as was clearly the case with Nestlé's purchase of Rowntree. The issue of separation from the brand is best illustrated by the words of John Stuart, former Chairman of Quaker Oats Ltd: 'If this business were to be split up, I would be glad to take the brands, trademarks, and goodwill, and you could have all the bricks and mortars. And I would fare better than you.' He is certainly right! However, a brand's value is not automatically transferable and the purchase of the brand could negatively affect its value. When the acquired brand becomes part of a new firm, it is divorced from the previous firm's management, culture and systems and, without the flair and networks that the previous owners had, it may lose its consumer base. Any sales are strongly influenced by promotions and shelf visibility, but more importantly there is also the goodwill from the corporation. In new hands, with a different corporate halo, the brand might not be as strong. When Ford acquired Jaguar they did not rebrand the firm, in part due to the strong corporate halo of Jaguar. Brands with unique functional qualities may not be manufactured in the same way by the purchaser of the brand, who might be seeking cost-saving initiatives.

Valuing brands is not an objective process as it is based on various assumptions and estimations. For example, the valuation of a marketing consultancy at 8 in the morning when few staff are there is different from its valuation at 11 in the morning. How do you account for a consultant working out his notice, who was particularly successful at winning new business? In view of the difficulties in valuing what are essentially clusters of mental associations recognisable through a name, some question the usefulness of valuing brands. Nevertheless, many companies believe that there are benefits in valuing their brands and have accepted this challenge. For example, an organisation learns much about the drivers of brand success through the brand auditing that is required.

It has been argued that valuing brands is a worthwhile exercise because it draws attention to the long-term implications of brand strategies. Moreover, being forced to consider long-term effects counterbalances the pressure that usually drives managers who focus on policies to achieve short-term profits, but which pay lip service to brand building. Brand valuation therefore encourages managers to think more about building brands than market share. Where managers' performance is evaluated on an annual basis by changes in their brand's equity, they are more likely to emphasise decisions that are beneficial to the long-term growth of their brand and are less inclined to accept quick-fix solutions, such as price-off promotions or brand extensions which become too remote from the core brand and may undermine the value of the parent brand. The brand represents a major marketing investment that it would be unwise not to evaluate, despite the fact that the assumptions underpinning the brand valuation process affect the resulting figure.

The value of the brand also differs according to the perspective it is considered from. From a firm's point of view, a brand's value is derived from the future incremental cash flow resulting from associating the brand with a product. For example, in a television factory once jointly owned by Hitachi and General Electric, Hitachi was able to sell the same product as GE but labelled Hitachi, with a £50 premium and at twice the volume. A brand brings numerous competitive advantages to the firm. For example, it provides a platform from which to launch new products and licenses; it builds resilience in times of crisis as seen by the quick sales recovery following the incident when Tylenol was tampered with; and it creates a barrier to entry, for instance, formidable barriers are present through names such as Chanel. From a trader's perspective, the value of a brand lies in its ability to attract consumers into their stores. From a consumer's point of view, the brand has value since it distinguishes the offering, reduces their perceptions of risk and reduces their effort in making a choice.

To manufacturers, retailers' and consumers' brands have value and therefore it is right that some attempt be made to quantify this. While one might argue whether Coca Cola's 2009 valuation of $67 billion (Brand Finance) is precisely correct, the issue really is that this is a multi-billion dollar asset and regular tracking is needed to assess how different branding activities are affecting its value.

The debacle at Saatchi & Saatchi illustrates how the value of a brand is heavily dependent on the intangible goodwill inherent in the brand's associations, which can fluctuate over time. In 1994, Maurice Saatchi was ousted as chairman of this famous advertising agency after the share price had fallen from £50 in 1987 to £1.50. At the time of his departure, the company rebranded itself as Cordiant. As a direct result of his leaving the company, it lost business worth £50 million and during the following six months its market value decreased by another third. However, the Saatchi brand came to life again in the new company founded by Maurice Saatchi, called M&C Saatchi, which benefited from the intangible 'Saatchi' assets, such as their creative employees and the clients he had taken with him.

METHODS OF MEASURING THE FINANCIAL VALUE OF A BRAND

There are a number of recognised methods for valuing trademarks or brands as defined here.

One can look at historic costs – what did it cost to create? In the case of a brand, one can look at what it cost to design, register and promote the trademarks and associated rights. Alternatively, one can address what they might cost to replace. Both the historic cost method and the replacement cost method are subjective, but we are often asked to value this way because courts may want to know what a brand might cost to create.

It is also possible to consider market value, though frequently there is no market value for intangibles, particularly trademarks and brands.

Generally speaking, the most productive approach to brand valuation is to employ an 'economic use' valuation method, of which there are a number.

First there is the price premium or gross margin approach, which considers price premiums or superior margins versus a "generic" business as the metric for quantifying the value that the 'brand' contributes. However, the rise of private label means that it is often hard to identify a 'generic' against which the price or margin differential should be measured.

Economic substitution analysis is another approach – if we didn't have that trademark or brand, what would the financial performance of the branded business be? How would the volumes, values and costs change? The problem with this approach is that it relies on subjective judgments as to what the alternative substitute might be.

The difficulties associated with these two approaches mean that the two most useful 'economic use' approaches are the 'earnings split' and 'royalty relief' approaches.

Under a 'royalty relief' approach one imagines that the business does not own its trademarks but licenses them from another business at a market rate. The royalty rate is usually expressed as a percentage of sales. This is the most frequently used method of valuation because it is highly regarded by tax authorities and courts, largely because there are a lot of comparable licensing agreements in the public domain. It is relatively easy to calculate a specific percentage that might be paid to the trademark or 'brand' owner.

Under an 'earnings split' approach one attributes earnings above a break-even economic return to the intangible capital. This involves four principal steps. The first is an appropriate segmentation of the market to ensure that we study the brand within its relevant competitive framework. The second step is to forecast the economic earnings of the branded business earnings within each of the identified segments. These are the excess earnings attributable to all the intangible assets of the business. The third step is to analyse the business drivers, research to determine what proportion of total branded business earnings may be attributed specifically to the brand. The final step is to determine an appropriate discount rate based on the quality and security of the brand franchise with both trade customers and end consumers.

Regardless of which method is used, the valuation usually will require a sensitivity analysis in which one flexes each of the assumptions made in the analysis, one at a time, to demonstrate the impact changes that each variable has on the overall valuation. However, this is a simple mechanical exercise intended to show which assumptions the valuation is most sensitive to. The valuer's dilemma lies in trying to determine which of the key assumptions is most likely to change and how, which is where all the brand audit data and brand equity measures becomes significant.

The 2009 Brand Finance Brand Tracker Table, for example, differs significantly from those of Interbrand and Milward Brown. The value of Coca Cola ranges from $33 billion to $67 billion. Google is valued by Brand Finance at $29 billion, by Interbrand at $26 billion and by Milward Brown at $100 billion.

In our experience, therefore, it is very important to express the final valuation number in context. This means explaining exactly what has been valued, using what method and what the key insights are as to the influence of the brand on the key operating variables of the business. This emphasises the importance of developing a valuation model that is presented in a user-friendly manner to help management make crucial decisions around marketing and branding strategy, objectively and with a high degree of financial rigour.

One way in which one can effectively express a valuation model in a simple format to help answer key marketing and branding investment decisions is a brand scorecard.

Brand Scorecards

Marketers are increasingly being challenged by their Boards and Chief Financial Officers to answer key questions:

- How much should we invest in marketing and branding?
- Which markets, customers, brands and channels will generate the highest return?

- Which strategy will generate the greatest value?
- How are our brands performing relative to competitors and targets?

In order to answer these vital investment decisions, brand managers should consider developing brand scorecards to inform brand management decisions before the finance department does it for them.

The success of a brand scorecard relies on the synergy between financial, market and customer analytical data. This integration of data allows one to gain greater commercial insight, improvements in the collection and utilisation of data and helps overcome understanding data in silos.

All relevant data collected from the brand audit, brand equity analysis, value mapping analysis and brand valuation are fed into the building of the scorecard.

How the scorecard is developed is dependent on who will be predominantly using the scorecard. As can be seen from Fig. 12.6, different management levels will use the scorecard for different purposes. As a result, it becomes imperative that all relevant stakeholders who will be involved in using the scorecard are identified to ensure that it is tailored suitably and includes the relevant brand metrics. In addition, one will need to identify what resources

FIGURE 12.6

Users of the Brand Scorecard. *Source: Brand Finance plc 2008.*

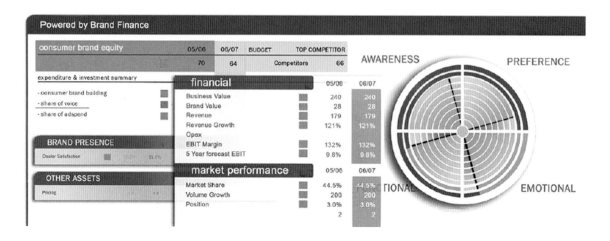

FIGURE 12.7
An example of a Brand Scorecard. *Source: Brand Finance plc 2008.*

are required to build and maintain the scorecard and which audiences will be tracked.

An example of a scorecard is given in Fig. 12.7.

Another effective manner in which brand/marketing managers can determine the relevant amount of investment required to achieve optimal marketing return on investment is through a statistical analysis called marketing mix modeling.

Marketing mix modelling

Marketing mix modeling is a statistical analysis to isolate and quantify the impact of different marketing activities. It is a model that guides the mix and combination of future marketing activities, making it a highly complex analysis that cannot be fully covered in-depth in this chapter.

The model can only be developed if an understanding of what the sales and profit objectives to be achieved by the marketing department in the financial year are known. These objectives should be based in terms of customers, average transaction value, frequency, penetration rates and product range. The objectives will form the basis of determining the optimal marketing budget using a demand driver approach.

The demand driver approach requires one to:

1. Know the product sales potential by customer segment;
2. Know the bottlenecks preventing increased product penetration;

3. Know the impacts and returns from differing marketing levers;
4. Use the most cost-effective communication channels for the task.

All these four steps will require a thorough econometric analysis into the most effective marketing activities to achieve the required objectives.

In order to determine the relevant payback of each activity, the model will identify the difference between strategic and tactical marketing spend. Strategic marketing spend is geared to support long term brand values such as image and awareness and covers a payback period of approximately 3 years, which gives a reason for consumers to buy a brand — a 'pull' strategy. In contrast, a tactical marketing spend is targeted to stimulate immediate sales promotions and distribution drives with a payback period of around 12 months, which gives the consumer an incentive to buy — a 'push' strategy. Often, not all marketing spend is strategic, but all of it needs to pay back at some point.

On completion of the model, it will help improve brand and marketing managers' understanding of factors that influence short term sales. It should also provide them with strategic insights that will shape overall business planning as the external environment changes and evolves, as well as providing key insights and understanding of why campaigns perform better in some situations than others.

It is often useful to flex all valuation models in order to determine the effect of varying brand strategies on financial performance.

Dynamic Scenario analysis

The word 'scenario' is a literary term meaning 'an outline of the plot of a dramatic or literary work'. While a literary scenario might be credible it is inevitably imaginary and there may be alternative plots or sub-plots. Brand planning can be much the same.

The original literary term has been adapted by commerce to describe 'a possible set of future events', 'an outline or model of an expected or supposed sequence of events', 'a postulated sequence of possible events'. Because all scenarios describe the future, they are inevitably hypothetical and uncertain. Some postulated scenarios may be quite implausible. To be taken seriously, they must be logical, internally consistent and credible.

In Michael Porter's words a scenario is: 'an internally consistent view of what the future might turn out to be - not a forecast, but one possible future'.

Therefore, when one talks about scenario analysis, it involves an in-depth evaluation into the future possibilities of a brand, which can be derived from flexing a valuation model in order to estimate the impact of alternative strategies on business value and brand value. As a result of such an analysis, a brand or marketing manager will be able to assess the short-term profit and long-term value implications of a range of potential strategies.

Scenario analysis forms a pertinent part of brand managers' responsibility. The credibility of a brand scenario needs to be explored and tested, using consumer and trade research, market, marketing and financial due diligence. If deemed consistent and credible, the value impact can be tested with a brand valuation model. If the new scenario implies that greater value will be created than the pre-existing scenario, then it should be actively pursued. This can be demonstrated in a case study of the brand 'Courvoisier.'

CASE STUDY – THE COURVOISIER SCENARIO DEBATE

In 1999, there was a debate within the global brand management group within Allied Domecq as to whether the Courvoisier brand should be classified as a 'core' brand or whether it should be divested given its poor financial performance and declining net present value (NPV).

The problem was that the previously preferred scenario for the Courvoisier brand, (which was marketed as a high-end cognac product, sold to high net worth individuals in Asia, Europe and East Coast USA, using traditional marketing and distribution, at a premium price) wasn't working. Given the high cost of production, high stock holding costs and low rate of sale, the brand had a low and declining NPV.

Market research indicated that the marketing of Courvoisier as a lower-end cognac, sold in different bottle sizes to mid-market ethnic consumers in heartland USA with contemporary marketing techniques, would radically change the value of the brand.

The two scenarios were modelled and the latter approach indicated a higher level of profits and cash flow, with lower capital investment and a faster rate of sale.

Variables were flexed but even allowing for changed assumptions, the NPVs of the two scenarios were quite different. This led to a change in brand strategy and a rapid improvement in the financial performance of the Courvoisier brand. In the recent acquisition of Allied Domecq's brand portfolio by Pernod Ricard, Courvoisier was referred to as a key brand.

In this example the two scenarios for Courvoisier implied radically different capital values for the brand, the higher of which was actually realised when the altered scenario was implemented.

CONCLUSION: FINANCIAL IMPLICATIONS FOR BRANDS

This chapter has described the process for identifying branded business value and the specific value of intangible assets, including trademarks or brands. While it is important to know the financial value of a trademark or brand within a branded business, the most important thing to know is the value of the branded business as a whole and how it can be maximised.

The 'brand valuation' framework described, indicates how to understand the impact of each audience on the financial model, how to map value, track brand equity, report performance to managers via scorecards and then plan business and brand value enhancement strategies using all the information. The approach is holistic because it incorporates both marketing and financial measures, all stakeholder audiences and both short and long term perspectives. It is both a historical measure of performance and a prediction of future performance.

In our view this all-encompassing framework is a vital tool for brand managers. It empowers them to manage their brands, just as CEOs manage the wider business. In fact, brand managers who have trained and operated with such accountable and strategic measurement frameworks have a higher than average propensity to become the CEO!

MARKETING ACTION CHECKLIST

To help clarify the direction of future brand measurement activity, it is recommended that the following exercises are undertaken:

1. With your marketing team, review the stages in the life cycle of your brands. Then assess how successful your brands have been at each stage. If you have not devised any system to measure the brand's success, consider which milestones marked the growth of each brand. Now consider the future of the brands. How do you plan to keep track of their development? Establish some benchmark figures to help you assess the growth of the brand. Do your short-term and long-term strategies support the achievement of these goals?

2. With your colleagues, select one of your firm's brands and go through the four factors that Young and Rubicam argue contribute to brand building (differentiation, relevance, esteem and familiarity), assigning a score on each of them. Compare your responses with those of your consumers, which may necessitate a customer survey. Any discrepancy between your view and consumers' views is an indicator of a possible performance gap and corrective action should be considered. With the results of the consumer survey, determine the brand score on vitality and stature. Analyse, using the same procedure, the brands of your competitors. Finally, plot your brand on the matrix shown in Fig. 12.3, together with the brands it is competing against. What conclusion can you draw from this comparison? Does your marketing programme protect your brand's position? Does the programme reflect your long-term objectives or does it primarily focus on achieving short-term results?

3. Using the model in Fig. 12.4, profile your brand against competitors. What strategies have led to these profiles? Does your marketing programme take the lower layers for granted? What marketing activities could you devise to improve your brand's profile?

4. From the list of attributes shown in the first box of brand equity in Fig. 12.5, identify those that would best apply to your brands, making explicit the reasons for including or disregarding any attribute. What market research

data do you have to assess these attributes both for your brand and your competitors? If your data is based on research undertaken more than 12 months ago, it is advisable to consider commissioning new market research.

5. For each attribute you have selected in the previous exercise (brand awareness, image, perceived quality, perceived value, personality and organisational associations), how is your brand performing? What improvements will help the brand on each attribute?

6. Selecting the brand strength dimensions from Fig. 12.5 most appropriate to evaluating your brand, how strong is your brand? By considering its weaknesses, consider possible routes to improvement.

7. Assess how appropriate and realistic it would be to value your brands solely on historic costs. With your colleagues, discuss how to attribute costs to that specific brand and identify those intangible elements that were important for the development of the brand. How could these intangible elements be valued?

8. From the material covered on brand valuation techniques, consider with your colleagues the valuation process that would be most suitable for your brands.

9. For one of your brands, devise an outline Brand Scorecard and discuss with your colleagues how it could be used by the different levels of management in your organisation.

References

Aaker, D. (1996). *Building Strong Brands*. New York: The Free Press.

Aaker, D. (1996). Measuring brand equity across products and markets. *California Management Review, 38*(3), 102–120.

Aaker, D., & Biel, L. (1993). *Brand Equity and Advertising*. Hillsdale, New Jersey: Lawrence Erlbaum Associates.

Ambler, T. (2000). *Marketing and the Bottom Line*. Harlow: Pearson Education Ltd.

Farquhar, P. (Sept. 1989). Managing brand equity. *Marketing Research, 1*, 24–33.

Feldwick, P. (2002). *What is Brand Equity Anyway?* Henley on Thames: World Advertising Research Centre.

Gordon, G., di Benedetto, A., & Calantone, R. (1994). Brand equity as an evolutionary process. *The Journal of Brand Management, 2*(1), 47–56.

Keller, K. (Jan. 1993). Conceptualizing, measuring and managing customer-based brand equity. *Journal of Marketing, 57*, 1–22.

Millward Brown International. (1996). *The Good Health Guide*. Warwick: Millward Brown International.

Simon, C., & Sullivan, M. (1993). The measurement and determinants of brand equity: a financial approach. *Marketing Science, 12*(1), 28–52.

Srivastava, R., & Shocker, A. (1991). *Brand equity: a perspective on its meaning and measurement* (pp. 91–124). Cambridge, Massachusetts: Marketing Science Institute Working Paper.

Vazquez, R., del Rio, A., & Inglesias, V. (2002). Consumer based brand equity: development and validation of a measurement instrument. *Journal of Marketing, 18*(1-2), 27—48.

Young, & Rubicam. (1994). *Brand Asset Valuation*. London: Young & Rubicam.

Further Reading

Aaker, D. (1991). *Managing Brand Equity*. New York: The Free Press.

Agarwal, M., & Rao, V. (1996). An empirical comparison of consumer-based measures of brand equity. *Marketing Letters, 7*(3), 237—247.

Barwise, P., Higson, A., Likierman, A., & Marsh, P. (1989). *Accounting for Brands*. London: London Business School and the Institute of Chartered Accountants.

Barwise, P. (1993). Brand equity: snark or boojum. *International Journal of Research in Marketing, 10* (1), 93—104.

Birkin, M. (1994). Assessing brand value. In P. Stebart (Ed.), *Brand Power*. Basingstoke: Macmillan.

de Chernatony, L. (1996). Integrated brand building using brand taxonomies. *Marketing Intelligence & Planning, 14*(7), 40—45.

de Chernatony, L., Dall'Olmo Riley, F., & Harris, F. (1998). Criteria to assess brand success. *Journal of Marketing Management, 14*(7), 765—781.

Dyson, P., Farr, A., & Hollis, N. (1996). Understanding, measuring and using brand equity. *Journal of Advertising Research, 36*(6), 9—21.

Gordon, W. (1999). *Good thinking*. Henley on Thames: Admap Publications.

Haig, D. (1996). *Brand Valuation*. London: Institute of Practitioners in Advertising.

Kamakura, W., & Russell, G. (1993). Measuring brand value with scanner data. *International Journal of Research in Marketing, 10*(1), 9—22.

Kapferer, J-N (1992). *Strategic Brand Management*. London: Kogan Page.

Mackay, M., Romaniuk, J., & Sharp, B. (1998). A classification of brand equity research endeavours. *Journal of Brand Management, 5*(6), 415—429.

Mackay, M. (2001). Evaluation of brand equity measures: further empirical results. *Journal of Product and Brand Management, 10*(1), 38—51.

Murphy, J. (Ed.). (1991). *Brand Evaluation*. London: Business Books.

Park, C., & Srinivasan, V. (May 1994). A survey-based method for measuring and understanding brand equity, and its extendibility. *Journal of Marketing Research, 31*, 271—288.

Perrier, R. (Ed.). (1997). *Brand Valuation*. London: Premier Books.

Pitta, D., & Katsanis, L. (1995). Understanding brand equity for successful brand extension. *Journal of Consumer Marketing, 12*(4), 51—64.

Srivastava, R., Shervani, T., & Fahey, L. (Jan 1998). Market based assets and shareholder value: a framework for analysis. *Journal of Marketing, 62*, 2—18.

Yoo, B., & Donthu, N. (2001). Developing and validating a multi-dimensional consumer based equity scale. *Journal of Business Research, 52*, 1—14.

Index

471